COMPREHENSIVE CHEMICAL KINETICS

COMPREHENSIVE

CHEMICAL KINETICS

EDITED BY

R.G. COMPTON

M.A., D.Phil. (Oxon.)
*University Lecturer in Physical Chemistry
and Fellow, St. John's College, Oxford*

VOLUME 28

REACTIONS AT THE
LIQUID–SOLID INTERFACE

ELSEVIER
AMSTERDAM–OXFORD–NEW YORK–TOKYO
1989

ELSEVIER SCIENCE PUBLISHERS B.V.
Sara Burgerhartstraat 25
P.O. Box 211, 1000 AE Amsterdam, The Netherlands

Distributors for the United States and Canada:

ELSEVIER SCIENCE PUBLISHING COMPANY INC.
655, Avenue of the Americas
New York, NY 10010, U.S.A.

ISBN 0-444-41631-5 (Series)
ISBN 0-444-87363-5 (Vol. 28)

with 114 illustrations and 8 tables

Printed in The Netherlands

COMPREHENSIVE CHEMICAL KINETICS

ADVISORY BOARD

Volumes in the Series

Contributors to Volume 28

P. HAMMONDS   Ciba – Geigy Industrial Chemicals,
Tenax Road,
Trafford Park,
Manchester M17 1WT, Gt. Britain

Present address

Baker Oil Treating
Tofts Farm West Industrial Estate,
Brenda Road,
Hartlepool,
Cleveland TS25 2BQ, Gt. Britain

W.A. HOUSE   Freshwater Biological Association,
River Laboratory,
East Stoke, Wareham,
Dorset BH20 6BB, Gt. Britain

A.T. HUBBARD   Department of Chemistry,
University of Cincinnati,
Cincinnati,
OH 45221-0172, U.S.A.

M. SPIRO   Department of Chemistry,
Imperial College of Science, Technology and Medicine,
Imperial College Road,
London SW7 2AY, Gt. Britain

Preface

Volume 28 is concerned with reactions occurring at the solid–liquid interface. Volumes 26 and 27 have dealt with electrochemical reactions. Thus the present volume covers those interfacial processes which involve no net passage of current.

The topics covered are as follows. The structure of the interfacial region and its experimental investigation are covered in Chapter 1. The following chapter reviews the mechanisms by which heterogeneous catalysis of solution reactions can take place. The third chapter is concerned with the mechanism and kinetics of crystal growth from solution and the final contribution deals with corrosion processes at the metal–solution interface.

The editor thanks Mr. A.J. Barwise for his assistance in compiling the index.

Oxford   R.G. Compton
January 1989

Contents

Chapter 3 (W.A. House)

Kinetics of crystallisation of solids from aqueous solution 167

Chapter 4 (P. Hammonds)

An introduction to corrosion and its prevention. 233

Chapter 1

Structure of the Solid–Liquid Interface

ARTHUR T. HUBBARD

1. Introduction

1.1 INVESTIGATION OF SOLID–LIQUID INTERFACES

A period of rapid progress in the study of solid–liquid interfacial structure is under way, although much further work will be required to arrive at a thorough understanding of this interesting area. This chapter offers a brief progress report. Emphasis is on the structure of the solid side of the interface, including adsorbed species, since progress is mainly in that area at present. Structural phenomena in the interfacial liquid as influenced by the nearby solid and adsorbed layer is an equally important topic about which it is hoped there will be more to report in a future article.

This chapter is arranged as follows: experimental strategies for direct determination of surface structure are discussed. Experimental findings are then presented in the areas of ionic adsorption, electrodeposition, alloy surface oxidation, and organic molecular adsorption.

Low-energy electron diffraction (LEED) was first described by Germer and Davisson [1] in experiments which demonstrated the wave-particle duality of electrons. LEED began to be recognized as an important surface characterization tool more than 30 years later when, in 1960, Germer and Hartman [2] described results obtained with an improved apparatus which incorporated the visual display system for "slow" electrons developed by Ehrenberg [3]. The discovery of LEED captured the imagination of the surface science community: the structure of surfaces at the atomic level could now be directly determined. Development of other techniques occurred in response to the desire to know the composition and other properties of the surface. Auger spectroscopy revealed surface elemental composition and cleanliness [4]. Electron energy-loss spectroscopy displayed surface vibrational bands [5]. Photoelectron spectroscopy yielded clues as to electron structure [6]. Efficient, metal ultra-high vacuum systems were developed specifically for experiments with clean surfaces.

The term "ultra-high vacuum" (UHV) has a special meaning in the context of surface investigations: the UHV environment is one in which no significant contamination of the surface occurs. The principal experimental requirements for achieving UHV conditions are: the absence of pump fluid, grease, volatile materials of construction, and other possible contaminants; the absence of sliding seals and other sources of leakage; and bake-out of the entire vacuum enclosure at 200°C or above to eliminate "virtual leaks", sources of contamination due to desorption or effusion from inside surfaces [7]. In any event, cleanliness of the sample is the essential criterion, not system pressure, as the acceptable residual gas pressure for a given experiment depends upon the nature of the species present in the residual gas and upon whether the sample surface is interactive with those species.

During the same period, surface studies by means of thin-layer electrodes [8] revealed that very many of the species adsorbed from solutions at electrode surfaces were covalently bonded to the surface sufficiently strongly that they would be expected to remain attached when the surface was transferred from solution to vacuum. This expectation was confirmed in practice, as discussed below. All was in readiness for surface electrochemistry to be part of the progress in the study of surfaces at the atomic level. Progess has been rapid, as noted in earlier reviews [9, 10].

Surfaces shown to be clean in UHV by Auger spectroscopy and to have a specific single-crystal surface structure by LEED are termed "well-defined surfaces". With apologies for stating the obvious, surfaces not directly characterized as to surface structure and composition are not well-defined surfaces, even though prepared from an oriented single-crystal. This is because the experience of experimenters around the world is that specific crystal planes do not emerge automatically at the surfaces of oriented single-crystals. Contamination, structural reconstruction, and disorder are the general rule. Only by repeated attempts, guided by direct experimental investigation, is the surface brought to a clean, ordered, "well-defined" state. Looking on the bright side, when faced with the considerable investment of time and money to design and build an apparatus for experimenting with surfaces in well-defined states, the same experimental tools required to bring a surface to a well-defined, reproducible initial state can then be employed to reveal the various interesting changes which occur at the surface as a result of virtually any subsequent chemical or electrochemical treatment or use of the surface. Indeed, the diagnostic power, generality, and directness of the multi-technical experiments perfected originally for the study of clean single-crystal surfaces are turning out to be similarly advantageous for exploring very complicated samples in a wide variety of other fundamental or practical investigations.

One group of surface analysis techniques derives its surface sensitivity from the limited free path or escape depth of "slow" electrons: LEED [11], Auger spectroscopy [12], photoelectron spectroscopy [13] (particularly ultraviolet photoemission spectroscopy, UPS), and X-ray absorption fine-structure experiments based upon the detection of emitted Auger electrons or photoelectrons (EXAFS) belong to this group. Other techniques take advantage of a similar limited free path of low-energy ions or atoms in solids; included in this group are secondary ion mass spectroscopy (SIMS) [14], electron-stimulated desorption (ESD) [15], laser-stimulated desorption mass spectroscopy, surface ion beam scattering, and surface molecular beam scattering experiments. A single monoatomic layer scatters about 30% of the electrons passing through it, as discussed below. Electrochemical methods derive their surface sensitivity from the separability of signals due to adsorbed species from those of the bulk solution by means of either thin-layer electrode or fast perturbation techniques. Other methods are based upon techniques which, while not inherently sensitive to the surface, are made somewhat more so by modulation. An example is infrared reflection absorption spectroscopy (IRRAS), which has recently been combined with optical polarization modulation or electrode potential modulation. Examination of a surface in the presence of a bulk fluid phase involves compromises in sensitivity, resolution, signal-to-noise ratio, or other aspects. In any event, no one technique measures enough of the properties of a surface to bring it to a well-defined state. Accordingly, there is no substitute for a balanced approach to surface characterization, a multi-technique approach including LEED, Auger spectroscopy, EELS, and IRRAS measurements. A useful exposition of developments in surface methodology is given by Somorjai [16].

Before proceeding to methods and applications of surface structure determination, perhaps we should remind ourselves of the motivations for investigating surface structure and composition. Chemical and electrochemical reactions of adsorbed atoms, ions, and molecules depend upon surface structure and composition, as discussed below. They also depend upon the mode of surface bonding of the adsorbed species. Most of the surface-catalyzed processes studied to date have proved to be sensitive to surface structure; most of the contradictions on this point in the earlier literature can now be understood in terms of missing or erroneous surface structural and compositional information. Adsorption processes are often highly selective and specific to a particular adsorbate and substrate, such that a species (for instance, sulfide or iodide) which might be only a trace ingredient of the fluid phase frequently turns out to be the principal ingredient of the surface. For this reason, such trace elements can usually be detected more efficiently at the surface than in the fluid. In any event, the substrate element or compound usually proves to be a "minority species" of an actual ordinary solid–liquid interface. The smallest traces of an adsorbate can strongly influence surface chemical and electrochemical processes. Surface stability, electronic properties, photochemical behavior, corrosion, passivation, wetability, adhesion, mechanical characteristics, and lubricity, to name a few, often depend strongly upon surface structure and composition at the atomic level. Accordingly, since surface structure and composition commonly affect the fundamental and practical properties of solid materials, electrodes, and catalysts, we concern ourselves in this chapter with the structure of surfaces in solution and the related matters of surface composition and chemical bonding.

1.2 PREPARATION OF SOLID SURFACES

Preparation of the surface of a pure element, alloy, or compound generally begins with cutting and polishing [17]. Single-crystal materials are oriented along with cutting and polishing by means of X-ray reflection (Laue) photography [18]. A small area of the sample is polished along some convenient, arbitrary direction for use in taking the initial X-ray reflection photographs. One can begin by taking photographs about ten degrees apart. One of the first few photographs usually reveals a zone, a row of spots related to a single crystal axis. Azimuthal and equatorial angles are then adjusted, keeping the zone near the center of the film, until an intersection with a second zone is brought to the center of the photographs. A low-index plane is identifiable from the symmetry of spots at that intersection point. Following one of those zones in a similar manner to another major intersection locates a second low-index plane. Identification of two low-index planes is normally sufficient to define the orientation of a crystal. The goniometer is then adjusted to put the crystal in the desired orientation for final polishing. A two-angle goniometer should be constructed for this purpose which is sufficiently rugged mechanically that the polishing procedures can be carried out without removing the crystal from the goniometer, preserving the accuracy of orientation. An angle error of less than $0.5°$ is necessary if atomic steps in the surface plane are to occur less frequently than one per hundred rows of surface atoms. Because the major planes are as much as $90°$ apart, it is advisable to provide a range of travel from -30 to $+120°$ on both axes of the goniometer. The entire surface area of the crystal should be cut and polished to one crystallographically equivalent orientation so as to expose only a single type of surface to the liquid upon immersion.

Crystal polishing for surface studies begins with metallographic procedures [17]. That is, a sequence of abrasives is used, from coarse to fine, in such a manner that the first stage removes the damage from the spark cutter or string saw and each successive finer abrasive removes the plowed material left by the previous stage. The force per unit area applied between the crystal and the abrasive is an important parameter. Measurements are made of the amounts of material removed at each stage. Sufficient pressure should be applied to remove the previously plowed material within about $2\,min$. The finer abrasives require the higher pressure (1–$4\,p.s.i.$). The plow depth of each abrasive is about one-tenth of the grit size. For example, the plow depth of 400 grit emery paper is approximately $1/400 \times 1/10 = 2.5 \times 10^{-4}\,in$. The procedure is usually begun with 240 grit emery paper. Kerosene is a convenient lubricant. While electric discharge machining (spark cutting) or string sawing is preferred for parting a wafer from the main crystal or boule, angle adjustments are more effectively made by means of coarse emery paper.

The series of grits typically employed is 240, 320, 400, and 600; $6\,\mu m$ diamond paste on nylon polishing cloth; and $1\,\mu m$, followed by $0.25\,\mu m$ diamond paste on napped cloth., The polishing resin in which the crystal is mounted should be at least $6\,mm$ wider than the crystal so as to minimize rounding of the edges of the crystal.

Malleable crystal materials should be mounted by attaching fine wires to the uppermost small edge, minimizing foreign or contaminated surfaces in contact with the solution. Wire of 0.5 mm diameter is usually appropriate for support and resistance heating of crystals (about 6 mm × 6 mm × 1.5 mm). Brittle materials can usually be supported by mechanical contact or solder on an upper face while essentially only the lower face parallel to the surface of the liquid is immersed.

The final polishing stage actually occurs after installation of the crystal into the vacuum system. Bombardment of the crystal with Ar^+ ions ($5 \times 10^{-6} A\,cm^{-2}$ at about 700 eV energy) [19] removes surface atoms at a rate of about one monoatomic layer per minute. Under the most favorable conditions, the metallographic mechanical polishing procedure leaves the surface with a selvedge no thicker than the plow depth of the final abrasive, about 100 Å. Accordingly, when combined with annealing in UHV and LEED examination of the surface, a LEED pattern indicative of an ordered single-crystal surface beings to appear after as little as 2 h of bombardment. Several days of bombardment are usually required before optimum results begin to appear and the variation of surface reactivity and spectroscopic properties is substantial during this final "polishing" process. Annealing temperatures equal to about 2/3 the melting point of the material usually give the best result. Although annealing in ultra-high vacuum is appropriate to elemental crystals and alloys, a chemical vapor atmosphere is often required to retain uniform surface composition of compound crystals. Frequent application of Auger electron spectroscopy during the final polishing–cleaning–annealing stage (and afterward) defines the cleanliness and composition of the surface and, indirectly, of the vacuum system.

1.3 IMMERSION OF SOLID SURFACES

Having formed a well-defined surface in UHV, the next task is to put the surface in contact with vapor and solution without accidental contamination. The inside surfaces of the apparatus employed to contain the crystal, solution, and gaseous environment must themselves be ultra-clean. UHV-compatible materials and operating procedures must be employed for all interior components and manifolds. A corollary of surface cleanliness in UHV is that the inert gases and reagents must be free of chemisorption agents to better than about 10^{-10} Torr. This criterion is not entirely practical as it requires that the inert atmosphere, solvent, and other principal reagents be free of chemisorptive impurities to better than one part in 10^{13}. In practice, the actual requirements have often proved to be less severe than this because the various contents of the liquid phase, especially the intended adsorbates, compete with the impurities for adsorption at the surface so as to contribute to the cleanliness of the surface.

A schematic diagram of a vacuum system for liquid–solid surface chemical and electrochemical studies, including structural investigations, is shown in Fig. 1. The sample surface under investigation is shuttled back and forth between UHV and various solutions at ambient pressure without

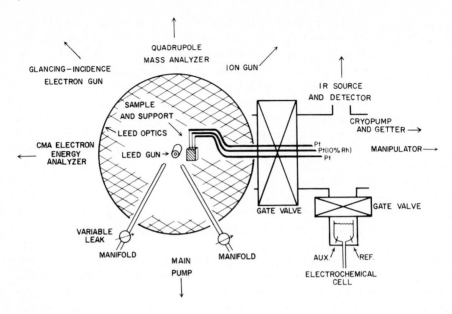

Fig. 1. Diagram of a surface electrochemistry instrument.

delays or accidental contamination. The apparatus consists of three vacuum chambers: a surface spectroscopy chamber (LEED, Auger, XPS, EELS, TDMS, and related procedures); a plenum joining the several subsystems of pumps; and a compact chamber which houses the solution cell/electrochemical cell, a part of the infrared reflection absorption experiment, and the mechanism for transferring the sample. Motion such as translation of the sample between chambers is introduced by means of stainless steel "welded" bellows which have the advantage of a large difference in expanded and compressed lengths (stroke) and remain leak-free throughout many years of use. If the fluid phases to be studied are typically very volatile or must be excluded from the solution-cell to an unusual degree, then it is advisable to design the transfer apparatus such that the sample can be detached and isolated from the solution-cell chamber; otherwise, a design in which the sample is translated/rotated from place to place without having to be disconnected is much more convenient.

Four pumping subsystems are required to manage the wide swings in pressure from atmospheric to 10^{-10} Torr while at the same time avoiding the introduction of possible contaminants such as pump oil or lubricants: two sorption pumps (zeolite beads at liquid nitrogen temperature) are used sequentially to lower the pressure from atmospheric (preferably nitrogen) to about 10^{-3} Torr; a cryogenic pump (activated charcoal at 14 K) lowers the pressure from 10^{-3} to about 10^{-7} Torr; ionization pumps are then employed from 10^{-7} Torr to about 10^{-9} Torr; titanium sublimation "getter" pumping diminishes residual gas levels such as CO during ion bombardment of the sample surface with Ar^+ ions, the ion pump being de-energized while Ar is present at elevated pressure (about 5×10^{-5} Torr).

Vacuum valves for UHV experiments are bellows-actuated. Where samples and other objects must be transferred through an opening, gate valves are preferred. Metal sealed valves are employed otherwise.

The entire vacuum system, including all components and accessories contacting the vacuum, must be baked at a uniform temperature of 200°C in order to remove adsorbed impurities from the inside surfaces.

A function of the vacuum chamber for surface spectroscopy is convenient placement of the sample surface at the focal points of the various spectrometers and at appropriate points for ion bombardment, immersion, and electrolysis. A sample manipulator for this purpose typically provides rotation about the axis of the cylindrical vacuum chamber with the sample offset 2.5–6 in. from the axis. By arranging the focal points of the spectrometers (LEED, Auger, XPS, etc.) on a circle of radius equal to the offset, the sample reaches the focal points by means of this single rotation. Short translations (± 0.5 in.) in Cartesian coordinates (X, Y, Z) permit fine adjustment of sample position. A coaxial rotation about an axis parallel to the sample surface allows exact to normal or other angles of incidence or emission, as well as alteration between front and back surfaces of the sample. All motions are bellows-activated. Flexible (braided) electrical connections to the sample allow electrical heating of the sample, and measurement of particle beam currents as well as electrolytic current.

A separate long-translation axis allows the sample to be unplugged from the spectroscopy manipulator and reconnected to a rotational feedthrough in the solution-cell chamber. This translation is typically about 12 in., actuated by an automatic mechanism through a small-diameter bellows. To maintain satisfactory cleanliness, this cannot involve any type of sliding seal.

The solution-handling subsystem must exclude dissolved contaminants and air. Inclusion of electrodes for measurement and control of potential and current is advisable because solid–liquid surface processes depend upon interfacial potentials, regardless of whether the objectives of the study are electrochemical. Potential-dependent interfacial properties include oxidation state, electron density, structure, chemical composition, ionic distribution, and hydrophilicity. Electrode potential is measured with respect to a reference electrode consisting of a short section of 3 mm glass tubing loosely capped on the ends with Teflon; a silver wire coated with silver chloride immersed in chloride solution is a convenient reference half-cell. The auxiliary electrode consists of an open section of 3 mm Teflon tubing containing a platinum wire. A conventional electronic potentiostat applies current between the sample surface and the auxiliary electrode to control the interfacial potential, which is measured between the sample and the reference electrode. Sample, reference, and auxiliary are contained in a Pyrex or Teflon vessel of about 10 ml volume, filled and drained through a tube at the bottom. Liquids are contained in deaerated (N_2) Pyrex bottles. Air is the most difficult impurity to exclude from the solutions. Diffusion of O_2, CO_2, and trace ingredients of air through the interconnecting Teflon tubing and into solutions and inert atmosphere is decreased to acceptable levels by a double-wall construction in which the Teflon tubes are enclosed by Teflon bellows continually purged with inert gas. Vacuum valves are used (rather than stopcocks) to avoid contact with the air. Short lengths of Pyrex glass tubing are used at points where visibility is required. Teflon compression fittings are employed for tubing interconnection. All parts are soaked in chromic acid solution until needed. Final rinsing is with distilled water pyrolyzed through a Pt gauze catalyst in pure O_2 at about 800°C and distilled once more. This pyrolytically distilled water [20] is also employed to prepare the aqueous solutions.

1.4 LOW-ENERGY ELECTRON DIFFRACTION (LEED)

The deBroglie wavelengths, λ, of electrons traveling with kinetic energies, KE, in the range from 20 to 200 eV are slightly smaller than typical interatomic distances and thus appropriate for diffraction from surfaces

$$\lambda = \left(\frac{150}{KE}\right)^{1/2} \tag{1}$$

where the wavelength, λ, is expressed in Ångstroms. Likewise, the mean free path of these "low-energy" electrons is about two atomic layers, such that low-energy electron diffraction is very surface-sensitive. The theory and practice of LEED are discussed in a number of recent books and reviews [11]. A diagram of the LEED experiment appears in Fig. 2. Electrons emitted from

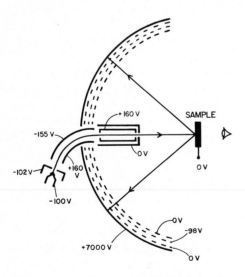

Fig. 2. Diagram of the LEED experiment.

an off-axis filament (to minimize stray light) are collimated to a beam about 1 mm in diameter. The beam strikes the sample at near normal incidence. About 1% of the incident electrons are reflected without exchange of kinetic energy and are diffracted. All reflected electrons travel in a field-free region toward the first concentric grid, which is grounded. Careful cancellation of stray magnetic fields by means of three orthogonal pairs of magnetic coils, or by magnetic shielding of the optics and sample, greatly improves the quality of LEED observations, particularly in the important 20–60 eV range; the improvement in beam current and lessened pattern distortion are very noticeable.

The second and third grids are polarized to repel all electrons not reflected elastically. The final grid is grounded. Finally, the diffracted electrons are accelerated to the concentric phosphor display screen, which is polarized at an adjustable positive high voltage, converting the electrons to visible light. When the principal objective is to identify the symmetry of the surface from the positions of beams at various energies, the LEED patterns can simply be recorded on fast black and white film. Patterns should be photographed at several energies, including an energy near 15 eV to bring out beams which might otherwise be lost near the center of the display; an energy near 60 eV to show the integral-index beams; an energy near 120 eV to display the second-order integral beams; and a few intermediate energies at which the various functional-index beams reach optimum intensity. On the other hand, if the uses to be made of the data include quantitative comparison of measured and calculated intensities as a function of energy, then the LEED pattern is recorded at 1 eV intervals by means of a video camera and digital data system.

Surface structures are deduced from LEED data by comparing the data with calculations based upon plausible model structures. Comparatively simple cases such as a monolayer of atoms, ions, or small molecules of a single type of an elemental single-crystal substrate can often be solved from the symmetry of the LEED pattern and the packing density of the monolayer, provided that the unit mesh (the surface unit cell) contains only a few molecules. More complicated cases require intensity–energy measurements and calculations as described in refs. 11(a)–(c). Vectors $(\mathbf{A}, \mathbf{B})$ defining the locations of LEED spots on the film plane in relation to the specular (direct-reflection) spot are related to the vectors defining the unit mesh of the surface layer structure as follows.

$$\mathbf{A} \cdot \mathbf{a} = \cos\theta_{\mathrm{Ii}} = |A||a| \cos\theta_{\mathrm{Aa}} \tag{2}$$

$$\mathbf{B} \cdot \mathbf{b} = \cos\theta_{\mathrm{Ii}} = |B||b| \cos\theta_{\mathrm{Bb}} \tag{3}$$

$$\mathbf{A} \cdot \mathbf{b} = 0 = \mathbf{B} \cdot \mathbf{a} \tag{4}$$

where $\mathbf{I}$ and $\mathbf{i}$ are the unit cell vectors of the substrate LEED pattern and substrate structure, respectively. That is, the lengths $|A|$ and $|B|$ of the LEED vectors are inversely proportional to the structure vector lengths $|a|$ and $|b|$.

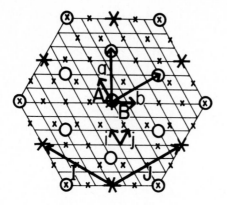

Fig. 3. Relationship between unit cell (mesh) vectors of a LEED pattern and unit mesh vectors of the corresponding surface structure. × = Integral-index LEED spots (due to substrate); × = fractional-index LEED spots (due to overlap); ○ = $(2\sqrt{3} \times 2\sqrt{3})R30°$ overlayer lattice (mesh); Grid = substrate mesh; **A, B** = unit cell (mesh vectors of LEED pattern; **a, b** = unit mesh vectors of overlayer structure; **I** = unit mesh vectors of substrate LEED pattern; and **i** = unit mesh vectors of substrate surface.

LEED vector **A** is perpendicular to structure vector **b**; likewise, **B** is perpendicular to **a**. These relationships are illustrated with a typical example in Fig. 3. Distances are commonly expressed in units of the substrate unit mesh length. The directions of **a** and **b** are referred to the directions of the unit mesh vectors. For the structure in Fig. 3, $|a| = |b| = 2\sqrt{3}$ unit mesh and the angle between **a** (or **b**) and a nearby unit mesh vector is 30°.

This structure is commonly denoted $(2\sqrt{3} \times 2\sqrt{3})R30°$ [11]. Alternately, this structure can be expressed in matrix notation, as illustrated with reference to Fig. 3, by the following equations, where **i** and **j** are the unit mesh vectors of the substrate.

$$\mathbf{a} = \mathbf{a}_1\mathbf{i} + \mathbf{a}_2\mathbf{j} = 2\mathbf{i} + 2\mathbf{j} \tag{5}$$

$$\mathbf{b} = \mathbf{b}_1\mathbf{i} + \mathbf{b}_2\mathbf{j} = 2\mathbf{i} + 4\mathbf{j} \tag{6}$$

$$\begin{pmatrix} \mathbf{a} \\ \mathbf{b} \end{pmatrix} = \begin{pmatrix} \mathbf{a}_1 & \mathbf{a}_2 \\ \mathbf{b}_1 & \mathbf{b}_2 \end{pmatrix} \begin{pmatrix} \mathbf{i} \\ \mathbf{j} \end{pmatrix} = \begin{pmatrix} 2 & 2 \\ -2 & 4 \end{pmatrix} \begin{pmatrix} \mathbf{i} \\ \mathbf{j} \end{pmatrix} \tag{7}$$

That is, $(2\sqrt{3} \times 2\sqrt{3})R30°$ is also denoted as $\begin{pmatrix} 2 & 4 \\ -2 & 4 \end{pmatrix}$. Matrix notation is particularly convenient for describing relatively complicated structures.

The four common types of unit mesh are hexagonal, square, rectangular, and oblique. Figure 4 illustrates the manner in which typical overlayer unit meshes up to (6 × 6) originate on an hexagonal substrate mesh.

There are numerous possible combinations, since unit meshes can be formed from pairs of vectors of unequal length (such as $2\sqrt{3} \times \sqrt{13}$) and

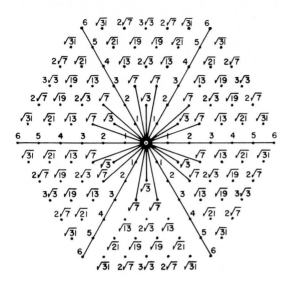

Fig. 4. Lengths and directions of unit mesh vectors at a hexagonal substrate surface (to 6 × 6).

inequivalent direction. There are multiple possible combinations (of directions) even of vectors having the same length. For example, there are three possible ($\sqrt{7} \times \sqrt{7}$) unit meshes, namely, one hexagonal mesh and two oblique meshes, as illustrated in Fig. 5. In addition, there are multiple ways in which the individual atoms or molecules can be placed within the unit mesh, the only requirements being that they must be located in such a way that the environment of each mesh point is equivalent and the locations are consistent with the size requirements of the atoms or molecules. This situation is illustrated for typical structures of a monoatomic adsorbate in a ($\sqrt{7} \times \sqrt{7}$)$R19.1°$ unit mesh (Fig. 6). Included are four hexagonal structures at packing densities of $\Theta = 1/7$, $3/7$, $4/7$, and $9/7$ adsorbed atoms per substrate surface atom and one oblique structure at $\Theta = 6/7$. All of these structures except the primitive lattice ($\Theta = 1/7$) contain adsorbed atoms in two or more types of sites. Structures for which $\Theta = 2/7$, $5/7$, or $8/7$ are

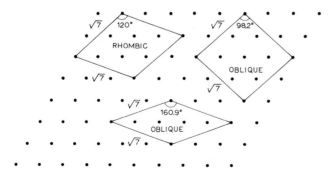

Fig. 5. Three possible ($\sqrt{7} \times \sqrt{7}$)$R19.1°$ meshes at a hexagonal substrate surface.

References pp. 64–67

12

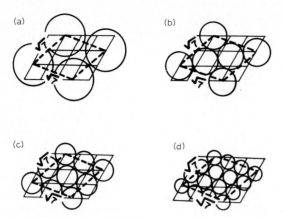

Fig. 6. Four hexagonally close-packed ($\sqrt{7}$ × $\sqrt{7}$)$R19.1°$ structures at a hexagonal substrate surface. (a) Packing density, Θ = 1/7; (b) Θ = 3/7; (c) Θ = 4/7; (d) Θ = 9/7.

considered unlikely because they would require a large departure from hexagonal packing and very dissimilar interatomic distances. At Θ = 7/7 = 1, the structure would be essentially (1 × 1). Structures at Θ = 9/7 and higher packing densities are not common as there are very few adsorbates for which the size is sufficiently small and the strength of adsorption is sufficiently large to permit a packing density greater than that of the substrate surface, while $\sqrt{7}$ structures having Θ = 3/7 or 4/7 are the most common.

The practice of expressing packing density in terms of adsorbed particles per substrate surface atom will be followed throughout this article.

$$\Theta_X \quad = \quad \frac{\Gamma_X(\text{moles X cm}^{-2})}{\Gamma_s(\text{moles substrate surface atoms cm}^{-2})} \tag{8}$$

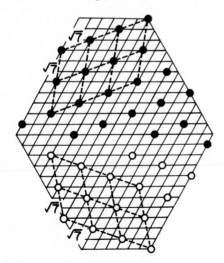

Fig. 7. Two non-superimposable (7 × 7)$R19.1°$ structural domains.

This will avoid the ambiguities which can exist in the classical coverage, θ, which is referenced to an assumed or measured limiting packing density, $\theta = \Gamma_X/\Gamma_X$ ("limiting").

For many of the possible unit meshes there exist alternative meshes which, although equivalent energetically are not superimposable, as illustrated in Fig. 7. Each of these non-superimposable structures yields a set of non-superimposable LEED beams. All possible LEED beams of this type are generally seen; that is, the fact that LEED patterns are generally observed to contain all beams due to the analogous alternative relational alignments of a given overlayer structure is a direct experimental indication that the surface layer tends to contain (within the 1 mm diameter of the incident beam) patches having each of these rotational alignments. Since the alternative alignments do not differ in energy, the choice of alignment of each patch is arbitrary and a random mixture of alignments is to be expected provided that layer growth begins independently at numerous points on the surface.

In some cases, there exist alternative meshes which differ only slightly in energy, giving rise to antiphase domains separated by phase shifts between non-equivalent, energetically similar binding sites. Figure 20(b), below, gives an example in which the phase shift is between the two three-fold sites of a rhombic substrate mesh. On the surfaces of close-packed solids, these sites differ due to the presence or absence beneath those sites of an atom of the next layer. Antiphase domains give rise to multiple LEED spots in the vicinity of beams characteristic of the local (short-range) coordinates of the surface layer (see Fig. 19, below). Antiphase domains also give rise to regions of uneven interatomic distances, as illsutrated by Fig. 7. Therefore, it is not surprising that substantial evidence has accumulated that surface processes tend to occur preferentially at domain boundaries: electrodeposition and dissolution; electrochemical oxidation–reduction of adsorbed molecules; and surface oxide layer formation and dissolution are examples, as described below.

1.5 AUGER ELECTRON SPECTROSCOPY

Auger spectroscopy is an important technique in structural investigations [12]: all elements except H and He can be detected, usually with limits of detection well below the lowest levels which significantly influence surface behavior; elemental Auger signals are readily measured from which atomic, ionic, and molecular packing densities can be obtained.

Auger spectroscopy of surfaces is illustrated in Fig. 8. A collimated electron beam is directed against the surface at 2000–10 000 eV kinetic energy. Selection of the incident energy is often made on the basis that the cross-section for ionization of a given atomic level of the sample is a maximum when the incident energy is about four times the binding energy [21]. In order to avoid beam damage to molecular layers and other non-metallic samples, the incident beam current must be less than about 10^{-7} A mm^{-2} and

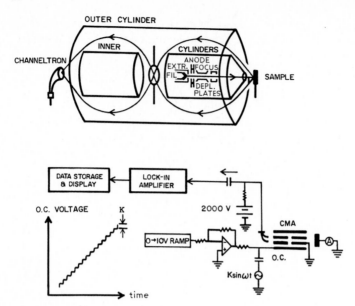

Fig. 8. Diagram of the Auger spectroscopy experiment.

for quantitative purposes regulated to within $\pm 1\%$; this will require modification of some conventional electron gun electronic controls. While some advocate the use of grazing incidence to increase Auger signal from the surface, this is not advisable in our experience; measurement of the beam current actually striking the sample surface is more difficult at non-normal angles, while the signal/noise ratio (which is the actual limiting consideration) is not affected appreciably by the incidence angle in most instances. The Auger process involves initial ionization of a core level, followed by capture of an outer electron and ejection of another outer electron. The kinetic energy corresponds sufficiently closely for most purposes to the simple equation

$$KE = \frac{E_1 - [E_2(Z) + E_2(Z + 1)]}{2} - \frac{[E_3(Z) + E_3(Z + 1)]}{2} \qquad (9)$$

where E_1 is the binding energy of the initially ionized level, $E_2(Z)$ and $E_3(Z)$ are the binding energies of the outer levels in the transition (Z identifies the atomic number), and $E_2(Z + 1)$ and $E_3(Z + 1)$ are the binding energies of the same atomic levels of the next higher element ($Z + 1$). Coghlan and Clausing [22] have prepared a very useful tabulation of atomic levels, Auger electron kinetic energies, and relative intensities (very approximate).

Packing densities, $\Gamma_X(\text{mol cm}^{-2})$, of elements, X, in a molecular layer can be found from the elemental Auger signals, I_X, by means of an equation of the form

$$\frac{I_X}{I°} = B_X \Gamma(f_i L_i + \ldots + f_N L_N) \tag{10}$$

where L_i is the fraction of atoms of element X located at level i in the surface layer ($i = 1$ is adjacent to the solid and N is outermost); f_i is the attenuation factor for element X in layer i (f_i is approximately 0.7 for C, N, and O); B_X is a constant which can be conveniently determined using a reference compound of element X; and $I°$ is an Auger signal from the clean substrate which serves as a convenient normalization constant to eliminate the need for frequent recalibration of the spectrometer. However, care must be taken to reproduce electronic control parameters and the sample position in relation to the analyzer. The usefulness of Auger spectroscopy for packing density measurements is further strengthened by an alternative method for the determination of packing density based upon attenuation of a *substrate* Auger signal, I, by an adsorbed layer:

$$\frac{I}{I°} = (1 - J_1 K \Gamma) \ldots (1 - J_i K \Gamma) \ldots (1 - J_N K \Gamma) \tag{11}$$

where J_i is the number of non-hydrogen atoms located at each level in the surface layer (N atoms per molecule) and K is an attenuation factor ($cm^2 mol^{-1}$). For Pt Auger signal at 161 eV attenuated by a hydrocarbon layer, $K = 1.60 \times 10^8 \, cm^2 mol^{-1}$. Examples and references are given in subsequent pages.

1.6 ELECTRON ENERGY-LOSS SPECTROSCOPY

Electron energy-loss spectroscopy (EELS) is an essential tool for surface investigations [23]: vibrational modes of all adsorbed species can be observed; detection limits are excellent and are usually well below the lowest levels encountered in typical surface processes.

The experimental aspects of EELS [24] are illustrated in Fig. 9. A carefully monochromated and focused beam is directed to the sample surface at an angle near 62° from the surface normal. The reflected electrons are energy-analyzed to detect energy losses resulting from interaction with the surface

$$KE = KE° - h\nu \tag{12}$$

Typical surface vibrational modes occur in the far- and mid-infrared region from 100 to 4000 cm^{-1} at an incident beam energy of 4 eV (80 cm^{-1}). Accordingly, all vibrational transitions above about 100 cm^{-1} are readily detected. Although this routinely achievable 80 cm^{-1} instrumental line width is about a factor of five greater than what is desired, it is sufficient to yield richly informative vibrational spectra of typical surface molecules. As such, EELS is emerging as a very important surface molecular structure probe. Along with its other interesting features, EELS differs in an interesting way from infrared spectroscopy: while IR responds most strongly to polar functional

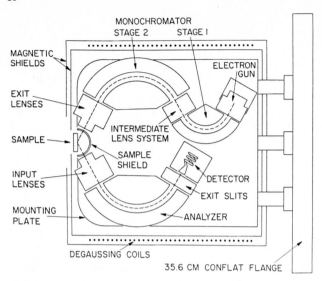

Fig. 9. Schematic of a high resolution electron energy loss spectrometer. Reprinted from ref. 24.

groups such as OH and NH, EELS operates primarily by a short-range impact scattering mechanism such that EELS tends to respond to all vibrational bands about equally. Comparison of EELS spectra detected at specular and off-specular angles provides some further delineation. Study of a given sample by EELS and surface IR will generally prove to be particularly definitive of the nature of surface molecular vibrations, molecular constitution, intramolecular interactions, and mode of attachment to the surface.

2. Structure of layers formed from ionic solutions

2.1 HALIDES

Immersion of Pt(111) and Ag(111) surfaces into aqueous ionic solutions at controlled pH and electrode potential results in the formation of a highly ordered chemisorbed layer, an adlattice [25, 26]. For example, the Pt(111) surface was examined by LEED and Auger spectroscopy after immersion into aqueous KBr and CaBr$_2$ solutions [27]. The pH and electrode potentials are both important variables in regard to surface layer structure and composition: pH was controlled at 4, 6, 8 and 10; potential was controlled at 50 mV increments from the negative limit due to H$^+$ reduction to the positive limit of O$_2$ evolution. Contrary to the traditional suppositions [28], the adsorbed particles are present primarily in the form of *neutral atoms* (Br) rather than as anions (Br$^-$). The adsorption process is

$$\text{Br}^- \underset{}{\overset{\text{Pt surface}}{\rightleftarrows}} \text{Br(adsorbed)} + \text{e}^- \tag{13}$$

That is, adsorption of Br$^-$ at Pt surfaces is a redox process. The reverse of eqn. (13), reductive desorption of Br as Br$^-$ anions, occurs at potentials more negative than -0.1 V vs. the Ag/AgCl (1 M KCl) reference). The onset of Br$^-$ desorption during a negative-going potential scan was signaled by a sharp, pH-independent current spike due to a structural transition within the Br layer.

$$\text{Pt(111)}(3 \times 3)\text{--Br}[\Theta_{\text{Br}} = 4/9] + \chi \text{e}^- \tag{14}$$

$$\rightleftarrows \text{Pt(111)}(4 \times 4)\text{--Br}[\Theta_{\text{Br}} = 7/16] + \chi \text{Br}^-$$

where Θ is the packing density of Br (atoms per surface Pt atom) and $\chi = 4/9 - 7/16 = 0.007$ electrons and Br$^-$ anions per surface Pt atoms, a very small proportion. [For the purpose of relating solid–liquid surface structure to the corresponding gas–solid behavior, it should be noted that the Pt(111)(3 × 3)–Br adlattice was also formed by saturation of Pt(111) with HBr vapor in vacuum [29].] Adsorption of Br was weaker in alkaline than in acidic solutions due, at least in part, to competing adsorption of OH.

$$2\,\text{Pt} + \text{OH} \rightleftarrows \text{Pt}_2\text{OH} \cdot \text{e}^- \tag{15}$$

$$\text{Pt}_2\text{OH} + 3\,\text{OH}^- \longrightarrow 2\,\text{Pt(OH)}_2 + 3\,\text{e}^- \tag{16}$$

The first stage of surface oxidation was virtually reversible and retained the ordered structure of the Pt(111) substrate to $\Theta_O = 1/2$. In the second stage of surface oxidation, both the O layer and the Pt surface were disorderd irreversibly. The limit was approximately $\Theta_O = 2$. An isotherm showing Br packing density versus pH and electrode potential is shown in Fig. 10. In contrast, cation packing densities displayed a *minimum* where Br packing density was a maximum and cation densities were 15–30-fold smaller than the Br density at that point (Fig. 11). No structural differences attributable to cations such as K$^+$, Cs$^+$, Ca^{2+}, or Ba^{2+} were observed by LEED. It is perhaps noteworthy in connection with the wetability of surfaces that, when Ca^{2+} ions were present, the surface retained water in vacuum to the extent of 5–15 water moloecules per Ca^{2+} ion, the upper value applying at pH $= 10$, the most alkaline studied. However, K$^+$ and Cs$^+$ ions, being less strongly hydrated, did not retain detectable amounts of water in vacuum. Due, at least in part, to the presence of these hydrated cations, the Pt surface was hydrophilic in Br$^-$ solutions under all conditions studied.

Immersion of the stepped Pt(s) [6(111) × (111)] surface into aqueous Br$^-$ solutions [30] and subsequent examination of the surface by Auger spectroscopy and LEED revealed that Br has no particular affinity for surface steps. Instead, the differences in adsorption behavior of the stepped and smooth (111) surfaces are due to the fact that different long-range adsorbed layer

18

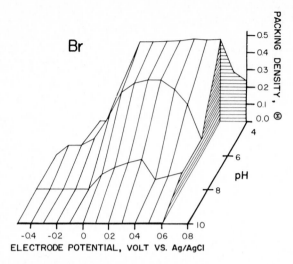

Fig. 10. Isometric projection of the adsorption profiles of Br (0.05 mM CaBr$_2$) at Pt(111). Reprinted from ref. 27.

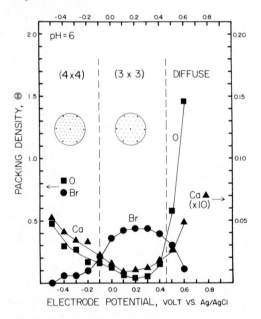

Fig. 11. Adsorption profiles of Br, Ca and O from 0.05 mM CaBr$_2$ (pH = 6) at Pt(111) as a function of electrode potential (Ag/AgCl/1 M KCl reference). Reprinted from ref. 27.

structures (different adlattices) are formed at the stepped and smooth surfaces.

When immersed in KI solutions [31] at controlled pH and electrode potential, Pt(111) acquires an ordered layer of iodine atoms.

$$I^- \underset{\text{Pt surface}}{\rightleftharpoons} I(\text{adsorbed}) + e^- \qquad (17)$$

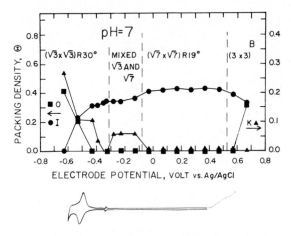

Fig. 12. Adsorption profiles of I, K and O from 0.1 mM KI (pH = 7) at Pt(111) as a function of electrode potential (Ag/AgCl/1 M KCl reference). Reprinted from ref. 31.

The structure of the adsorbed layer was potential-dependent, as shown in Fig. 12. Starting from positive potentials (but below the range where oxidation of the surface occurred) the adlattice structure was Pt(111)(3 × 3)–I. An hexagonal layer of iodine atoms containing $\Theta_I = 4/9$ I atoms per surface Pt atom. At potentials in mid-range, the adlattice structure was Pt(111)($\sqrt{7} \times \sqrt{7}$)$R19°$–I [$\Theta_I = 3/7$], while at moderately negative potentials, the layer reconstructed to Pt(111)($\sqrt{3} \times \sqrt{}$)$R30°$–I [$\Theta_I = 1/3$]. At extremely negative potentials, reductive desorption in the form of I⁻ ions occurred, the reverse of eqn. (17). The pH dependence of iodine adsorption was relatively slight, probably due to the ability of iodine to suppress adsorption of OH⁻ and related oxygen species.

A remarkable characteristic of iodide-treated Pt(111) surfaces is that the Pt(111)(3 × 3)–I and Pt(111)($\sqrt{7} \times \sqrt{7}$)$R19°$–I adlattices are strongly *hydrophobic* while the Pt(111)($\sqrt{3} \times \sqrt{3}$)$R30°$–I adlattice is hydrophilic. Evidently, the close-packed (3 × 3) and ($\sqrt{7} \times \sqrt{7}$) atomic layers are incompatible with water molecules, while the ($\sqrt{3} \times \sqrt{3}$) adlattice does not fully block the Pt surface. LEED patterns and structures for adlattices formed by immersion of Pt(111) in aqueous KI solutions are shown in Figs. 13 and 14.

The limiting packing density of I atoms at Pt(111) under gas–solid conditions ($\Theta_I = 0.62$) corresponded to a Pt(111)($9\sqrt{3} \times 3\sqrt{3}$)$R30°$–I oblique adlattice which formed when Pt(111) was heated with solid I_2 [32].

Adsorption of I_2 vapor onto a stepped Pt surface, Pt(s) [6(111) × (111)], displayed no particular affinity of I atoms for step sites of the surface [33]. The steps resulted in iodine atomic layer structures having (3 × 3) or ($\sqrt{3} \times \sqrt{3}$)$R30°$ local geometry and repeat distances spanning one or several terraces. Examples are Pt(s)[6(111) × (111)](3 × T)–I; Pt(s)[6(111) × (111)](8 × T:)–I; and Pt(s)[6(111) × (111)](20 × $2T$)–I.

When Pt(111) was immersed at controlled potential in solutions containing less than 10 mM Cl⁻, an ordered layer did not form [34]. Only when the

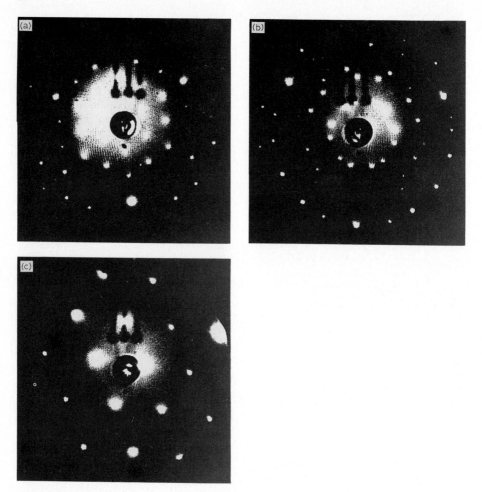

Fig. 13. LEED patterns after immersion of Pt(111) into 0.1 mM HI. (a) Pt(111)(3 × 3)–I. Electrode potential = 0.4 V vs. Ag/AgCl; beam energy = 83 eV; (b) Pt(111)($\sqrt{7}$ × $\sqrt{7}$)R19.1°–I. Electrode potential = −0.2 V; beam energy = 73 eV; (c) Pt(111)($\sqrt{3}$ × $\sqrt{3}$)R30°–I. Electrode potential = 0.34 V; beam energy = 67 eV. Reprinted from ref. 31.

Cl^- concentration exceeded 10 mM was the adsorption sufficiently strong to form the Pt(111)(3 × 3)–Cl (θ_{Cl} × 0.25) layer (Fig. 15).

$$Cl^- \underset{}{\overset{Pt\ surface}{\rightleftharpoons}} Cl(adsorbed) + e^- \tag{18}$$

HCl did not form an ordered layer at Pt(111) in vacuum, although it did so at Pt(100) [29]. Evidently, Cl adsorption is much weaker at Pt(111) than that of Br or I (Fig. 16). This can also be seen in that the packing density of Cl was only about half that for a hexagonally close-packed adlattice (Θ_{Cl} = 0.5).

Retention of Ca^{2+} by the surface varied with potential and was two times

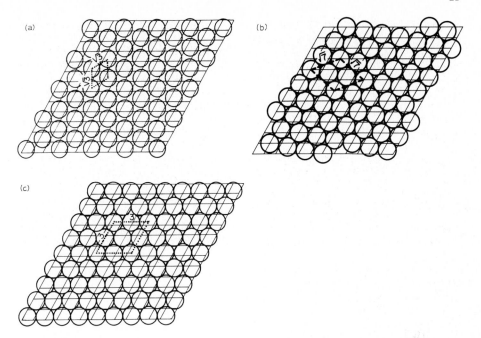

Fig. 14. Model structures. (a) Pt(111)(3 × 3)R30°–I, Θ_I = 1/3; (b) Pt(111)(7 × 7)R19°–I, Θ_I = 3/7; (c) Pt(111)(3 × 3)–I, Θ_I = 4/9. Reprinted from ref. 31.

more dependent on potential than was observed for $CaBr_2$ under similar conditions. The limiting coverage of Ca^{2+} depended on the amount that the Ca^{2+} was hydrated (from 10 to 20 H_2O molecules per Ca^{2+} ion) which varied with pH.

Fluorine is not strongly adsorbed at Pt [35], so surface properties were dominated primarily by H^+ and OH^- adsorption in F^- aqueous solutions.

A well-defined Ag(111) surface has been studied by Auger spectroscopy and LEED following immersion in aqueous halide solutions [26]. There was no adsorption of F^- or F detectable by Auger spectroscopy. Dissolution of Ag in aqueous KF/HF electrolytes resulted in a surface state only slightly less well-ordered than that produced by Ar^+ ion bombardment and annealing in UHV.

Strong adsorption of Cl, Br, or I occurred when Ag(111) was immersed in aqueous KCl, KBr, or KI solutions, respectively, at open circuit and throughout most of the useful range of electrode potentials. As for Pt surfaces, halide adsorption on Ag(111) was a redox process.

$$X^- \overset{\text{Ag(111) surface}}{\rightleftharpoons} X(\text{adsorbed}) + e^- \quad (X = Cl, Br, I) \tag{19}$$

Oxidative adsorption and reductive desorption of Cl, Br, or I [eqn. (19)] gave rise to a very broad feature spanning much of the accessible potential range. Iodide was removed only partially after several minutes at potentials where hydrogen is generated. At potentials in mid-range, Cl and Br formed simple

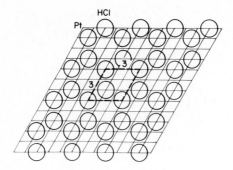

Fig. 15. Model structure Pt(111)(3 × 3)–Cl, Θ_{Cl} = 2/9. Reprinted from ref. 34.

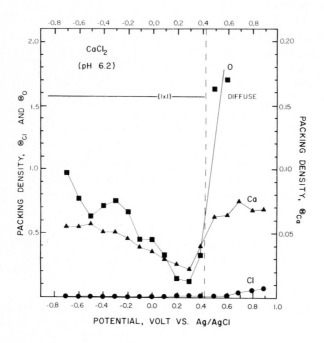

Fig. 16. Adsorption profiles of Cl, Ca, and Q from 0.05 mM $CaCl_2$ (pH = 6.2), 0.1 mM acetate buffer) at Pt(111) as a function of electrode potential (Ag/AgCl/1 M KCl reference). Reprinted from ref. 34.

Ag(111) ($\sqrt{3} \times \sqrt{3})R30°$–X adlattices ($\Theta_X$ = 1/3), while iodide yielded some of the same complex structures as for Ag electrodeposition at iodine-pretreated Pt(111) [see Fig. 30(b) and 20(c), below]. That is, the Ag(111) surface is reconstructed in iodide solutions over most or all of the accessible potential range. A polycrystalline separate AgX phase (X = Cl, Br, I) begins to form in X solutions as the potential approaches that of the corresponding Ag/AgX couple.

2.2 CYANIDE AND THIOCYANATE

Immersion of a well-defined Pt(111) surface in aqueous KCN or KSCN solutions formed an ordered layer of chemisorbed species of composition resembling that of the parent ions (CN^- or SCN^-). The structure of the CN layer was found to be potential-dependent [36, 37]: at potentials more positive than 0.45 V, the structure was Pt(111)$(2\sqrt{3} \times 2\sqrt{3})R30°$–CN with $\Theta_{CN} = 7/12$, while at potentials more negative than -0.45 V, the adlattice was Pt(111)$(\sqrt{13} \times \sqrt{13})R14°$–CN ($\Theta_{CN} = 7/13$). The SCN layer structure was Pt(111)(2×2)–SCN ($\Theta_{SCN} = 1/2$); the potential-dependence of the SCN layer structure has not been investigated.

The acid–base behavior of these ordered adsorbed CN and SCN layers was explored by measuring (by means of Auger spectroscopy) the Cs or Ca packing densities resulting from rinsing the coated surfaces with CsCl or CaCl$_2$ solutions adjusted at intervals of about 0.5 pH units from pH = 4 (0.1 mM H^+) to pH = 10 (0.1 mM OH^-). Rinsing ion-exchanged Cs^+ or Ca^{2+} ions for the original K^+ counterions, but did not change the LEED pattern. The cation packing density was independent of electrode potential, apart from potentials where oxidation of the surface took place. The Cs^+ packing density varied with pH (Fig. 17) in a manner reminiscent of a polybasic acid. That is, the first stage of ionization occurred readily, as for a strong acid. The Cs^+ packing density corresponded to an array of non-adjacent ion pairs [Fig. 18(a)]. The final stage of ionization proceeded readily above pH = 10 and corresponded in Cs packing density to adjacent ion pairs. In contrast, Ca^{2+} ions gave rise to a single plateau above pH = 3 (Fig. 17) corresponding to one Ca^{2+} ion per $(2\sqrt{3} \times 2\sqrt{3})R30°$ unit mesh [Fig. 18(b)]. This is evidently due to the large effective radius of the strongly hydrated Ca^{2+} ion [27].

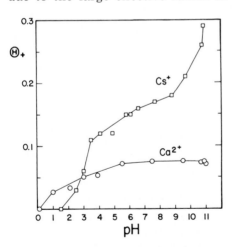

Fig. 17. Cation packing density vs. pH at Pt(111)$(2\sqrt{3} \times \sqrt{3})R30°$–CN. Experimental conditions: the CN layer was immersed into 0.1 mM Cs^+ or Ca^{2+}; pH was adjusted by means of HCl, CsOH, or Ca(OH)$_2$; the electrode potential was 0.1 V (Ag/AgCl/1 M KCl reference). Reprinted from ref. 36.

(a)

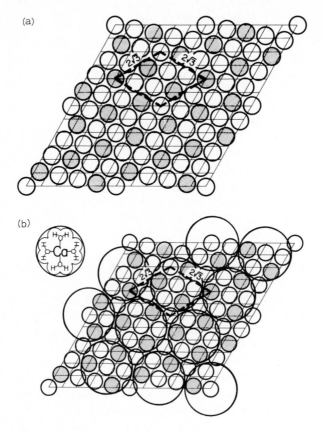

(b)

Fig. 18. Surface structures. (a) Pt(111)(2$\sqrt{3}$ × 2$\sqrt{2}$)$R30°$–CN, Cs; (b) Pt(111)(2$\sqrt{3}$ × 2$\sqrt{3}$)$R30°$–CN, Ca. Shaded circles: CN$^-$ anions, open circles, neutral CNH. Reprinted from ref. 36.

Consistent with this explanation, exactly analogous acid–base behavior was manifested by the Pt(111)(2 × 2)–SCN adlattice.

3. Structure of electrodeposited layers

3.1 ELECTRODEPOSITION AND LEED

Experiments have been performed in which a well-characterized single-crystal surface was immersed in electroplating solution, plated with a fraction of a monolayer to several hundred monolayers of metal, and then examined by LEED, Auger spectroscopy, and related techniques [32, 38–49]. The initial results have been very revealing and work is continuing.

Electrodeposition of Ag has been studied at Pt(111) surfaces pretreated with I$_2$ vapor to form a Pt(111)($\sqrt{ }$ × $\sqrt{7}$)$R19°$–I adlattice of iodine atoms which protected the Pt and Ag surfaces from attack by the electrolyte and possible residual gases. The results are illustrated by Fig. 19. Electrodeposi-

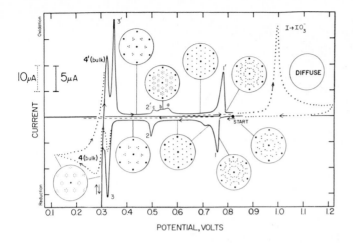

Fig. 19. Cyclic voltammogram and LEED patterns. Experimental conditions: 0.1 mM Ag$^+$ in 1 m HClO$_4$ at the Pt(111)($\sqrt{7} \times \sqrt{7}$)$R19°$–I surface; scan rate, 2 mV s^{-1}. Reprinted from ref. 39.

tion took place in four separate regions of electrode potential: three process-es at potentials more *positive* than the equilibrium potential of the Ag$^+$/Ag half-cell (underpotential deposition, UPD), followed by electrodeposition of bulk Ag [40]. It is noteworthy that the iodine layer remained attached to the surface during multiple cycles of electrodeposition and re-dissolution of Ag. Indeed, the iodine atomic layer remained *atop*: the Auger signal due to I did not decrease significantly as the amount of deposited Ag increased, while the Pt signal eventually vanished [39]; the I/Ag and I/Pt Auger signal ratios increased sharply with decreasing angle of incidence of the primary electron beam [32]. Evidently, the Ag atoms being electrodeposited are able to pass through the iodine layer during the deposition process due to covalent interatomic distances (I–Ag = 2.67 Å), which are smaller than van der Waals distances (I–I = 4.3 Å).

LEED patterns obtained at each stage of the electrodeposition process revealed interesting facts of the mechanisms of deposition: LEED patterns observed in the midst of the first UPD peak, Fig. 19, exhibited all the beams of the original Pt(111)($\sqrt{7} \times \sqrt{7}$)$R19°$–I pattern plus all the new beams of the Pt(111)(3 × 3)–Ag,I pattern from the first UPD process. Therefore, elec-trodeposition formed the Ag monolayer in patches (domains), leaving re-gions of the $\sqrt{7}$ iodine layer temporarily unaffected, until eventually the (3 × 3) Ag domains engulfed the entire surface. LEED patterns obtained during the second and third UPD processes likewise displayed this behavior. The latter two LEED patterns contained beams clustered near the (1/3, 1/3) and equivalent positions, indicating ($\sqrt{3} \times \sqrt{3}$)$R30°$ local structure with characteristic splittings of beams signaling the pressure of antiphase do-mains. Very close agreement was found between the amount of Ag elec-trodeposited and the packing density detected by Auger spectroscopy. Com-bining the LEED, Auger, and voltammetric information with theoreteical

References pp. 64–67

(a)

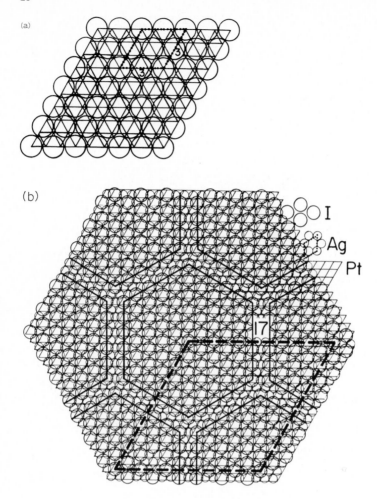

(b)

Fig. 20. Surface structures of Ag electrodeposited at iodine-pretreated Pt(111). (a) Pt(111)(3 × 3)–Ag, I, $\Theta_{Ag} = \Theta_1 = 4/9$; (b) Pt(111)($\sqrt{3} \times \sqrt{3}$)$R30°$–Ag, I with 17-unit domains, Θ_{Ag} 2/3, $\Theta_1 = 1/3$; (c) Pt(111)($\sqrt{3} \times \sqrt{3}$)$R30°$–Ag, I with 11-unit (middle layer) and 17-unit (lower layer) domains, $\Theta_{Ag} = 5/3$, $\Theta_1 = 1/3$. Reprinted from refs. 32 and 40.

prediction of the LEED patterns revealed the specific structures associated with each stage of the UPD process (Fig. 20). It can be noted that the Pt(111)(3 × 3)–Ag,I adlattice parallel to the (111) plane is the usual zinc blende structure of AgI.

Surface disorder can be removed from ion-bombarded or redox-cycled Pt single-crystal surfaces by annealing at ambient pressure in argon containing I_2 vapor [32, 41]. Familiar surface structures can be readily identified from their distinctive voltammetric behavior in the electrodeposition of Ag [32, 41].

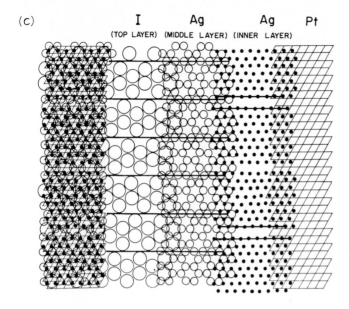

3.2 INFLUENCE OF SUBSTRATE STRUCTURE

Substrate surface structure exerts a controlling influence upon monolayer electrodeposition, as can be seen by comparing deposition of Ag at I_2-pretreated Pt(111) and Pt(100), all else being equal [41, 45] (Fig. 21). The UPD process at Pt(100)($\sqrt{2} \times \sqrt{8}$)R45°–I took place in two stages, rather than three, and the electrodeposit structures were related to the square mesh of the Pt(100) substrate in contrast to the hexagonal structures observed at Pt(111).

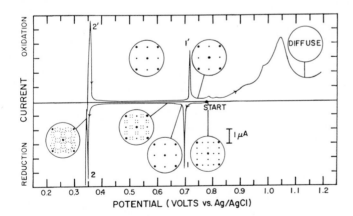

Fig. 21. Cyclic voltammogram and LEED patterns for deposition of Ag at the Pt(100) Pt(100)[$c(\sqrt{2} \times \sqrt{8})$]R45°–I adlattice. Experimental conditions: 1.0 mM Ag$^+$ in 1 M HClO$_4$; scan rate, 2 mV s^{-1}. Reprinted from ref. 45.

References pp. 64–67

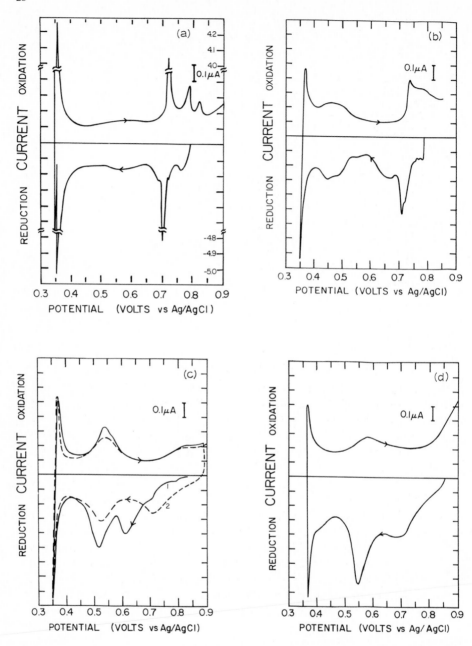

Fig. 22. Cyclic voltammograms for deposition of Ag onto iodine-pretreated, (100)-oriented Pt surfaces prepared in various ways. (a) Pt(100)[c($\sqrt{2}$ × $\sqrt{8}$)]$R45°$–I; (b) Pt(100)(1.92$\sqrt{2}$ × $\sqrt{2}$)$R45°$–I (incommensurate); (c) Pt(100)[c(5$\sqrt{2}$ × $\sqrt{2}$)$R45°$–I, first (——) and second (– – –) cycle; (d) Pt(oxidized and reduced)–I; (e) Pt(ion-bombardment)–I. Experimental conditions: 1.1 mM Ag$^+$ in 1 M HClO$_4$; scan rate, 2 mV s^{-1}. Reprinted from ref. 45.

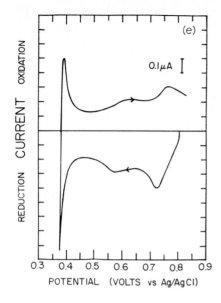

3.3 INFLUENCE OF ADSORBED LAYERS

Adsorbed layer structure and composition also control the structure of electrodeposited metallic monolayers [41, 42], as illustrated by Fig. 22. For example, the voltammogram shown in Fig. 22(b) involves a slight variation of the conditions used in Fig. 22(a): the packing density of I atoms was increased by only $\Delta\Theta_I = 0.02$ such that the I layer formed an incommensurate structure (not regularly coinciding with the Pt surface mesh) [50]. The effect of this incommensuracy was that the Ag electrodeposition process no longer produced a stable, ordered deposit. Similarly, electrodeposition of Ag into the Pt(100)(5$\sqrt{2}$ × $\sqrt{2}$)$R45°$–I adlattice produced very different electrodeposits [Fig. 22(c)] from deposition into the Pt(100)(2 × 8)$R45°$–I adlattice [Figs. 21 and 22(a)]. Slight roughening of the Pt surface also had a profound effect on electrodeposition of Ag [32, 41] [Fig. 22(d)], as did surface roughness introduced by Ar$^+$ ion bombardment [Fig. 22(e)].

A completely analogous series of variations in electrodeposition behavior was observed for the electrodeposition of Ag at Pt(111) surfaces [31, 39, 40].

3.4 INFLUENCE OF DEPOSITED ELEMENT

The nature of the deposited element and the nature of the supporting electrolyte anion strongly influence monolayer electrodeposition behavior as described below.

Electrodeposition of Cu onto I$_2$-pretreated Pt surfaces proceeded rather

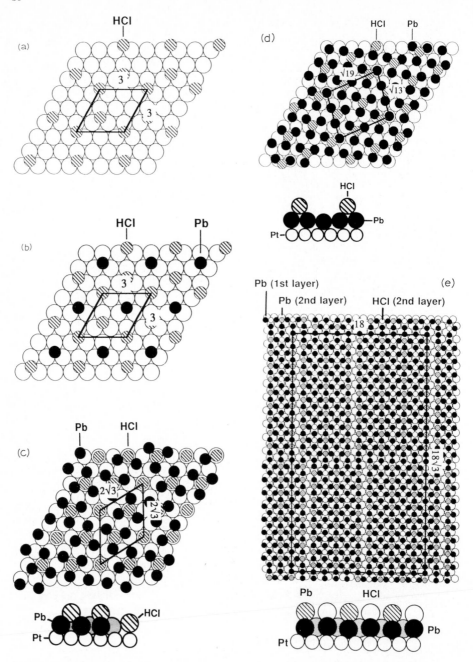

Fig. 23. Surface structures of Pb electrodeposited at Pt(111) from Cl⁻ media. (a) Pt(111)(3 × 3)–
Cl, $\Theta_{Cl} = 2/9$; (b) Pt(111)(3 × 3)–Pb, Cl, $\Theta_{Cl} = 2/9$; (c) Pt(111)(2$\sqrt{3}$ × 2$\sqrt{3}$)R30°–Pb, Cl,
$\Theta_{Pb} = 6/12$, $\Theta_{Cl} = 2/9$; (d) Pt(111)($\sqrt{13}$ × $\sqrt{19}$)–Pb, Cl, $\Theta_{Pb} = 2/3$, $\Theta_{Cl} = 2/9$; (e) Pt(111)(3 × 3)–
Pb, Cl, $\Theta_{Pb} = 0.85$, $\Theta_{Cl} = 0.26$; (f) Pt(111)[$c(5\sqrt{3}$ × $\sqrt{3})$]–Pb, Cl, $\Theta_{Pb} = 0.93$, $\Theta_{Cl} = 1/3$; (g)
Pt(111)(3 × $\sqrt{3}$)–Pb, Cl, $\Theta_{Pb} = 4/3$, $\Theta_{Cl} = 1/3$. Reprinted from ref. 44.

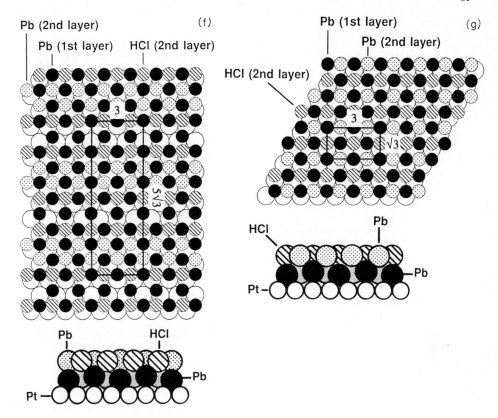

differently from deposition of Ag under otherwise identical conditions [42, 43]. While the first UPD process for Cu again yielded a (3 × 3) structure Pt(111)(3 × 3)–Cu,I, for which $\Theta_{Cu} = 4/9$, the second UPD process produced a Pt(111)(10 × 10)–Cu,I superlattice. Furthermore, electrodeposition of Cu at I_2-pretreated Pt formed isolated grains of Cu rather than uninterrupted monolayers as for Ag. The LEED patttern of Pt(111)(10 × 10)–Cu,I and the Auger signals for Pt remained in evidence due to the intervening thinly covered surface area, even after deposition of the equivalent of hundreds of Cu monolayers. The copper granules were readily visible at that stage by optical microscopy.

Electrodeposition behavior of Pb at Pt(111) illustrates the influence both of the nature of the supporting electrolyte anion and of the nature of the depositing metal [44, 46, 47]. The Pt(111) surface was not pretreated with I_2 in these studies. At the onset, comparison of the Pb Auger signal with the coulometric data for the amount of Pb deposited revealed that the Pb deposit was not stable at open circuit when the packing density of deposited Pb exceeded about one monoatomic layer. Accordingly, the structural investigations of Pb electrodeposition at Pb(111) have thus far been limited to monolayer and submonolayer deposits.

Electrodeposition of Pb from aqueous $PbCl_2/HCl$ solutions was characterized by a large number of UPD processes (at least 12 voltammetric peaks were observed) [44]. Consideration of coulometric data, Auger spectra, LEED patterns, and theory as discussed in refs. 32 and 38–49 indicated the structures shown in Fig. 23. An interesting aspect of Pb electrodeposition onto Pt(111) from Cl^- solutions is the formation of a variety of complicated oblique LEED patterns and structures, although hexagonal structures also occurred at some stages. Rectangular structures were also found. Thus far, none of the other electrodeposited metals studied (Ag, Cu, Bi, Tl, Sn) has been shown to form oblique monolayer structures; perhaps the large and therefore comparatively diffuse size of the Pb atom makes possible these complex, distorted-hexagonal structures (Fig. 23). Electrodeposition of Pb from Br^- formed a different series of monolayer structures than was found in Cl^- media [46], although the structures were hexagonal and oblique types analogous to those in Cl^-.

The fact that immersion of Pt(111) in HI/HBr mixtures formed at $Pt(111)(\sqrt{7} \times \sqrt{7})R19°$–I adlattice instead of the $Pt(111)(3 \times 3)$–I adlattice found in pure I^- solutions at the same potential allowed studies to be made of the electrodeposition of Pb into the $\sqrt{7}$ layer for comparison with electrodeposition of Ag into the same adlattice [47]. No detectable Br was present in the chemisorbed layer. Pb, unlike Ag, formed a $Pt(111)(4 \times \sqrt{3})$–Pb,I rectangular superlattice and a $Pt(111)(\sqrt{13} \times \sqrt{13})R14°$–Pb,I superlattice. Pb electrodeposited into the $Pt(111)(3 \times 3)$–I adlattice (in the absence of Br^-) was unstable.

Tin, in contrast to the other metals, deposited spontaneously onto Pt (that is, at open circuit, without the need for current flow in an external circuit) [51]. Auger spectra following spontaneous deposition showed a strong oxygen signal. Anodic electrolysis (oxidation) increased the oxidation state of the surface layer somewhat and rendered the surface passive except for evolution of H_2 at very negative potentials and O_2 at very positive potentials. Once immersion of Pt(111) in Sn^{2+} (Cl^- or Br^-) solution had taken place, the Sn deposit could not be removed from the surface by electrolysis in the same electrolyte. LEED patterns of the Sn layer were diffuse, indicating that the tin oxide layer was disordered. The pathway of spontaneous Sn deposition probably involves disproportionation, followed by oxidation of the metallic tin.

$$2\,Sn^{2+} \longrightarrow Sn + Sn^{4+} \tag{20}$$

$$Sn + 2\,H_2O \longrightarrow Sn(OH)_2 + H_2 \tag{21}$$

The overall process would then be

$$2\,Sn^{2+} + 2\,H_2O \longrightarrow Sn(OH)_2 + Sn^{4+} + H_2 \tag{22}$$

Electrodeposition of Ag onto Ag(111) from aqueous F^- solutions [26] occurred epitaxially, maintaining the simple (1 × 1) structure of the surface.

Electrodeposition of Bi, Pb, Tl, or Cu at Ag(111) occurred with noticeably smaller underpotentials than for the deposition of the same or similar metals at Pt(111) [32, 38–48]. Indeed, only Bi possessed sufficient stability in the absence of external potential control to permit determination of layer structure in vacuum.

Electrodeposition of Bi onto Ag(111) from aqueous acetate electroyte yielded an hexagonally close-packed superlattice, $Ag(111)(2\sqrt{3} \times 2\sqrt{3})R30°$–Bi for which $\Theta_{Bi} = 7/12$. The $(2\sqrt{3} \times 2\sqrt{3})R30°$ cells were present in parallel, strip-shaped domains averaging 8.5 silver unit meshes in width and unlimited length. In contrast, electrodeposition of Cu at Ag(111) yielded no UPD peaks. While the Cu electrodeposit at Ag was stable at open circuit and was present as monoatomic layers as judged by attenuation of the Ag substrate Auger signal, it was somewhat disordered as judged by LEED.

Pb and Tl electrodeposited onto Ag were unstable at open circuit in a variety of electrolytes tested representing various anions (ClO_4^-, F^-, Cl^-, Br^-, I^-) and alkaline, neutral, or acid pH [49]. Ordered layers of PbO or TlOH were formed during evacuation. Oxide layer structures were determined in each case and were found to vary with the electrolyte anion.

4. Structure of oxide films on single-crystal stainless steel

Alloys of Fe, Cr, and Ni crystallise in the face-centered cubic system provided that the Ni content is greater than about 10% [52–55]. The constituent elements are distributed statistically among the f.c.c. lattice positions. Studies have been reported in which an alloy single-crystal of composition (70 at.% Fe, 18 at.% Cr, 12 at.% Ni) resembling that of type 304 stainless steel was oriented, polished, cleaned by Ar^+ ion-bombardment, and annealed in UHV as described in Sect. 1.2 and refs. 52–55. The elemental ratios of the clean alloy surfaces were simlar to the bulk. The LEED patterns of the clean alloy (111) surface [52–54] closely resembled those for Pt(111), Ag(111), or Ni(111). That is, the clean alloy (111) surface is not reconstructed and the LEED technique is not element-specific. On the other hand, the clean alloy (100) surface reconstructed to a (2 × 2) structure consisting of atomic rows spaced 2 apart at 45° to the unit cell vector of the bulk (100) plane [55]. Heating these surfaces during or after exposure to O_2, H_2O vapor, or aqueous solutions led to Cr enrichment and formation of various ordered oxide films. Prior to heating, the films were amorphous, hydrated, and of mixed (Fe/Cr/Ni) composition as shown by LEED, TDMS, and Auger spectroscopy, respectively.

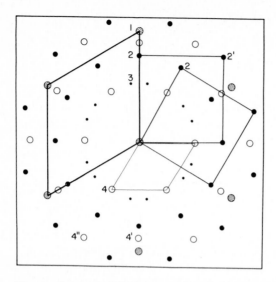

Fig. 24. Composite LEED pattern for the FeCrNi(111) surface exposed to electrolytes and annealed. Type 1: integral-index beams due to alloy. Type 2: pattern-of-twelve beams due to square CrO mesh. Type 3: beams due to initial growth phase of oxide with (2×2) local structure in (11×11) domains. Type 4: beams due to $Cr_2O_3(001)$ film. The alloy contained Fe (70 at.%), Cr (18 at.%), and Ni (12 at.%). Reprinted from ref. 54.

Films formed in acidic solution were enriched in Cr_2O_3 even prior to heating due to selective dissolution of Fe and Cr. Anodic dissolution likewise led to Cr_2O_3 enrichment of the surface layer. Cyclic voltammetry experiments confirmed that the single-crystal alloy surface passivated rapidly in alkaline borate with relatively little current flow, but more slowly in H_2SO_4 or $HClO_4$ after an active–passive transition involving relatively high current and never fully passivated in HCl [54].

When the room temperature amorphous oxide film was heated in UHV, water was evolved from the surface in separate processes at 300 and 700°C. Reduction of Fe and Ni oxides by Cr occurred to give an oxide layer of nearly pure Cr_2O_3.

The annealed alloy (111) oxide films yielded distinctive LEED patterns summarized in Fig. 24. Integral index beams, Type 1 in Fig. 24, were not seen when the oxide film was present, demonstrating that the film was continuous and was thicker than a few atomic layers. Thicknesses of about 2 Å were typical. Beams were produced by the oxide film, indicative of a hexagonal superlattice mesh having a lattice constant of about 4.9 Å and rotated 30° with respect to the mesh of the clean alloy surface (Type 4 of Fig. 24). These are the correct dimensions for the (001) plane of Cr_2O_3 in which the oxygen–oxygen vectors are parallel to the interatomic vectors of the alloy. Beams of Type 2 were also present, which were indicative of a square mesh; these

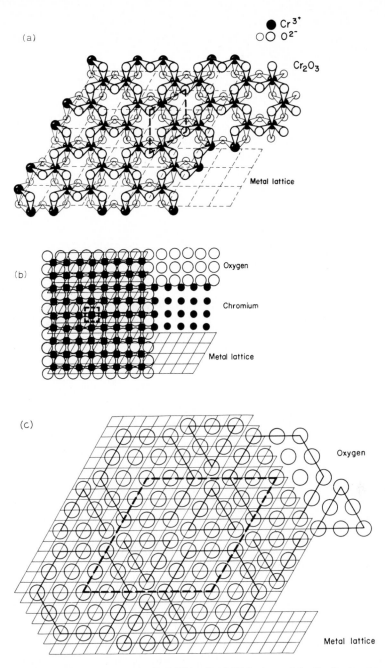

Fig. 25. Surface structures at FeCrNi(111). (a) (001) plane of the Cr_2O_3 structure situated on the alloy substrate as experimentally determined. (For clarity, only a few layers of the oxide are shown.); (b) Square CrO structure; (c) Growth phase structure (11 × 11); (d) Alternative growth phase structure (11 × $\sqrt{3}$). Reprinted from ref. 54.

(d)

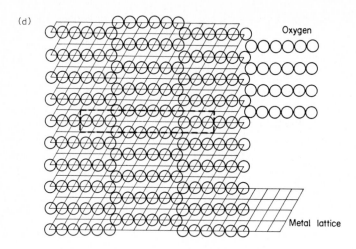

latter beams were most noticeable under conditions where the oxide film was thin. Beams of Type 3, due to (11 × 11) domains of (2 × 2) local structure, were present during the initial stages of film growth, after heating the surface in H_2O vapor, or after treatment of the surface with HCl solutions. Structures consistent with these findings are shown in Fig. 25.

Re-immersion of the ordered oxide films into $HClO_4$ or HCl solutions led to the disappearance of the LEED beams of Type 2. That is, the CrO phase was not stable in acid solutions. This is an indication that acidic electrolytes, particularly HCl, attacked the passive layer at the comparatively thin CrO regions, replacing or covering those regions with a thin, hydrated amorphous iron oxide layer.

In contrast, the alloy (100) surface formed predominantly a CrO film upon annealing, which was relatively thin and not particularly stable in acidic solutions [55].

5. Structure of adsorbed monolayers

Studies of molecular adsorption from solution at well-defined solid surfaces is yielding important results. Well-defined surfaces have a simplifying effect on such studies by eliminating many of the structural imperfections which would otherwise complicate the results with a mixutre of adsorption states. Surface analysis methods such as LEED, Auger spectroscopy, EELS, XPS and voltammetry are very well suited to the characterization of surface molecular structure, composition, and bonding. As a result, clear correlations between adsorbed state and surface chemical or electrochemical reactivity are beginning to emerge.

5.1 ADSORPTION AT POLYCRYSTALLINE PLATINUM THIN-LAYER ELECTRODES

5.1.1 Influence of adsorbate concentration

While polycrystalline Pt surfaces are not, as yet, fully defined as to structure and related properties and thin-layer electrodes (TLE) do not fully define surface molecular structure and reactivity, nevertheless the extensive results recently obtained provide important clues as to the type of behavior which can occur, promising areas for more thorough investigation, possible pitfalls, and so forth. Accordingly, we will begin this section by surveying recent TLE results. The distinction between TLE and other electrochemical investigations is that the solution under investigation is located in a thin film (20 μm) near an electrode surface. Surface cleanliness is easier to maintain because only a tiny volume of liquid contacts a large electrode surface area. As a result, surface adsorbed material constitutes a relatively large proportion of the total material. And the theoretical difficulty of separating chemical or electrochemical kinetics from mass transfer rates can be greatly simplified by proper choice of experimental conditions [56]. Limitations of TLE and the fact that, in general, only compounds exhibiting electrochemical reactivity readily distinguishable from background reactions of the electrode and electrolyte can be efficiently studied. Precision TLEs have not been conveniently adapted to surface characterization experiments in UHV. On balance, although the structural information yielded by TLE is indirect, it is highly reliable, quantitative, and readily available. Therefore, TLE experiments are a suitable starting point when exploring a new area.

Numerous compounds related to hydroquinone have been studied with regard to their adsorption and adsorbed state reactivity at polycrystalline Pt TLE. These compounds were chosen because their electrochemical reactivity makes them an inviting target and because this large family of compounds offers a wide range of properties for investigation.

$$\text{(quinone)} + 2 \text{ H}^+ + 2 \text{ e}^- \rightleftharpoons \text{(hydroquinone)} \tag{23}$$

Immediately it was noticed that the electrochemical reactivities of the adsorbed and unadsorbed forms of these compounds are very different, as illustrated by Fig. 26. This separability of adsorbed and unadsorbed reactivity is useful for the measurement of packing densities (T, mol cm^{-2}) from which adsorbate molecular orientation and certain other structural characteristics can be obtained.

Determination of Γ by means of TLE is accomplished as follows: the electrolyte surface is cleaned by electrochemical oxidation/reduction in pure supporting electrolyte; annealing can be carried out as described in refs. 32 and 41. If adsorption happens to be strong and the solution is very

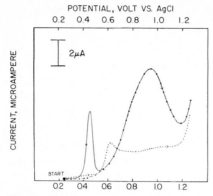

Fig. 26. Thin-layer current potential curves for hydroquinone (HQ) at a polycrystalline Pt electrode. ———, HQ solution at presaturated surface; · · ·, presaturated surface rinsed to remove unadsorbed HQ; · · · ·, clean surface in pure electrolyte. Experimental conditions: electrolyte was 1 M $HClO_4$; HQ concentration, 0.15 mM; thin layer volume, 4.08 μl; electrode area, 1.18 cm^2; potential sweep rate, 2 mV s^{-1}; temperature, 292 K. Reprinted from ref. 58.

dilute, then multiple filling of the TLE is required in order to supply sufficient adsorbate (solute peaks were absent from the voltammetric scans). The amount of material taken up by the surface during the first fillings was

$$\Gamma_1 + \Gamma_2 + \ldots + \Gamma_k = \frac{kVC^\circ}{A} \tag{24}$$

If the next filling $(k + 1)$ leads to a dissolved excess of surfactant (solute peak present in the voltammetric scan), then the amount is detected by thin-layer coulometry

$$\Gamma_{k+1} = \frac{(Q_{k+2} - Q_{k+1})}{nFA} \tag{25}$$

where n is the number of Faradays per mol electrolyzed and F the Faraday constant. The final $(k + 2)$ filling then leads to no further adsorption and is used to calibrate the bulk solute concentration, C°:

$$C^\circ = \frac{Q_{k+2} - Q_b}{nFV} \tag{26}$$

where Q_b is the coulometric charge due to background reactions occurring under conditions identical to $k + 2$ except that dissolved surfactant had been rinsed from the TLE with pure electrolyte. The packing density, Γ, is the sum of these individual terms.

$$\Gamma = \sum_{j=1}^{k+1} \Gamma_j \tag{27}$$

Details of this procedure are given in ref. 57.

Packing densities (nmol cm^{-2}) measured as described above for 28 com-

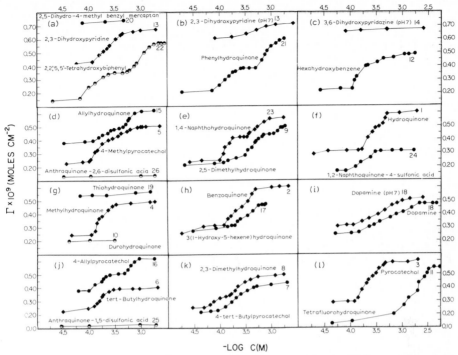

Fig. 27. Packing density, (mol cm^{-2}) versus adsorbate concentration (mol l^{-1}) at a polycrystalline Pt thin-layer electrode. The sizes of points represent the average experimental deviation. Experimental conditions: as Fig. 14, except as noted. Reprinted from ref. 57.

pounds at polycrystalline Pt TLE are shown in Fig. 27. Adsorbate concentration was varied systematically and the electrode was cleaned before each trial. Each compound displayed one or more packing density plateaus versus adsorbate concentration (at constant pH and electrode potential). It was discovered that each plateau correlates with a plausible molecular orientation [58]. The appropriate orientation can be deduced quite readily and

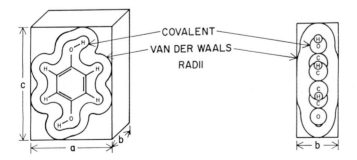

Fig. 28. Molecular unit cell assumed in theoretical estimates of packing density. Reprinted from ref. 58.

References pp. 64–67

reliably from calculations of adsorbed molecular area using van der Waals radii as described by Pauling [59] (Fig. 28).

Compound 20, Fig. 27(a), is representative of a large class of adsorbates which attach to the Pt surface in a single predominant orientation as a result of a surface-active functional group.

$$(28)$$

Incidentally, the hydroquinone pendant moiety in eqn. (28) is reversibly electroactive and will be discussed further in Sect. 5.2.1, below.

Compound 1, Fig. 27(f), is representative of another large class of compounds for which two adsorption plateaus are observed, corresponding to horizontal (low coverage) and edgewise (high coverage) orientations

$$(29)$$

It is also noteworthy that, as implied by eqn. (29), adsorption from hydroquinone (HQ) solutions leads to double dehydrogenation of the adsorbate in either final orientation.

Compound 22, Fig. 27(a), and related compounds exhibit three adsorbed orientations

$$(30)$$

The pendant HQ moiety on the right-hand side of eqn. (30) is reversibly electroactive, as discussed in Sect. 5.2.2, below.

Compound 13, Fig. 27(b), represents a class of acids and bases the orientations of which depend upon pH.

$$(31)$$

Compounds heavily halogenated at the aromatic ring, such as those shown in Fig. 27(l), tend not to adsorb through the ring; attachment occurs through functional groups where present.

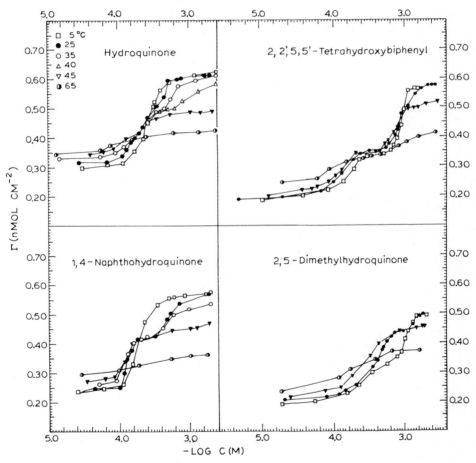

(32)

Although the details of such surface molecular states await further investigation, nevertheless the available evidence has already revealed the existence of very important and interesting surface molecular behavior, which we will attempt to summarize in the following.

Fig. 29. Packing density, Γ (nmol cm^{-2}), versus adsorbate concentration at various temperatures. Experimental conditions: polycrystalline Pt thin-layer electrode; 1 M HClO$_4$ electrolyte. Reprinted from ref. 60.

5.1.2 Influence of temperature

Adsorbates such as HQ which are capable of more than one surface orientation can undergo a transition from uniformly oriented adsorption to mixed or disordered orientation [60] as the temperature rises about 45°C, as shown in Fig. 29.

On the other hand, thiophenols and other compounds which have a single predominant surface orientational state, eqn. (28), are less noticeably influenced by temperature changes.

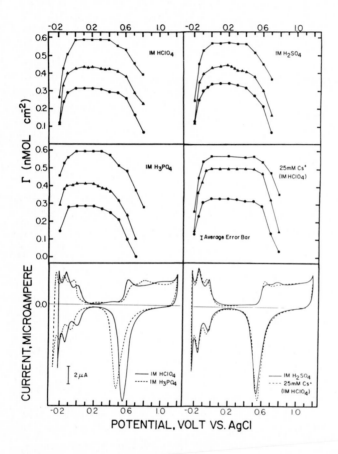

Fig. 30. Packing density versus electrode potential at a given orientation in various supporting electrolytes. ■, n^2-orientation; ▲, transition region (HQ concentrations were 0.36, 0.36, 0.25, and 0.37 mM in $HClO_4$, H_2SO_4, H_3PO_4, and $CsClO_4$, respectively); ●, n^6-orientation. The solid lines simply connect experimental points and do not assume any theoretical curve. Bottom figures: thin-layer cyclic current potential curves for clean Pt electrode in the respective supporting electrolytes. Thin-layer volume, $V = 3.29 \mu l$; electrode surface area, $A = 1.16 \, cm^2$; rate of potential sweep, $r = 2.00 \, mV \, s^{-1}$; temperature, $T = 23 \pm 1°C$. The average standard deviation in Γ was $\pm 3\%$ below 1 mM and $\pm 6\%$ above 1 mM. Reprinted from ref. 61.

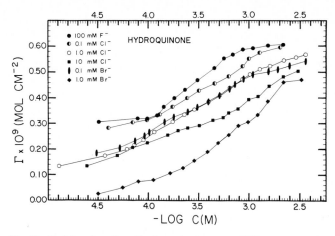

Fig. 31. Packing density of hydroquinone versus HQ concentration at polycrystalline Pt electrodes. Experimental conditions: electrode potential, 0.2 V (vs. Ag/AgCl) at pH = 0 or − 0.2 V at pH = 7 (F⁻ only); electrolyte was 1 M HClO₄ or 1 M NaClO₄ (F⁻ only); temperature was 23 ± 1°C. Reprinted from ref. 63.

5.1.3 Influence of electrode potential

Electrode potential influences organic adsorption at Pt surfaces from aqueous solutions primarily by bringing to the surface competing adsorbates such as OH or H [61] (Fig. 30) and by oxidation or reduction of the adsorbate. Based upon the limited reliable data available, it appears that these redox and chemisorption effects greatly outweigh the electrostatic effects.

5.1.4 Influence of anions

Aromatic adsorbates are adsorbed sufficiently strongly at Pt that weakly adsorbing anions such as F^-, ClO_4^-, PF_6^-, and SO_4^{2-} or cations such as K^+, Ca^{2+}, and La^{3+} have no appreciable influence [62]. However, I^-, Br^-, and Cl^- compete strongly with aromatics at Pt surfaces [63] (Fig. 31). Presumably, CN^-, SCN^-, S^{2-}, and other strongly adsorbing anions also would have a strong effect, although their influence has not yet been systematically investigated. For example, iodide ions compete strongly with aromatic adsorption and displace pre-adsorbed aromatics within minutes [64]. Also, iodide evidently causes reorientation of horizontally oriented adsorbed aromatics to other more compact orientations [64].

$$\text{(structure)} + I^- + H^+ \longrightarrow \text{(structure)} + \frac{1}{2} H_2 \qquad (33)$$

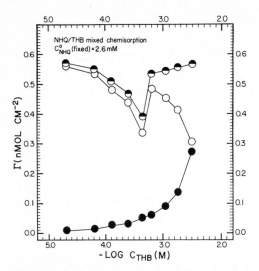

Fig. 32. Individual and total packing densities for 1,4-naphthohydroquinone (NHQ) and 2,5,2',5'-tetrahydroxybiphenyl (THBP) at a polycrystalline Pt electrode. ○, NHQ; ●, THBP; ◓, total. Experimental conditions: NHQ concentration, 2.5 mM; electrolyte, 1 M HClO₄; temperature, 23 ± 1°C. Reprinted from ref. 65.

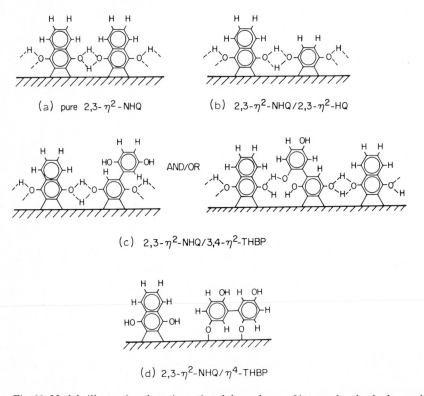

Fig. 33. Models illustrating the orientational dependence of intermolecular hydrogen-bonding in vertically oriented chemisorbed polyphenols (by analogy with crystalline diphenols). Reprinted from ref. 65.

5.1.5 Mixtures of aromatic adsorbates

Studies of competitive adsorption from mixtures of aromatic compounds leads frequently to packing densities that are not simply additive [65, 66]. This situation is illustrated by competition between 1,4-naphthohydroquinone (NHQ) and 2,2′,5,5′-tetrahydroxybiphenyl (THBP) (Fig. 32). Disordered (inefficient) packing was signaled by a deep minimum in the total packing density (primarily at the expense of NHQ) under conditions where THBP constituted about 10% of the layer. Structural models which account for this effect are shown in Fig. 33. According to these models, at certain proportions of THBP in the layer, the molecules were present in an orientation which is not conducive to intermolecular hydrogen bonding. These models were patterned after intermolecular hydrogen bonding in crystalline diphenols [67].

5.1.6 Influence of non-aqueous solvents

Typical non-aqueous solvents chemisorb at Pt surfaces. Included in the strongly adsorbing group are benzene [68], acetonitrile [69], pyridine, dimethylsulfoxide [70], sulfolane, and dimethylacetamide. Moderately strong adsorption at Pt is manifested by acetic acid, ethyl acetate, and tetrahydrofuran [29, 71]. Even when diluted in aqueous solutions, these materials strongly interfered with the adsorption of typical aromatic compounds [68–70]. Accordingly, solvent chemisorption is expected to dominate electrode surface chemistry in non-aqueous electrolytes, particularly with solvents in the strongly adsorbing group above.

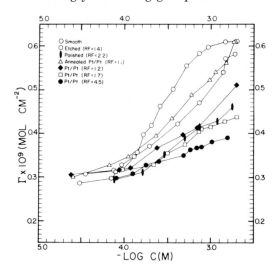

Fig. 34. Packing density of hydroquinone versus HQ concentration at polycrystalline Pt thin-layer electrodes roughened by various methods. Experimental conditions: 1 M $HClO_4$; electrode potential, 0.2 V (Ag/AgCl reference); temperature 23 ± 1°C; RF = real area/geometric area. Reprinted from ref. 72.

References pp. 64–67

5.1.7 Influence of surface roughness

Surfaces roughened to a moderate extent by slight platinization or mechanical abrasion were discovered to be unsuitable substrates for oriented aromatic adsorption [72]. Roughness (real area/geometric area) from 1.02 to 4.50 were examined (Fig. 34). The effect of surface roughness near 4.5 on the adsorption of HQ was to eliminate the horizontal-to-edgewise packing density transition. Corresponding differences in adsorbed molecule reactivity were also observed, as described in the next section.

5.1.8 Electrochemical reactions of adsorbed molecules

2,5-Dihydroxy-4-methylbenzylmercaptan (DMBM) [eqn. (28)], above, is representative of the behavior of a special group of adsorbates for which a reversibly electroactive adsorbed pendant is observed [57].

$$+ 2 H^+ + 2 e^- \tag{34}$$

The electroactivity of the adsorbed states of this group of compounds is evidently associated with the presence on the adsorbed layer of a pendant functionality which is electroactive in the unadsorbed molecule and is attached to the surface in such a way that the pendant is virtually unperturbed structurally or electronically by the surface. In contrast, a molecule such as hydroquinone which is electroactive prior to adsorption but is strongly perturbed by adsorption (that is, by direct covalent attachment to the Pt surface) is not reversibly electroactive in the adsorbed state (Fig. 26). Accordingly, adsorbed DMBM is reversibly electroactive under almost all conditions, while adsorbed HQ is not reversibly electroactive under any conditions thus far studied. In between these two extremes is THBP, which is reversibly electroactive in one of its adsorbed states but not in the other two.

$$+ 2 H^+ + 2 e^- \tag{35}$$

No reaction $\tag{36}$

$$\text{(structure)} \quad \longrightarrow \quad \text{No reaction} \qquad (37)$$

All chemisorbed compounds are electroactive to at least some degree at sufficiently extreme applied potentials. The most common behavior is rapid, irreversible oxidation at relatively positive potentials [61, 72–76]. In particular, experiments have been performed in which a surface containing an adsorbed layer is oxidized in pure supporting electrolyte at electrode potentials shown to cause complete oxidative desorption of the layer. Measurement of the coulometric charge, Q_{ox}, and the background charge observed in the absence of an adsorbed layer, Q_b, permitted calculation of the number of electrons removed in the oxidation of the adsorbed molecule, n_{ox}

$$n_{ox} = \frac{Q_{ox} - Q_b'}{FA\Gamma} \qquad (38)$$

where the other symbols have their usual meaning. Measured values of n_{ox} are graphed in Fig. 35. Similar results were obtained for numerous other

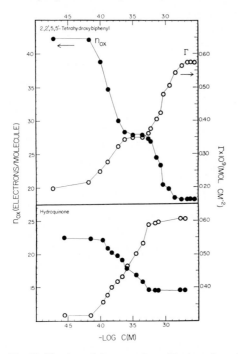

Fig. 35. Number of electrons for oxidation of an adsorbed molecule of 2,2′,5,5′-tetrahydroxybiphenyl (THBP) or hydroquinone (HQ). Experimental conditions: polycrystalline Pt thin-layer electrode in 1 M HClO$_4$ electrolyte at 23 ± 1°C. Reprinted from ref. 73.

compounds [74–76]. Exact correspondence is observed between the extent of oxidation, n_{ox}, and the molecular orientation of the adsorbate. Horizontal orientations became oxidized to CO_2.

$$+ \quad 10\ H_2O \longrightarrow 6\ CO_2 + 24\ H^+ + 24\ e^- \tag{39}$$

Edgewise orientation led to maleic acid or analogous products.

$$+ 6\ H_2O \longrightarrow \qquad + 2\ CO_2 + 12\ H^+ + 12\ e^- \tag{40}$$

These results indicate that, by proper control of adsorbed intermediate structure, it should be possible to employ such adsorbates to produce highly specific electrochemical products. Also, there is every reason to expect highly specific *chemical* reactions of well-defined adsorbed intermediates reacting with solutions.

Electrochemical reduction of specific oriented chemisorbed intermediates is also very selective [77–79]. For example, reduction of chemisorbed thiophenols (and mercaptans) result in selective scission of the carbon–sulfur bond to yield an unadsorbed hydrocarbon and an adsorbed sulfur atom.

$$+ \ H^+ + \ e^- \longrightarrow \qquad + \tag{41}$$

5.2 ADSORPTION OF WELL-DEFINED Pt(111) SURFACES

Studies of molecular adsorption from solution at well-defined solid surfaces have begun [80–84]. The simplifications inherent in the use of well-defined surfaces for such studies and the great usefulness of surface-sensitive spectroscopic methods provide ample motivation for the studies.

5.2.1 Thiophenols and mercaptans

Adsorption at Pt(111) from aqueous solutions of a series of compounds related to thiophenol has been studied by means of voltammetry assisted by Auger spectroscopy, EELS, and LEED [81], e.g. thiophenol (TP), penta-

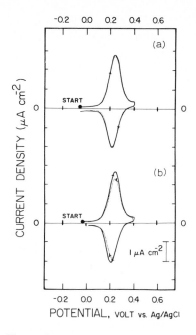

Fig. 36. Cyclic voltammograms of 2,5-dihydroxy-4-methylbenzylmercaptan (DMBM) adsorbed at Pt(111). (a) ——, After immersion into 0.7 mM DMBM and rinsing with pure electrolyte; · · · ·, as above, then 1 h in vacuum prior to voltammetry. (b) ——, First scan; · · · ·, second scan. Reprinted from ref. 81.

fluorothiophenol (PFTP), 2,3,5,6-tetrafluorothiophenol (TFTP4), 2,3,4,5-tetrafluorothiophenol (TFTP2), 2,5-dihydroxythiophenol (DHTP), 2,5-dihydroxy-4-methylbenzylmercaptan (DMBM), and benzylmercaptan (BM).

Two of these compounds, DMBM and DHTP, undergo reversible oxidation/reduction in the adsorbed state. Shown in Fig. 36(a) (solid curve) is the cyclic voltammogram of a DMBM adsorbed layer at Pt(111). Adsorbed DMBM reacts as shown in eqn. (34). The dotted curve in Fig. 36(a) is particularly noteworthy: it was obtained under the same conditions as the solid curve except that the electrode was held for 1 h in UHV prior to the voltammetric scan. As can be seen, the results obtained before and after evacuation are essentially identical. That is, the adsorbed layer of DMBM is stable in vacuum (undergoes no irreversible change as a result of evacua-

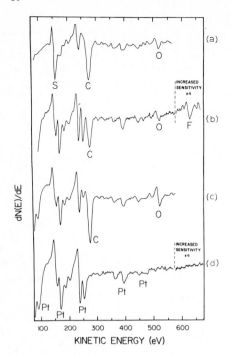

Fig. 37. Auger spectra of adsorbed thiophenol and related compounds at Pt(111). (a) Clean Pt(111); (b) 2,5-dihydroxy-4-methylbenzylmercaptan (DMBM); (c) Thiophenol (TP); (d) Penta-fluorothiophenol (PFTP). Experimental conditions: adsorbate concentrations 0.7 mM; incident beam 100 nA, 2000 eV at normal incidence. Reprinted from ref. 81.

tion). The results in Fig. 36(a) are quantitatively consistent with those obtained with thin-layer electrodes (entirely in solution). Since LEED, Auger, and EELS were employed to examine the surface layer during the hour in UHV, there was evidently no appreciable beam damage to the layer. In other words, the DMBM layer exhibits ample stability to allow its full characterization by surface-sensitive spectroscopies in UHV. However, in contrast to the UHV results, cyclic voltammetry caused appreciable damage to the DMBM layer, primarily scission of the carbon–sulfur bond [78], as can be seen by comparing the first and second voltammetric cycles [Fig. 36(b)]. Similar stability in UHV and damge due to cyclic voltammetry has been observed for all of the reversibly electroactive adsorbed compounds studied thus far [80–84].

Auger spectra of thiophenol and related compounds are shown in Fig. 37. Packing densities of individual elements, X, of the adsorbed layer were determined from the elemental Auger signals, I_X, by means of the following equations and were in good agreement with calculated values.

$$\frac{I_X}{I_{Pt}^\circ} = B_X \Gamma_X (L_1 f_1 + \ldots + L_i f_i + \ldots + L_N f_N) \tag{42}$$

where L_i is the fraction of atoms of element X located in layer i ($i = 1$ is adjacent to the solid and N is the outermost layer) and f_i is the scattering factor at the Auger electron kinetic energy of element X in layer i. The molecular packing densities were determined independently from the positive lobe of the Pt Auger signal at $235\,\mathrm{eV}$ by means of the equation

$$\frac{I_{\mathrm{Pt}}}{I^{\circ}_{\mathrm{Pt}}} = (1 - J_1 kT) \ldots (1 - J_i kT) \ldots (1 - J_N kT) \tag{43}$$

where J_i represents the number of atoms located at each level in the layer. In particular, the packing densities of C, O, and S in DMBM are given by

$$\Gamma_{\mathrm{C}} = \frac{I_{\mathrm{C}}/I^{\circ}_{\mathrm{Pt}}}{B_{\mathrm{C}}[(5f/8) + (3/8)]} \tag{44}$$

where I°_{Pt} was measured at the positive lobe of the Pt signal at $235\,\mathrm{eV}$ to minimize overlap with other peaks in the spectrum, $B_{\mathrm{C}} = 8.48 \times 10^8\,\mathrm{cm^2\,mol^{-1}}$, and $f = 0.70$ [from the Pt(111)(3 × 3)–HQ reference structure] [80].

$$\Gamma_0 = \frac{I_0/I^{\circ}_{\mathrm{Pt}}}{B_0[(f^3_0/2) + (1/2)]} \tag{45}$$

$$\Gamma_{\mathrm{S}} = \frac{I_{\mathrm{S}}/I^{\circ}_{\mathrm{Pt}}}{B_{\mathrm{S}} F^2_{\mathrm{S}}} \tag{46}$$

where $B_0 = 1.272 \times 10^9\,\mathrm{cm^2\,mol^{-1}}$, $B_{\mathrm{X}} = 30.0 \times 10^{10}\,\mathrm{cm^2\,mol^{-1}}$, $f_0 = 0.70$, and $f_{\mathrm{S}} = 0.60$. The molecular packing density of DMBM was determined independently from Pt signal attenuation due to the adsorbed layer by use of

$$\frac{I_{\mathrm{Pt}}}{I^{\circ}_{\mathrm{Pt}}} = (1 - kT)(1 - 5kT)^2 \tag{47}$$

where $k = 1.53 \times 10^8\,\mathrm{cm^2\,mol^{-1}}$ [82]. The results are $\Gamma_{\mathrm{C}} = 3.07\,\mathrm{nmol\,cm^{-2}}$, $\Gamma_0 = 0.63\,\mathrm{nmol\,cm^{-2}}$, $\Gamma_{\mathrm{S}} = 0.39\,\mathrm{nmol\,cm^{-2}}$, Γ (molecular packing density from $\Gamma_{\mathrm{C}}/11$) $= 0.384\,\mathrm{nmol\,cm^{-2}}$. The ideal molecular packing density of the model structure shown in Fig. 38 is $\Gamma = 0.39\,\mathrm{nmol\,cm^{-2}}$.

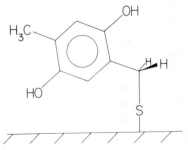

Fig. 38. Structural model of DMBM adsorbed on Pt(111) at saturation coverage. Reprinted from ref. 81.

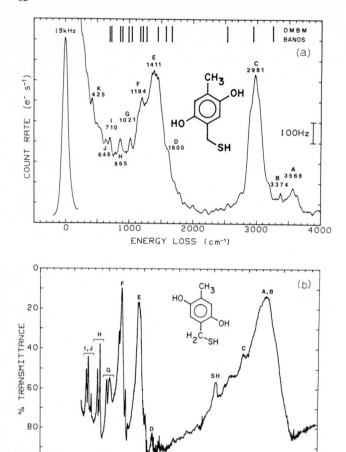

Fig. 39. Vibrational spectra of DMBM. (a) EELS spectrum of adsorbed DMBM at Pt(111); (b) IR spectrum of solid DMBM in KBr. Experimental conditions: DMBM solution concentration, 0.1 mM in 10 mM KF/HF electrolyte (pH = 4); EELS incidence and detection angles, 62° from surface normal; beam energy, 4 eV; beam current, 15 nA; EELS resolution, 10 meV (80 cm^{-1}) FWHM; IR resolution, 4 cm^{-1} FWHM. Reprinted from ref. 81.

The EELS spectrum of DMBM [Fig. 39(a)] and the IR spectrum of solid DMBM in KBR [Fig. 39(b)] are very similar except for the absence of the sulfhydryl SH stretch in the EELS spectrum (2500 cm^{-1}). [The difference in appearance of the EELS and IR spectra results from several causes: the resolution of EELS (80 cm^{-1}) was poorer than that of IR (4 cm^{-1}); EELS emphasizes electron impact scattering while IR emphasizes dipole scattering, an experimentally valuable distinction; and the intrinsic line widths may be broader for the adsorbed layer than for the solid compound.] Assignments of the EELS bands are proposed in ref. 81. Absence of the SH stretch from EELS spectra of DMBM is evidence for removal of the sulfhydryl hydrogen during adsorption

$$\text{(structure)} \longrightarrow \text{(structure)} + \text{H}^+ + \text{e}^- \tag{48}$$

DMBM formed an ordered layer at a packing density near $\Gamma = 0.2\,\text{nmol cm}^{-2}$ ($\Theta = 1/12$), $\text{Pt}(111)(2\sqrt{3} \times 2\sqrt{3})R30°$–DMBM, a hexagonal adlattice having a primitive $2\sqrt{3}$ unit mesh. At higher packing densities, registry with the surface was evidently lost and the LEED pattern was again diffuse. The limiting packing density was $\Gamma = 0.38\,\text{nmol cm}^{-2}$ ($\Theta = 0.15$). Based upon EELS spectra, packing densities from Auger spectroscopy with coulometry, and LEED structures, the model structure shown in Fig. 38 was proposed.

5.2.2 Phenols and quinones

Adsorption and electrochemical oxidation of a series of compounds related to hydroquinone (HQ) at Pt(111) in aqueous solutions has been studied [81, 82], including benzoquinone (BQ), phenol (PL), perdeuterophenol (PDPL), tetrafluorohydroquinone (TFHQ), 2,2′,5,5′-tetrahydroxygiphenyl (THBP), and 2,2′,5,5′-tetraketodicyclohexadiene (TKCD).

HQ   BQ   PL   PDPL   TFHQ

THBP   TKCD

Two of these compounds, THBP and TKCD, displayed reversible electroactivity in the edge-pendant adsorbed state [eqn. (25)]. The electroactive species were stable in vacuum [82]. However, reversible electrochemical redox hycyclic voltammetry led to gradual removal of the adsorbed layer.

Packing densities of THBP measured at Pt(111) by Auger spectroscopy [eqns. (42) and (43)] as well as n_{ox} measurements [eqn. (38)] revealed that the horizontal orientation of THBP [eqn. (36)] is more stable at Pt(111) than at polycrystalline Pt relative to the other orientations.

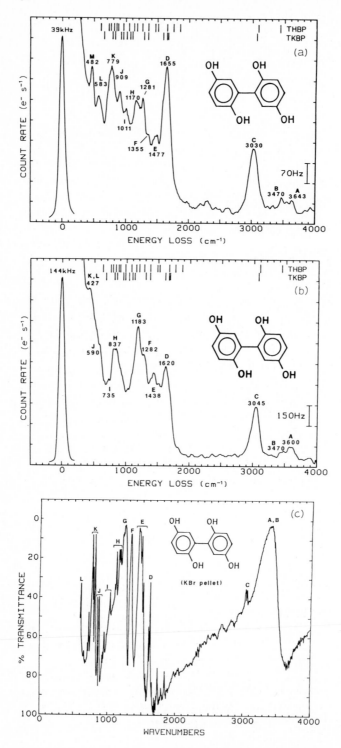

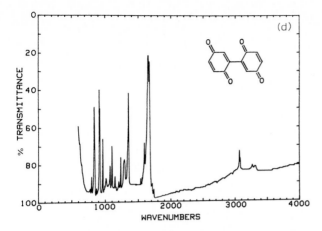

Fig. 40. Vibrational spectra of THBP and TKCD. (a) EELS spectrum of Pt(111) after immersion into 0.1 mM THBP; (b) EELS spectrum of Pt(111) after immersion into 5 mM THBP; (c) IR spectrum of solid THBP in KBr; (d) IR spectrum of solid TKCD in KBr. Experimental conditions: as in Fig. 27. Reprinted from ref. 82.

EELS spectra are shown in Fig. 40(a) (0.1 mM THBP) and (b) (5 mM THBP). Shown for reference are the mid-IR spectra of solid THBP in KBr [Fig. 40(c)] and of the solid quinone form, TKCD [Fig. 40(d)]. Assignment of the EELS bands is discussed in ref. 82. The EELS OH stretching peaks (3470 and 3600 cm^{-1}) and the phenolic C–O stretching peak (1183 cm^{-1}) increased with increasing THBP concentraiton. Evidently, when adsorption was carried out at concentrations below 0.1 mM, the phenolic hydrogens were lost from the horizontal orientation during adsorption while, at higher concentrations, an edgewise-pendant orientation adsorbed state was formed which resulted in retention of the phenolic hydrogens [eqn. (30)].

Measurement of packing densities from HQ solutions by Auger spectro-

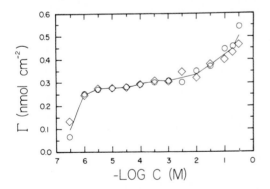

Fig. 41. Packing density, Γ, of HQ/BQ at Pt(111). (a) Γ, based upon I_c/I_{Pt}° [Eqn. (34)] (○); (b) Γ, based upon I_{Pt}/I_{Pt}° [Eqn. (35)] (◇). Experimental conditions: supporting electrolyte, 10 mM KF adjusted to pH = 4 with HF; electrode potential, 0.2 V vs. Ag/AgCl (1 M KCl) reference; temperature, 23 ± 1°C. Reprinted from ref. 80.

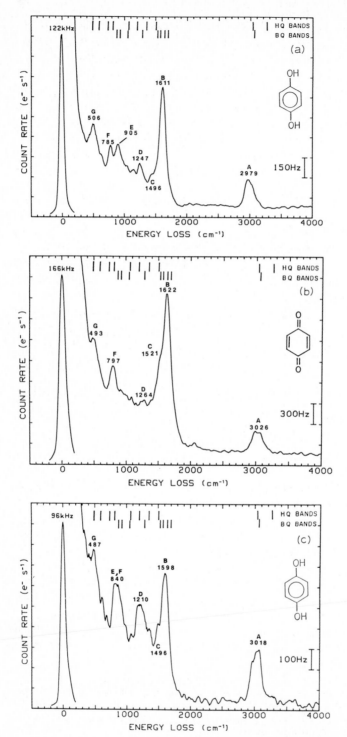

scopy [80] (Fig. 41) revealed an orientational transition from horizontal to edgewise with increasing HQ concentration, although the transition occurred at higher HQ concentrations at Pt(111) than at polycrystalline surfaces (Figs. 27 and 29). Measurements of n_{ox} for the full range of concentrations in Fig. 27 by coulometry ($Q_{ox} - Q'_b$) combined with Auger spectroscopy [80] [eqn. (38)] confirmed the existence of this transition [eqns. (39) and (40)] and once again evidenced the direct dependence of solid–liquid catalytic reactivity upon adsorbed intermediate molecular orientation and mode of attachment to the surface. Measurements of n_{ox} after subjecting the sample to 1 h in UHV yielded identical values of n_{ox}, demonstrating the stability of adsorbed HQ in UHV.

EELS spectra of the adsorbed layers formed from HQ or BQ solutions were very similar [Figs. 42(a) and (b)], indicating formation of a common horizontally oriented adsorbed state. Also shown in Fig. 42 are the locations of the principal mid-IR bands of solid HQ and BQ. The virtual absence of the OH stretches (3260 cm^{-1}) from the EELS spectra of adsorbed layers formed at HQ concentrations below 1 mM indicates that the phenolic hydrogens are lost during adsorption [eqn. (33)].

LEED patterns observed after HQ adsorption to packing densities slightly lower than the plateau value were Pt(111)(3 × 3). Based upon the EELS, Auger, LEED and n_{ox} data, the structure shown in Fig. 43 has been proposed. Phenol (PL) displayed a similar (3 × 3) LEED pattern under these same conditions [80].

Further evidence leading to the assignment of EELS bands of adsorbed phenols is provided by the EELS spectrum of adsorbed perdeuterophenol (phenol-d_6, PDPL) [Fig. 44(b)]. As expected, the CH stretching band of PL (3020 cm^{-1}) was replaced by a CD stretching band at 2258 cm^{-1}. Peaks F–I (CD modes) were shifted, while peaks B–E remained relatively constant (C–C and ring modes) [80].

In the initial PDPL experiments, it was noticed that traces of hydrocarbon contaminants were present in the surface layer, giving rise to a distinct CH stretching band in the EELS spectrum near 3000 cm^{-1}. Evidently, PL, like other aromatic adsorbates, is vulnerable to competition from a variety of chemisorbable organic compounds [79] present in typical reagents and in the environment. Accordingly, the presence or absence of a CH stretching band in the EELS spectrum of adsorbed PDPL is persuasive evidence as to whether the cleanliness of the reagents and apparatus is sufficient to yield the intended adsorbed layer. It is recommended that others attempting to obtain EELS, IR, or related data for organic materials adsorbed at electrode surfaces first obtain a spectrum of adsorbed PDPL as a procedural test before

Fig. 42. EELS spectra of adsorbed HQ/BQ at Pt(111). (a) HQ (1.0 mM); (b) BQ (1.0 mM); (c) HQ (500 mM). Experimental conditions: supporting electrolyte, 10 mM KF adjusted to pH = 4 with HF; electrode potential, 0.2 V vs. Ag/AgCl (1 M KCl) reference; temperature, 23 ± 1°C; beam energy, 4 eV; beam current, 0.15 nA; reflected beam was detected at the specular angle (62° from the surface normal). Reprinted from ref. 80.

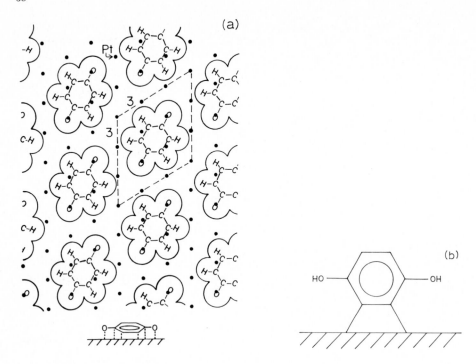

Fig. 43. Structure of BQ/HQ adsorbed at Pt(111). (a) Pt(111)(3 × 3)–BQ/HQ, Θ = 1/9 (Γ = 0.276 nmol cm^{-2}); (b) Vertical orientation of adsorbed HQ (2,3-n^2). Reprinted from ref. 80.

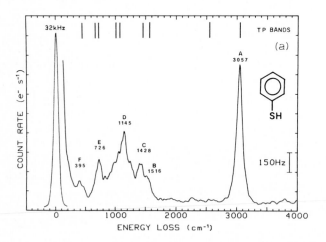

Fig. 44. EELS spectra of adsorbed PL and PDPL at Pt(111). (a) Phenol (PL); (b) Perdeuterophenol (PDPL, phenol-d_6). Experimental conditions: adsorbate concentration, 10 mM; other conditions as in Fig. 30. Reprinted from ref. 80.

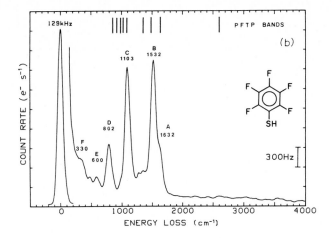

proceeding to other compounds. In our experience, this will save time and resources in the long run. Tetrafluorohydroquinone is also suitable for this purpose.

5.2.3 Amino acids and related compounds

Adsorption of a series of amino acids and other compounds related to L-dopa has been studied [84], including L-dopa (LD), L-tyrosine (TYR), L-cysteine (CYS), L-phenylatamine (PHE), L-alanine (ALA), dopamine (DA), catechol (CT), 3-mercaptopropionic acid (MPA), and 2-aminoethanethiol (AET).

LD   TYR   CYS   PHE   ALA

DA   CT   MPA   AET

60

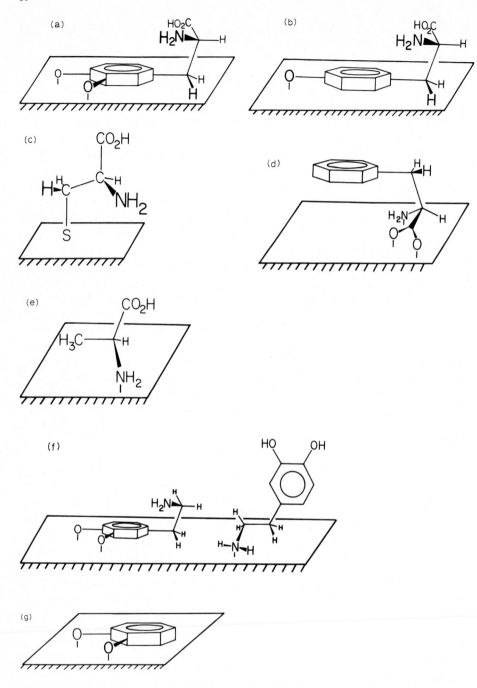

Fig. 45. Structural models of adsorbed molecules at Pt(100) and Pt(111) surfaces. (a) L-Dopa; (b) L-Tyrosine; (c) L-Cysteine; (d) L-Phenylalanine; (e) L-Alanine; (f) Dopamine; (g) Catechol. Reprinted from ref. 84.

As for the other compounds thus far studied, these amino acids and related adsorbates were not removed from Pt by evacuation, as judged by measurements of Q_{ox} or n_{ox}.

Packing densities measured by Auger spectroscopy were analogous to those shown in Fig. 41, although solubility limitations prevented measurement of amino acid behavior at concentrations above about 1 mM. Best agreement with experimental packing densities is found for the model structures shown in Fig. 45. Modes of bonding based upon packing densities and molecular models were entirely consistent with the EELS spectra and n_{ox} behavior of the adsorbed compounds. In particular, the phenolic amino acids such as LD and TYR are attached to the surface predominantly through the phenolic aromatic ring [Figs. 45(a) and (b)]. In contrast, PHE, which has no phenolic OH groups, adsorbs through the amino acid moiety [Fig. 45(d)]. CYS attaches through the sulfur atom, as expected [Fig. 45(c)]. The related compounds DA and CT also adsorb primarily through horizontally oriented attachment of the phenolic ring to the surface [Figs. 45(f) and (g)].

EELS spectra of PA and related adsorbed molecules are shown in Fig. 46. Also shown are the locations and relative intensities of IR spectral bands of the solid compounds in KBr. Assignment of the EELS bands is discussed in Ref. 84. The S–H stretch of CYS is not seen in the EELS spectrum [Fig. 46(c)] due to loss of the sulfhydryl hydrogen upon adsorption [Fig. 45(c)]. Likewise, the phenolic OH stretches of LD, TYR, DA, and CT were absent from the EELS spectra due to loss of the phenolic hydrogens during adsorption of those compounds [Fig. 45(a), (b), (f), and (g)]. However, the carboxylic OH stretches of LD, TYR, and CYS are present in Fig. 46(a)–(c); this indicates that the rate and/or strength of adsorption of the phenol or sulfhydryl moiety is sufficient to prevent reaction between the Pt surface and the carboxylate moiety for these compounds. The amine hydrogens are also retained, including those of ALA and PHE, in spite of attachment of the amino acid moiety to the surface in those cases (3200 and 3400 cm^{-1}). The CH stretching band is of a higher frequency for the aromatic amino acids (near 3070 cm^{-1}), evidence of the retention of aromatic character in the adsorbed state. The carboxylic $C = O$ stretch is present, as expected, for LD, TYR, and CYS but is relatively weak for PHE, perhaps due to the direct amino acid–Pt interaction of PHE. The striking resemblances among EELS spectra of LD, TYR, and CYS [Fig. 46(a)–(c)] is consistent with their mutual low symmetry (C_1), similar molecular constitution, and pendant-amino acid adsorbed structure.

References pp. 64–67

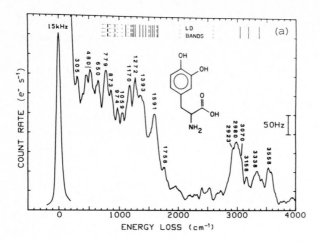

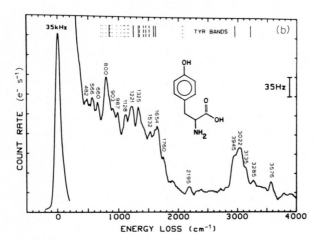

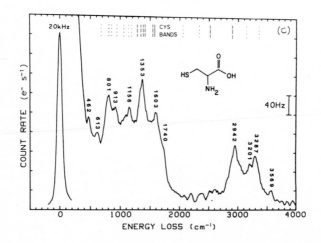

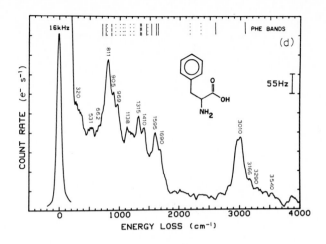

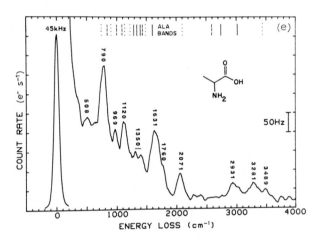

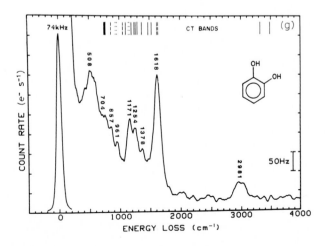

References pp. 64–67

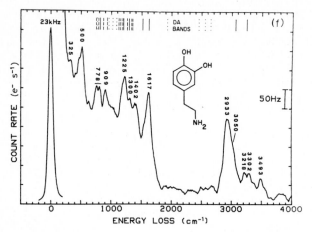

Fig. 46. EELS spectra of adsorbed amino acids and related compounds at Pt(111). (a) L-Dopa; (b) L-Tyrosine; (c) L-Cysteine; (d) L-Phenylalamine; (e) L-Alanine; (f) Dopamine; (g) Catechol. Experimental conditions: adsorbate concentrations (mM) were LD (5.0), TYR (1.0), CYS (0.5), PHE (1.0), ALA (1.0), DA (1.0), CT (1.0); other conditions were as in Fig. 20. Reprinted from ref. 84.

Acknowledgements

The author gratefully acknowledges research support from the Air Force Office of Scientific Research, the Department of Energy, the Gas Research Institute, the National Institute of Health, the National Science Foundation, and the Petroleum Research Fund administered by the American Chemical Society. Douglas G. Frank assisted with preparation of this article.

References

1   L.H. Germer and C. Davisson, Nature (London), 119 (1927) 558.
2   L.H. Germer and C.D. Hartman, Rev. Sci. Instrum., 31 (1960) 784.
3   W. Ehrenberg, Philos. Mag., 18 (1934) 878.
4   J.J. Lander, Phys. Rev., 91 (1953) 1382.
5   F.M. Propst and T.C. Piper, J. Vac. Sci. Technol., 4 (1967) 53.
6   K. Siegbahn, C. Nordling and A. Fahlman, ESCA: Atomic, Molecular and Solid State Structure Studied by Means of Electron Spectroscopy. Almqvist-Wikell, Uppsala, Sweden, 1967.
7   (a) S. Dushman, Scientific Formulations of Vacuum Technique, Wiley, New York, 2nd edn., 1962. (b) P.A. Redhead, J.P. Hobson and E.V. Kornelson, The Physical Basis of Ultrahigh Vacuum, Chapman and Hall, London, 1968. (c) R.W. Roberts and T.A. Vanderslice, Ultrahigh Vacuum and Its Applications, Prentice-Hall, Englewood, NJ, 1963.
8   A.T. Hubbard, Crit. Rev. Anal. Chem., 3 (1973) 201.
9   A.T. Hubbard, Acc. Chem. Res., 13 (1980) 177.
10 A.T. Hubbard, J.L. Stickney, M.P. Soriaga, V.K.F. Chia, S.D. Rosasio, B.C. Schardt, T. Solomun, D. Song, J.H. White and A. Wieckowski, J. Electroanal. Chem., 168 (1984) 43.

11 (a) C.B. Duke, Adv. Chem. Phys., 27 (1974) 215. (b) P.J. Estrup, in L.H. Lee (Ed.) Charac-
    terization of Metal and Polymer Surfaces, Vol. 1, Academic Press, New York, 1977, p. 187.
    (c) G.A. Somorjai and H.H. Farrell, Adv. Chem. Phys., 20 (1971) 215.

12 (a) C.C. Chang, Surf. Sci., 25 (1971) 53. (b) D.T. Hawkins, Auger Electron Spectroscopy, A
    Bibliography, 1927–1975, Plenum Press, New York, 1977. (c) G.A. Somorjai and F.J. Szal-
    kowski, Adv. High Temp. Chem., 4 (1971) 137. (d) N.J. Taylor, in R.F. Bunshah (Ed.),
    Techniques of Metals Research, Vol. 7, Wiley–Interscience, New York, 1971, p. 117. (e) M.
    Thompson, M.D. Baker, A. Christie, J.F. Tyson, Auger Electron Spectroscopy, Wiley, New
    York, 1985.

13 (a) D. Betteridge, Photoelectron Spectroscopy: Chemical and Analytical Aspects, Perga-
    mon Press, New York, 1972. (b) T.A. Carlson, Photoelectron and Auger Spectroscopy,
    Plenum Press, New York, 1975. (c) W.N. Delgas, T.R. Hughes and C.S. Fadley, Catal. Rev.,
    4 (1970) 179. (d) H.K. Herglotz and H.L. Suchan, Adv. Colloid Interface Sci., 5 (1975) 79.

14 R.E. Honig, Adv. Mass Spectrom., 6 (1974) 337.

15 (a) T.E. Madey and J.T. Yates, Jr., Surf. Sci., 63 (1977) 203. (b) T.E. Madey and J.T. Yates,
    Jr., J. Vac. Sci. Technol., 8 (1971) 525. (c) D. Menzel and R. Gomer, J. Chem. Phys., 41 (1964)
    3311.

16 G.A. Somorjai, Chemistry in Two Dimensions: Surfaces, Cornell University Press, Ithaca,
    New York, 1981.

17 L.E. Samuels, Metallographic Polishing by Mechanical Methods, Pitman, London, 1967.

18 E.A. Wood, Crystal Orientation Manual, Columbia University Press, New York, 1963.

19 G. Carter and J.S. Colligon, Ion Bombardment of Solids, Elsevier, Amsterdam, 1975, pp.
    223–273.

20 B.E. Conway, H. Angerstein-Kozlawska, W.B.A. Sharp and E.E. Criddle, Anal. Chem., 45
    (1973) 1331.

21 M. Gryzinski, Phys. Rev. A., 138 (1965) 305, 322, 336.

22 W.A. Coghlan and R.E. Clansing, At. Data, 5 (1973) 317.

23 (a) H. Ibach and D.L. Mills, Electron Energy Loss Spectroscopy and Surface Vibrations,
    Academic Press, New York, 1982. (b) J.E. Demuth, K. Christman and P.N. Sanda, Chem.
    Phys. Lett., 75 (1980) 201. (c) N.R. Avery, Appl. Surf. Sci., 13 (1982) 171. (d) B.A. Sexton, J.
    Vac. Sci. Technol., 16 (1979) 1033. (e) H. Froltzheim, H. Ibach and S. Lehwald, Rev. Sci.
    Instrum., 46 (1975) 1325.

24 L.L. Kesmodel, J. Vac. Sci. Technol. A, 1 (1983) 1456.

25 J.L. Stickney, S.D. Rosasco, G.N. Salaita and A.T. Hubbard, Langmuir, 1 (1985) 66.

26 G.N. Salaita, F. Lu, L. Laguren-Davidson and A.T. Hubbard, J. Electroanal. Chem., 229
    (1987) 1.

27 G.N. Salaita, D.A. Stern, F. Lu, H. Baltraschat, B.S. Schardt, J.L. Stickney, M.P. Soriaga,
    D.G. Frank and A.T. Hubbard, Langmuir, 2 (1986) 20.

28 P. Delahay, Double Layer and Electrode Kinetics, Wiley, New York, 1965.

29 G.A. Garwood, Jr. and A.T. Hubbard, Surf. Sci., 112 (1982) 281.

30 H. Baltruschat, M. Martinez, S.K. Lewis, F. Lu, D. Song, D.A. Stern, A. Datta and A.T.
    Hubbard, J. Electroanal. Chem., 217 (1987) 111.

31 F. Lu, G.N. Salaita, H. Baltruschat and A.T. Hubbard, J. Electroanal. Chem., 222 (1987)
    305.

32 A. Wieckowski, B.C. Schardt, S.D. Rosasco, J.L. Stickney and A.T. Hubbard, Surf. Sci., 146
    (1984) 115.

33 T. Solomun, A. Wieckowski, S.D. Rosasco and A.T. Hubbard, Surf. Sci., 147 (1984) 241.

34 D.A. Stern, H. Baltruschat, M. Martinez, J.L. Stickney, D. Song, S.K. Lewis, D.G. Frank
    and A.T. Hubbard, J. Electroanal. Chem., 217 (1987) 101.

35 M.P. Soriaga, V.K.F. Chia, J.H. White, D. Song and A.T. Hubbard, J. Electroanal. Chem.,
    162 (1984) 143.

36 D.G. Frank, J.Y. Katekaru, S.D. Rosasco, G.N. Salaita, B.C. Schardt, M.P. Soriaga, D.A.
    Stern, J.L. Stickney and A.T. Hubbard, Langmuir, 1 (1985) 587.

37 B.C. Schardt, J.L. Stickney, D.A. Stern, D.G. Frank, J.Y. Katekura, S.D. Rosasco, G.N. Salaita, M.P. Soriaga and A.T. Hubbard, Inorg. Chem., 24 (1985) 1419.

38 D.A. Stern, H. Baltruschat, M. Martinez, J.L. Stickney, D. Song, S.K. Lewis, D.G. Frank and A.T. Hubbard, J. Electroanal. Chem., 217 (1987) 101.

39 A.T. Hubbard, J.L. Stickney, S.D. Rosasco, M.P. Soriaga and D. Song, J. Electroanal. Chem., 150 (1983) 165.

40 J.L. Stickney, S.D. Rosasco, D. Song, M.P. Soriaga and A.T. Hubbard, Surf. Sci., 130 (1983) 326.

41 A. Wieckowski, S.D. Rosasco, B.C. Schardt, J.C. Stickney and A.T. Hubbard, Inorg. Chem., 23 (1984) 565.

42 J.L. Stickney, S.D. Rosasco, D. Song, M.P. Soriaga and A.T. Hubbard, Surf. Sci., 130 (1983) 326.

43 J.L. Stickney, S.D. Rosasco and A.T. Hubbard, J. Electrochem. Soc., 131 (1984) 260.

44 B.C. Schardt, J.L. Stickney, D.A. Stern, A. Wieckowski, D.C. Zapien and A.T. Hubbard, Surf. Sci., 175 (1986) 520.

45 J.L. Stickney, S.D. Rosasco, B.C. Schardt and A.T. Hubbard, J. Phys. Chem., 88 (1984) 251.

46 B.C. Schardt, J.L. Stickney, D.A. Stern, A. Wieckowski, D.C. Zapien and A.T. Hubbard, Langmuir, 3 (1987) 239.

47 J.L. Stickney, D.A. Stern, B.C. Schardt, D.C. Zapien, A. Wieckowski and A.T. Hubbard, J. Electroanal. Chem., 213 (1986) 293.

48 P.M. Ishikawa, J.Y. Katekaru and A.T. Hubbard, J. Electroanal. Chem., 86 (1978) 271.

49 L. Laguren-Davidson, F. Lu, G.N. Salaita and A.T. Hubbard, Langmuir, 4 (1988) 224.

50 T.E. Felter and A.T. Hubbard, J. Electroanal. Chem., 100 (1979) 473.

51 J.L. Stickney, B.C. Schardt, D.A. Stern, A. Wieckowski and A.T. Hubbard, J. Electrochem. Soc., 133 (1986) 648.

52 D.A. Harrington, A. Wieckowski, S.O. Rosasco, G.N. Salaita, A.T. Hubbard and J.B. Lumsden, Proceedings of the Purbaix Symposium, The Electrochemical Society, Princeton, NJ, 1984.

53 D.A. Harrington, A. Wieckowski, S.D. Rosasco, B.C. Schardt, G.N. Salaita, J.B. Lumsden and A.T. Hubbard, Corros. Sci., 25 (1985) 849.

54 D.A. Harrington, A. Wieckowski, S.D. Rosasco, G.N. Salaita and A.T. Hubbard, Langmuir, 1 (1985) 232.

55 D.G. Frank, M. Schneider, K. Werner and A.T. Hubbard, Langmuir, 3 (1987) 860.

56 A.T. Hubbard, Ph.D. Thesis, California Institute of Technology, 1967, University Microfilms, Ann Arbor, Michigan, Document 67-6066.

57 M.P. Soriaga and A.T. Hubbard, J. Am. Chem. Soc., 104 (1982) 3397.

58 M.P. Soriaga and A.T. Hubbard, J. Am. Chem. Soc., 104 (1982) 2735.

59 L.C. Pauling, Nature of the Chemical Bond, Cornell University Press, New York, 3rd edn., 1960, pp. 221-264.

60 M.P. Soriaga, J.H. White and A.T. Hubbard, J. Phys. Chem., 87 (1983) 3048.

61 V.K.F. Chia, M.P. Soriaga and A.T. Hubbard, J. Electroanal. Chem., 167 (1984) 97.

62 M.P. Soriaga, V.K.F. Chia, J.H. White, D. Song and A.T. Hubbard, J. Electroanal. Chem., 162 (1984) 143.

63 M.P. Soriaga, J.H. White, D. Song and A.T. Hubbard, J. Electroanal. Chem., 171 (1984) 359.

64 M.P. Soriaga and A.T. Hubbard, J. Am. Chem. Soc., 104 (1982) 2742.

65 D. Song, M.P. Soriaga and A.T. Hubbard, Langmuir, 2 (1986) 20.

66 D. Song, M.P. Soriaga and A.T. Hubbard, J. Electroanal. Chem., 193 (1985) 255.

67 K. Maartmann-Moe, Acta Crystallogr., 21 (1966) 979.

68 D. Song, M.P. Soriaga, K.L. Vieira, D.C. Zapien and A.T. Hubbard, J. Phys. Chem., 89 (1985) 3999.

69 D. Song, M.P. Soriaga and A.T. Hubbard, J. Electroanal. Chem., 201 (1986) 153.

70 D. Song, M.P. Soriaga and A.T. Hubbard, J. Electrochem. Soc., 134 (1987) 874.

71 J.Y. Katekaru, J. Hershberger, G.A. Gardwood Jr. and A.T. Hubbard, Surf. Sci., 121 (1982) 396.

72   J.H. White, M.P. Soriaga and A.T. Hubbard, J. Electroanal. Chem., 177 (1984) 89.
73   M.P. Soriaga, J.L. Stickney and A.T. Hubbard, J. Mol. Catal., 21 (1983) 211.
74   M.P. Soriaga, J.L. Stickney and A.T. Hubbard, J. Electroanal. Chem., 144 (1983) 207.
75   V.K.F. Chia, M.P. Soriaga, A.T. Hubbard and S.E. Anderson, J. Phys. Chem., 87 (1983) 232.
76   M.P. Soriaga and A.T. Hubbard, J. Phys. Chem., 88 (1984) 1758.
77   M.P. Soriaga and A.T. Hubbard, J. Electroanal. Chem., 159 (1983) 101.
78   K.L. Vieira, D.C. Zapien, M.P. Soriaga, A.T. Hubbard, K.P. Low and S.E. Anderson, Anal. Chem., 58 (1986) 2964.
79   J.L. Stickney, M.P. Soriaga, A.T. Hubbard and S.E. Anderson, J. Electroanal. Chem., 125 (1981) 73.
80   F. Lu, G.N. Salaita, L. Laguren-Davidson, D.A. Stern, E. Wellner, D.G. Frank, N. Batina, D.C. Zapien, N. Walton and A.T. Hubbard, Langmuir, 4 (1988) 637.
81   D.A. Stern, E. Wellner, G.N. Salaita, L. Laguren-Davidson, F. Lu, N. Batina, D.G. Frank, D.C. Zapien, N. Walton and A.T. Hubbard, J. Am. Chem. Soc., 110 (1988) 4885.
82   G.N. Salaita, L. Laguren-Davidson, F. Lu, N. Walton, E. Wellner, D.A. Stern, N. Batina, D.G. Frank, C.H. Lin, C.S. Benton and A.T. Hubbard, J. Electroanal. Chem., 245 (1988) 253.
83   N. Batina, D.G. Frank, J.Y. Gui, B.E. Kahn, C.H. Lin, F. Lu, J.W. McCargar, G.N. Salaita, D.A. Stern, D.C. Zapien and A.T. Hubbard, Electrochim, Acta, in press.
84   D.A. Stern, G.N. Salaita, F. Lu, J.W. McCargar, N. Batina, D.G. Frank, L.L. Davidson, C.H. Lin, N. Walton, J.Y. Gui and A.T. Hubbard, Langmuir, 4 (1988) 711.

Chapter 2

Heterogeneous Catalysis of Solution Reactions

MICHAEL SPIRO

1. General features

1.1 INTRODUCTION

The heterogeneous catalysis of gas reactions has been extensively studied and indeed forms the subject matter of three previous volumes (19–21) of Comprehensive Chemical Kinetics. The heterogeneous catalysis of solution ractions has received far less systematic attention. This is surprising since the phenomenon has been known and utilised sporadically for almost 150 years. As long ago as 1845, Millon [1] found that the oxidation of oxalic acid by iodate

$$2 \text{ HIO}_3 + 5 \text{ (COOH)}_2 \rightarrow \text{ I}_2 + 10 \text{ CO}_2 + 6 \text{ H}_2\text{O} \tag{I}$$

was catalysed by platinum sponge. The catalysis of the same reaction by platinum black was confirmed in 1921 by Lemoine [2] who, typically for this field, seems to have been unaware of Millon's earlier work. It was not until 1965 [3] that the many scattered literature reports of the catalysis by platinum metal of oxidation–reduction reactions were collected together and interpreted mechanistically.

The present review will begin by analysing various steps that can be rate-determining in heterogeneously catalysed solution reactions. These mechanisms can be distinguished in practice by the resulting kinetic behaviour and by other means that will be described. General stoichiometric and thermodynamic aspects will then be discussed. The later parts of this chapter will be devoted to a detailed survey of the specific types of catalysed reaction (substitution, isomerisation and redox) which have been studied in the literature.

The present chapter is concerned only with catalysis at the solid/liquid interface and will not deal with microheterogeneous catalysis by enzymes, micelles and polyelectrolytes even though the resulting kinetics are closely similar [4]. Moreover, little reference will be made to catalytic processes involving gases as these have been the subject of Vols. 19–21 of this series, nor to catalytic polymerisations which have been treated in Vols. 14, 14A, and 15.

1.2 CATALYTIC RATE

Let the overall reaction

$$v_\text{A} \text{ A} + v_\text{B} \text{ B} + \ldots \rightarrow v_\text{P} \text{ P} + v_\text{Q} \text{ Q} + \ldots \tag{II}$$

take place in a volume, V, of solution that contains a catalyst of mass m and surface area A. The catalytic rate per unit area of catalyst (mol $m^{-2} s^{-1}$) is then defined as

$$v_{cat} = -\frac{1}{Av_A}\frac{dn_A}{dt} = \frac{1}{Av_P}\frac{dn_P}{dt} = \ldots \tag{1}$$

where n_i represents the number of moles of i and where t is time. The quantity v_{cat} is sometimes termed the areal rate of reaction [5]. Division by m instead of by A gives a related quantity often called the specific rate of the reaction [5]. Chemical engineers often divide by the volume of catalyst pellets (cf. Sect. 1.6.2). Since the reaction is normally followed by periodically analysing the bulk solution for either reactant or product, eqn. (1) can be more usefully written in the form

$$v_{cat} = -\frac{V}{Av_A}\frac{dc_A}{dt} = \frac{V}{Av_P}\frac{dc_P}{dt} = \ldots \tag{2}$$

In practice, two other factors must also be taken into account. The first arises from the existence, in many cases, of a parallel homogeneous or bulk reaction whose kinetics can be determined in separate experiments. Its contribution can either be incorporated into the overall kinetic equation [e.g. eqn. (14), below] or else a correction must be applied to the overall measured reaction rate. A second correction is necessary whenever samples of solution (free from catalyst) are removed for analysis. This progressively decreases the ratio V/A and so creates artificially high changes in concentration. A formula incorporating both corrections was developed by Freund and Spiro [6] for the situation where only the early stages of the reaction are being followed. The velocities of the homogeneous reaction, v_{hom} (mol $dm^{-3} s^{-1}$) and of the heterogeneous reaction, v_{cat} (mol $m^{-2} s^{-1}$) can then be taken as constant. It follows that

$$c_j = c_o + v_{hom}\sum_1^j \Delta t_i + v_{cat}AS_j \tag{3}$$

where c_j is the concentration of the product being monitored at the time t_j when the jth aliquot is taken, Δt_i is the interval between sampling at times t_i and $t_{i-1'}$ and the function S_j is given by

$$S_j = \sum_1^j \frac{\Delta t_i}{V_o - (i-1)\Delta V} \tag{4}$$

in which V_o is the initial volume of the reaction mixture and ΔV the size of the aliquots being removed. Values of v_{cat} can then be derived from a plot of

$(c_j - c_o - v_{hom}\sum_1^j \Delta t_i)$ against S_j. This treatment has made no assumptions

about the rate law governing the catalytic reaction. However, if it is known that the latter is first order with a rate constant k, the sampling correction alone can be treated by Bradley's equation [7]

$$\sum_{0}^{j-1} V_i \ln \left(\frac{c_i}{c_{i+1}} \right) = kAt_j \tag{5}$$

The function on the left-hand side is plotted against t_j to yield a value for k. Other correction equations can easily be developed to fit different circumstances.

1.3 RATE-DETERMINING STEPS IN THE CATALYSIS

Heterogeneous catalysis must of necessity involve interaction between the surface and at least one of the reactants. The catalytic process therefore involves five distinct steps [8, 9]:

(i) mass transport of reactant(s) to the catalyst surface from the bulk solution;

(ii) adsorption of reactant(s) on the surface;

(iii) chemical reaction at the surface;

(iv) desorption of product(s) from the surface; and

(v) mass transport of product(s) away from the surface into the bulk solution.

Any of these steps could be rate-determining. We shall now consider them individually together with their kinetic consequences.

1.4 ADSORPTION AND COMPETITIVE ADSORPTION

1.4.1 Rates of adsorption and desorption

The kinetics of adsorption of solutes in solution have rarely been studied in detail and the kinetics of desorption have been studied even less. Some information is available on the contact times necessary for adsorption equilibrium to be attained: these vary from a few seconds or minutes to several days for porous adsorbents or for polymer solutes [10, 11]. However, contact times depend on both steps (i) and (ii), above. The mass transport step (i), whether involving "external" diffusion through the solution to the outer surface of the adsorbent or "internal" diffusion within its pores, is almost always rate-determining in adsorption. It will be discussed at greater length in Sect. 1.6. The intrinsic adsorption step (ii) has been found to require times of the order of minutes for the adsorption on to metal surfaces of molecular solutes such as linoleic and dilinoleic acids [12], dimethylsulphoxide [13], and dimethylformamide [14]. These experiments were carried out by using radioactively labelled solutes. For many other systems, step (ii) is too fast to be measured by conventional methods. Recent relaxation technique experiments by Yasunaga and his group have yielded very large rate constants for the adsorption/desorption of aqueous transition metal ions on γ-Al$_2$O$_3$ [15], IO$_3^-$ ions on TiO$_2$ [16], H$^+$ on γ-Fe$_2$O$_3$ and Fe$_3$O$_4$ [17], and acetic acid on silica–alumina [18]. This last molecular solute adsorbed more slowly than did the ions, many of which reached adsorption equilibrium in a fraction of a second. Hydroquinone also takes a few seconds to adsorb on to platinum [19]. It should be added that the desorption step (iv) is often slower than step (ii) and there is some evidence in the literature for slow desorption of solute species from carbon surfaces [20]. But in only one case, the very rapid catalysis by colloidal platinum of the reaction between H$^+$ ions and the methylviologen radical cation MV$^+$, has it been shown that adsorption of a reactant (MV$^+$) or desorption of a product (MV^{2+}) is rate-limiting [21]. On balance, therefore, we may conclude that steps (ii) and (iv) are usually too fast to be rate-controlling in solution catalysis. The greater rate of an overall adsorption process compared with that of a surface cata-lytic process is illustrated later in Fig. 2.

Before we leave this topic, it would be wise to note the results of some recent research on heterogeneously catalysed gas reactions. Here finite rates of adsorption and desorption had to be introduced into the reaction scheme in order to explain the occurrence of multiple steady states and oscillatory phenomena. This observed exotic behaviour could be reproduced by solving a set of coupled equations for the rates of adsorption/desorption, the rate of the surface reaction, and the mass balance relations [22, 23]. Adsorption steps (ii) and (iv) may therefore need to be invoked for any heterogeneously catalysed solution reactions that are found to exhibit sim-ilar dynamic behaviour.

1.4.2 Adsorption isotherms

The *extent* of adsorption can have a profound effect on the rate of the surface reaction. Equilibrium isotherms of many kinds have been reported for adsorption from solution and have been classified by Giles et al. [24–27]. The shapes of these adsorption curves often furnish qualitative information on the nature of the solute–surface interactions. Several of the types of isotherm observed in dilute solution are represented reasonably well by three simple and popular isotherm equations, those of Henry, Langmuir, and Freundlich. Their shapes are illustrated in Fig. 1. Each of these isotherms relates the surface concentrations c_{ads} (mol m^{-2}) to the bulk equilibrium concentration c of the solute species in question. When few surface sites are occupied, Henry's law adsorption

$$c_{ads} = hc \tag{6}$$

can apply, with h a parameter of the system. Under such low-coverage conditions the adsorption of any other species, whether reactant or product, takes place independently and is described by a similar equation with an appropriate value of h.

Many more cases of adsorption from solution lead to isotherms in which c_{ads} rises with increasing c in dilute solution and then levels off to a limiting

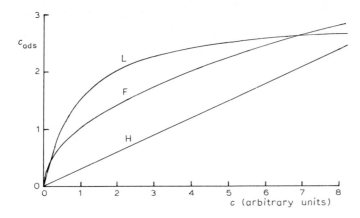

Fig. 1. Typical shapes of Henry (H), Langmuir (L) and Freundlich (F) isotherms. For the Freundlich isotherm, α has been taken as 0.5 ($n = 2$).

plateau value. This behaviour is often found with compounds that are solid at room temperature and only sparingly soluble in the solvent concerned [10, 28]. Such isotherms are usually well described by the Langmuir equation

$$c_{ads} = \theta c_{mono} = \frac{Kc\, c_{mono}}{1 + Kc} \tag{7}$$

where c_{mono} (mol m^{-2}) is the surface concentration corresponding to monolayer coverage, θ the fraction of the surface covered, and K the Langmuir adsorption coefficient. The value of K decreases with rising temperature and its dependence on substituent effects within adsorbate molecules is referred to in Sect. 1.7. The theoretical assumptions underlying the Langmuir model are well known [29, 30]. It is one of its major assumptions that the heat of adsorption on every surface site is the same, a postulate that has been shown to be untrue [29, 30]. However, not all sites will be equally effective in catalysis. Too strong an attachment of a reagent species to a site may immobilise it for reaction; too weak an attachment will not activate the species sufficiently to induce the necessary bond making or breaking. There is now good experimental evidence [31] that for kinetic purposes only sites with a relatively narrow range of intermediate adsorption energies are effective, a situation that corresponds to the Langmuir model. A further assumption of the model, that adsorption takes place only up to monolayer coverage and does not continue to the multilayer stage, matches the conditions of liquid-phase catalyses better than those of gas-phase ones because in solution the ubiquitous solvent molecules compete with reactant species for surface sites [28]. These considerations help to explain the success of the Langmuir isotherm. Of course co-reactants, product species, and other solutes may also compete for the surface sites. The appropriate Langmuir equation for a given species then becomes

References pp. 159–166

$$c_{ads} = \theta c_{mono} = \frac{Kc\,c_{mono}}{1 + \Sigma_i K_i c_i} \qquad (8)$$

Another common isotherm is that of Freundlich which may be written

$$c_{ads} = gc^\alpha \quad (\alpha < 1) \qquad (9)$$

although the exponent is often expressed as $1/n$ where $n > 1$. Both g and n usually decrease with increasing temperature. This isotherm can be derived from a model in which, as a result of surface heterogeneity, the heat of adsorption falls logarithmically as θ increases. Although the observed adsorption enthalpies usually decrease linearly with rising θ, this postulate is at least qualitatively in accord with the facts [29, 30].

1.4.3 Catalyst area

The area is an important surface parameter for catalytic studies. It is needed to evaluate the rate constant of the surface reaction from the kinetics as well as to allow a fair comparison to be made of the effectiveness of different catalysts. Areas are commonly determined by nitrogen or krypton gas adsorption interpreted by the Brunauer–Emmett–Teller (BET) isotherm [30, 32]. A number of other methods has been proposed and utilised including microscopy, isotopic exchange, chromatography, gas permeability, adsorption from solution, and negative adsorption (desorption) of co-ions [30, 33].

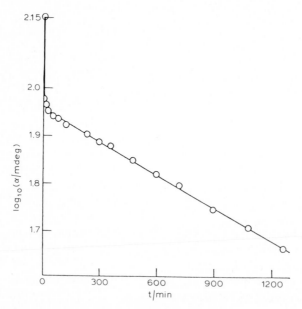

Fig. 2. Variation of the logarithm of the optical rotation amplitude (α) with time in the racemisation of 2×10^{-3} mol dm^{-3} $(+)_{589}$-[Co(en)$_3$]$^{3+}$ catalysed by Black Pearls carbon (0.25 g in 25 cm^3 solution) at 40°C. (After Mureinik and Spiro [35 and private communication].)

Van den Hul and Lyklema [33] have compared the results of several methods for three different solids. The agreement was generally satisfactory although for dispersed silver iodide the area obtained by "wet" methods was approximately 3 times as large as by "dry" methods. Adsorption from solution is therefore an attractive possibility for surface area measurements, being both quicker and simpler than a method requiring vacuum apparatus [34]. One should, however, heed the experimental precautions enjoined by Everett [11] and note Kipling's prudent comments on the interpretation of the results [34]. It is particularly important to realise that the area obtained depends upon the size and orientation of the adsorbate employed. The one recommended by Giles et al. [25, 27] is p-nitrophenol. This coloured molecule adsorbs end-on at polar surfaces and flat on benzenoid surfaces like graphite. Giles and Nakhwa [25] found reasonable agreement with BET (N_2) values although the extent of nitrogen adsorption was greater for solids with small pores that admitted only the smaller molecule. The sizes and shapes of pores in porous solids are briefly discussed in Sect. 1.6.2.

Ideally one would like to determine the effective surface area of a catalyst using as adsorbate the reactant species in question. This has indeed proved possible in a few cases where the adsorbent area is large. One such favourable situation occurred in the racemisation of $(+)_{589} - [Co(en)_3]^{3+}$ catalysed by a carbon black [35] where fast initial adsorption of the substrate was followed by its slow isomerisation, as shown in Fig. 2. The difference between the initial reading and the intercept of the first-order line was a measure of the extent of the substrate adsorption.

In principle one should even be able to determine the extent of substrate adsorption during the course of the reaction itself, provided an appreciable fraction of the reactant is adsorbed. If the experimenter knows the initial reactant concentration, a, and can measure both its bulk concentration, c, at a given time as well as the concentration, x, of a non-adsorbed product, then he can calculate the number of moles of reactant absorbed from the equation

$$Ac_{ads} = V([a - x] - c) \tag{10}$$

where A is the area of catalyst and V the volume of solution. This method was used by Mortimer and Spiro [36] in their study of the solvolysis of $PhCH_2CMe_2Cl$ in the presence of an activated carbon. Here the product H^+ remained non-adsorbed because its concentration was kept low and constant with a pH-stat. The c_{ads} versus c data obtained led to an isotherm from which the value of c_{mono} could be deduced and compared with that derived from BET (N_2) measurements. The figures showed that fewer sites on the surface were accessible to the large substrate molecules than to the much smaller N_2 molecules, a result in keeping with adsorption studies in the literature on non-reacting organic solutes and consonant with the presence of pores in the catalyst.

References pp. 159 166

1.4.4 Surface orientation and competitive adsorption

For insights into the catalytic mechanism it is important to know not only the number of adsorbed species on the surface but also their orientation. Configurations on the surface can, in certain cases, be inferred from infrared spectroscopy or more commonly from adsorption data and molecular dimensions [10, 28]. A table listing orientations deduced in this way has been given by Parfitt and Rochester [28] for various organic solutes on adsorbents such as TiO_2, SiO_2, Al_2O_3, and graphon. A further source of information comes from preferential or selective adsorption studies with liquid mixtures or solutions [37], a technique that has been extended to adsorption on electrodes [38]. Competitive adsorption has also proved to be a potent tool in catalytic research. Thus Pincock and his group [39, 40] discovered that the very strong catalysis by carbons of the racemisation of 1, 1'- binaphthyl in acetone

(III)

was inhibited by the addition of quite small quantities of higher molecular weight aromatics. Their effect increased in the sequence benzene ≪ naphthalene < anthracene < pyrene < perylene. Dramatic poisoning was produced by as little as 1×10^{-5} mol dm^{-3} of the larger polynuclear aromatics. Their inhibition of the catalysis was consistent with a mechanism in which the planar transition state of the 1, 1'- binaphthyl was adsorbed in a trans-planar arrangement onto the graphite-like carbon surface. A different example comes from a study by Barbosa et al. [41] of the solvolysis of t-BuBr catalysed by AgBr. These workers found that the rate of the catalysed reaction was not affected by adding the product t-BuOH but was markedly reduced by adding small concentrations of KBr. The presence of KNO_3, on the other hand, produced only a modest rate decrease. These facts can easily be explained if the t-BuBr molecule adsorbs on to the AgBr surface with its –Br end and not with its t-Bu– end. Bromide ions and t-BuBr molecules then compete for surface silver ion sites. The lesser effect of KNO_3 will have been due to the formation of an electrical double layer which partly shielded the surface.

1.5 SURFACE KINETICS

In surface-controlled catalyses, the rate-determining step involves the reaction on the surface of an adsorbed reactant or of a derived species. The resultant kinetics may be more or less complicated depending upon the circumstances, and treatments of various cases have been given in reviews dealing with the heterogeneous catalysis of gas reactions [9, 31, 42]. Several basic equations appropriate for the heterogeneous catalysis of solution reactions are given below. Except in Sect. 1.5.3, the back reactions have not been included although the theories can easily be extended to cover these.

It is convenient to divide surface reactions in solution into four main categories: unimolecular, racemisation and isotopic exchange, bimolecular and electron transfer.

1.5.1 Unimolecular reactions

In unimolecular surface reactions, a single reactant species is adsorbed and reacts on the surface. A good example is given by catalysed nucleophilic solvolyses discussed further in Sect. 2.1. The reaction scheme may then be written

$$
\begin{array}{ccc}
A & \xrightarrow{\ k_{\text{hom}}\ } & P & + & Q \\
\Big\updownarrow & & \Big\updownarrow & & \\
A_{\text{ads}} & \xrightarrow{\ k_{\text{het}}\ } & P_{\text{ads}} &
\end{array}
\tag{IV}
$$

If the initial concentration of reactant A is a, then at any given time t the bulk concentrations of A, P, and Q will be $a - x - (Ac_{A_{\text{ads}}}/V)$, $x - (Ac_{P_{\text{ads}}}/V)$, and x respectively. The corresponding differential rate equation can be shown to be [43]

$$
\frac{dx}{dt} = k_{\text{hom}}\left(a - x - \frac{A}{V}c_{A_{\text{ads}}}\right) + \frac{A}{V}k_{\text{het}}c_{A_{\text{ads}}}
\tag{11}
$$

so that

$$
v_{\text{cat}} = k_{\text{het}}c_{A\text{ads}}
\tag{12}
$$

If adsorption equilibrium is established rapidly and the adsorbed and bulk species remain in equilibrium throughout the reaction, $c_{A\text{ads}}$ can be expressed in terms of a suitable isotherm. This allows the differential kinetic equation to be integrated. For example, if Henry's law adsorption is presumed to apply [43]

$$
\frac{dx}{dt} = (a - x)\left(k_{\text{hom}} + \frac{Ah[k_{\text{het}} - k_{\text{hom}}]}{V + Ah}\right)
\tag{13}
$$

It is evident that the overall first-order rate constant will be greater than the homogeneous rate constant if there is appreciable heterogeneous catalysis ($k_{\text{het}} > k_{\text{hom}}$) and smaller if adsorption occurs without significant catalysis. Both cases have been observed in the laboratory [36, 41].

Should the adsorption follow the Langmuir format, with A and P adsorbing competitively on the same sites, then

$$\frac{dx}{dt} = (a - x)\left(k_{\text{hom}} + \frac{A}{V} \cdot \frac{k_{\text{het}} K_A c_{\text{mono}}}{1 + K_A(a - x) + K_P x}\right) \tag{14}$$

provided the extents of adsorption of both A and P are small [43]. Because of this restriction in the derivation, non-reactive adsorption of A would wrongly appear to bring about no diminution in the overall rate. The integrated form of eqn. (14) contains two first-order terms

$$(1 + \psi) k_{\text{hom}} t = \ln\left(\frac{a}{a - x}\right) + \psi \ln\left(\frac{a + \xi[1 + \psi]}{a + \xi[1 + \psi] - x}\right) \tag{15}$$

where

$$\xi = \frac{1 + K_P a}{K_A - K_P}$$

and

$$\psi = \frac{k_{\text{het}} A}{k_{\text{hom}} V} \cdot \frac{K_A c_{\text{mono}}}{1 + K_P a}$$

The treatment of Freundlich adsorption also becomes tractable only by making the assumption that the extents of adsorption are small. This leads to the kinetic equations

$$\frac{dx}{dt} = k_{\text{hom}}(a - x) + \frac{A}{V} k_{\text{het}} g(a - x)^\alpha \tag{16}$$

$$(1 - \alpha) k_{\text{hom}} t = \ln\left(\frac{\phi + a^{1-\alpha}}{\phi + [a - x]^{1-\alpha}}\right) \tag{17}$$

where $\phi = Agk_{\text{het}}/Vk_{\text{hom}}$. The way in which the various differential and integrated rate equations can be applied to experimental data has been discussed in detail in the literature [36, 41, 43].

1.5.2 Optical racemisation and isotopic exchange reactions

In all the reactions

$$\text{D-(A)} \rightleftharpoons \text{L-(A)} \tag{V}$$

$$\text{AX} + \text{BX*} \rightleftharpoons \text{AX*} + \text{BX} \tag{VI}$$

$$\text{Ox} + \text{Red*} \rightleftharpoons \text{Red} + \text{Ox*} \tag{VII}$$

(where Ox and Red are conjugate components of a given redox couple), there are no overall changes in the concentrations of the *chemical* components during the reaction. McKay [44, 45] was the first to recognize that, in consequence, any isotopic exchange run will always exhibit first-order kinetics, irrespective of the actual mechanism of the reaction. The mechanism can only be deduced by carrying out a series of runs with different starting concentrations of reagents. Many years later, Mureinik and Spiro [35]

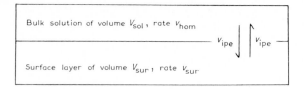

Fig. 3. Two-phase model used to develop the kinetic equations for catalysed racemisations and isotopic exchange reactions. The rate of inter-phase exchange (v_{ipe}) is much faster than the rate of the reaction in each phase.

pointed out that pure optical racemisations should obey the same rule. Totterdell and Spiro [46, 47] subsequently derived rate equations for racemisation and isotopic exchange reactions when they take place both homogeneously in bulk solution and heterogeneously on a catalyst surface. Only the bulk solution can be sampled and analysed whereas most of the reaction may occur at the interface. The problem was made tractable by regarding the reaction layer on the surface as a separate phase immiscible with the bulk solution phase (Fig.3). The exchange of solute species between the two phases was assumed to be rapid, much more rapid than the rates of racemisation or exchange within each phase. This model led to general kinetic equations that included the contributions from both homogeneous and heterogeneous processes. Applications to real systems will be discussed in Sects. 3 and 4.2.

Let us first consider the racemisation reaction (V). In purely homogeneous solution [35]

$$\frac{1}{t} \ln \left(\frac{c_D + c_L}{c_D - c_L} \right) = \frac{1}{t} \ln \left(\frac{\alpha_0}{\alpha_t} \right) = \frac{2v_{hom}}{c_0} \tag{18}$$

where the overall concentration $c_0 = c_D + c_L$ and where α is the optical rotation by which the reaction is followed. In any given run, c_0 and the rate v_{hom} (mol dm^{-3} s^{-1}) remain constant so that first-order kinetics are observed. Further experiments with other c_0 values are needed to reveal the functional dependence of v_{hom} on c_0 and so provide a guide to the mechanism. For example, if the racemisation is a truly first-order process then $v_{hom} = k_{hom} c_0$. When a catalyst is present, the model described above leads to the rate equation [46]

$$\frac{1}{t} \ln \left(\frac{c_D + c_L}{c_D - c_L} \right)_{sol} = \frac{1}{t} \ln \left(\frac{\alpha_0}{\alpha_t} \right)_{sol} = \frac{2(v_{hom} V_{sol} + v_{sur} V_{sur})}{(c_0)_{sol} V_{sol} + (c_0)_{sur} V_{sur}} \tag{19}$$

where subscript sol refers to the bulk solution, V_{sol} is its volume, and V_{sur} the volume of the thin surface layer (normally $V_{sol} \gg V_{sur}$). The rate in the surface layer, v_{sur} (mol dm^{-3} s^{-1}), is related to the catalytic rate v_{cat} (mol m^{-2} s^{-1}) by the relation

$$v_{cat} = \frac{v_{sur} V_{sur}}{A} \tag{20}$$

The significance of eqn. (19) is understood more easily by dividing both numerator and denominator of the right-hand side by $(V_{sol} + V_{sur})$. Comparison with eqn. (18) then shows that the appropriate rate in the presence of a catalyst is the sum of the volume-weighted rates in the two phases (i.e. in the bulk solution and in the surface layer). Similarly, the appropriate overall concentration has become the sum of the volume-weighted overall concentrations. This is equal to the global concentration of the system as a whole, $(c_0)_{sys}$.

Some corollaries are worthy of mention.

(1) The solid will act catalytically provided $v_{sur} > v_{hom}$. If $v_{sur} < v_{hom}$, introduction of the solid will slow down the reaction.

(2) Should the homogeneous rate v_{hom} be zero, so that the reaction proceeds only on the catalyst surface, then the right-hand side of eqn. (19) reduces to $2v_{sur}V_{sur}/(c_0)_{sys}V_{sol}$ if, as usual, $V_{sur} \ll V_{sol}$. The effective reaction rate (mol dm^{-3} s^{-1}) is thus $v_{sur}V_{sur}/V_{sol}$ or $v_{cat}A/V_{sol}$.

(3) For systems in which the surface reaction is first order

$$v_{sur} = k_{sur}(c_0)_{sur} = \frac{k_{sur}(c_0)_{ads}A}{V} = \frac{k_{sur}(c_0)_{ads}}{\delta_s} \tag{21}$$

where $(c_0)_{ads}$ is the sum of the adsorbed concentrations of the optically active forms and δ_s is the effective thickness of the surface layer. It is interesting to note that the form taken up by eqn. (19) when the bulk and surface reactions are both first order can equally well be derived by the procedure described in Sect. 1.5.1 using the reaction scheme

$$
\begin{array}{ccc}
\text{D-(A)} & \underset{k_{-1}}{\overset{k_1}{\rightleftarrows}} & \text{L-(A)} \\
\big\updownarrow & & \big\updownarrow \\
\text{D-(A)}_{ads} & \underset{k'_{-1}}{\overset{k'_1}{\rightleftarrows}} & \text{L-(A)}_{ads}
\end{array}
\tag{VIII}
$$

As long as the bulk and surface concentrations of both species are presumed to be in equilibrium throughout, the desired result can be obtained by inserting a suitable isotherm such as that of Henry or Langmuir [43].

Isotopic exchange reactions of type (VI) and (VII) follow the same pattern. The course of the reaction may be represented by the concentrations in the scheme below in which $a \gg x_\infty$ and $b \gg y_0$.

	AX	+	BX*	$\rightleftharpoons$	AX*	+	BX	
	Ox$_1$	+	Red$_1^*$	$\rightleftharpoons$	Ox$_1^*$	+	Red$_1$	
At time $t = 0$	a		y_0		0		$b - y_0$	(IX)
At time $t = t$			y		x			
At time $t = \infty$			y_∞		x_∞			

McKay [44, 45] proved that the homogeneous reaction for such systems invariably proceeds according to the first-order equation

$$\ln\left(\frac{x_\infty}{x_\infty - x}\right) = \ln\left(\frac{y_0 - y_\infty}{y - y_\infty}\right) = v\left(\frac{1}{a} + \frac{1}{b}\right)t \tag{22}$$

The overall rate of exchange, v, is a function of a and b and so remains constant throughout a given run. In order to establish the mechanism of the reaction, the functional dependence of v must be found by carrying out experiments with different concentrations a and b. If, for instance, the exchange process is first order in each reactant, $v = kab$ [45].

In the presence of a solid catalyst, the exchange reaction proceeds in the surface layer as well as in the bulk solution. Application of the two-phase model, together with the assumption of extremely fast inter-phase exchange, then leads to the equation [47]

$$\frac{1}{t}\ln\left(\frac{x_\infty}{x_\infty - x}\right)_{sol} = (v_{hom} V_{sol} + v_{sur} V_{sur})$$

$$\times \left(\frac{1}{a_{sol} V_{sol} + a_{sur} V_{sur}} + \frac{1}{b_{sol} V_{sol} + b_{sur} V_{sur}}\right) \tag{23a}$$

$$= \left(\frac{v_{hom} V_{sol} + v_{sur} V_{sur}}{V_{sol} + V_{sur}}\right)\left(\frac{1}{a_{sys}} + \frac{1}{b_{sys}}\right) \tag{23b}$$

where $a_{sys} = (a_{sol} V_{sol} + a_{sur} V_{sur})/(V_{sol} + V_{sur})$, and similarly for b_{sys}. This equation bears a marked resemblance to eqn. (22), with the velocity of isotopic exchange now being given by the sum of the volume-weighted exchange velocities in the two phases. If the homogeneous rate is zero and if, as usual, the volume of the surface layer is much smaller than that of the bulk solution, the effective rate of isotopic exchange in the system becomes simply $v_{sur} V_{sur}/V_{sol}$ or $v_{cat}A/V_{sol}$. A bimolecular reaction step on the surface would be revealed if further experiments showed that $v_{sur} \propto a_{sur}b_{sur}$ or $v_{cat} \propto a_{ads}b_{ads}$. The rate equation for the catalysed electron exchange reaction (VII) is discussed in more detail in Sect. 4.2.

1.5.3 Bimolecular reactions

In bimolecular surface reactions we must distinguish between 4 different types of rate-determining step. With all stoichiometric coefficients taken as unity, the first type can be written

$$A_{ads} + B \rightarrow P_{(ads)} + Q_{(ads)} \tag{X}$$

This is the Rideal–Eley mechanism in which adsorbed reactant A is attacked by reactant B from the bulk solution. The product(s) may or may not be adsorbed. Thus

$$v_{cat} = k_{het}c_{A_{ads}}c_B \tag{24}$$

If the adsorption of A follows the Langmuir form (7) or (8)

$$v_{cat} = \frac{k_{het}K_A c_{A_{mono}} c_A c_B}{1 + K_A c_A (+ K_P c_P + K_Q c_Q)} \tag{25}$$

This mechanism is fairly rare in the heterogeneous catalysis of gas reactions [9, 31, 42] and no clear case of it has yet been reported with heterogeneously catalysed solution reactions.

In the second type of bimolecular surface reaction, the rate-determining step takes place between adsorbed reactant A and absorbed reactant B sitting on adjacent sites

$$A_{ads} + B_{ads} \rightarrow P_{(ads)} + Q_{(ads)} \tag{XI}$$

This Langmuir–Hinshelwood mechanism is the one most commonly encountered in the heterogeneous catalysis of gas reactions and the appropriate rate expressions for various special cases are well known [9, 31, 42]. In general, we may write

$$v_{cat} = k_{het} c_{A_{ads}} c_{B_{ads}} \tag{26}$$

although some authors like Thomas and Thomas [42] prefer the alternative version

$$v_{cat} = \frac{k_{het} c_{A_{ads}} c_{B_{ads}}}{c_{mono}} \tag{27}$$

The adsorbed concentrations can now be related to the bulk concentrations by a suitable isotherm such as that of Langmuir. If the reactants (and possibly the products) adsorb competitively on the surface sites

$$v_{cat} = \frac{k_{het} K_A K_B c_{mono}^2 \, c_A c_B}{(1 + K_A c_A + K_B c_B [+ K_P c_P + K_Q c_Q])^2} \tag{28}$$

Equation (28) has been called the Langmuir rate law [5]. Certain special cases of this equation lead to a variety of different kinetic forms. For example, if all species are only slightly adsorbed, the denominator tends to unity and the reaction becomes simply first-order in each of A and B. On the other hand, if A (and P and Q) is weakly adsorbed and B strongly, the denominator reduces to $K_B^2 c_B^2$ which converts the reaction into one that is first order in A but inverse first order in B. Strong adsorption of one of the reactants thus denies surface sites to the other reactant and effectively stifles the catalytic process.

The third type of bimolecular surface reaction is also represented by eqns. (XI) and (26) but in this case the two reactants adsorb non-competitively on different kinds of surface site. In terms of the Langmuir model

$$v_{cat} = \frac{k_{het} K_A K_B c_{A_{mono}} c_{B_{mono}} c_A \, c_B}{(1 + K_A c_A)(1 + K_B c_B)} \tag{29}$$

No P and Q terms have been included in the denominator as the products could sit on either kind of site. Once more a range of kinetic forms emerges for special conditions. Weak adsorption by both reactants again yields straightforward second-order kinetics whereas strong adsorption by one reactant (say, B) leads to a rate equation that is simply first order in the other reactant (A). Situations of non-competitive adsorption arise more frequently in the catalysis of solution reactions than they do in the catalysis of gas reactions. Examples are given in Sects. 2.2 and 4.2.

In all these three types of surface-controlled catalysis, it is advisable to check the adsorption inferences reached by interpretation of the experimental kinetic law. Wherever possible, independent adsorption experiments should be carried out with the individual substances concerned. Alternatively, or in addition, infrared and preferably FTIR measurements will often reveal useful information on the presence and state of adsorbed species.

The mechanism of the fourth category of bimolecular surface steps is peculiar to redox reactions catalysed by metals and semiconductors. Here both reactants sit on the surface, not necessarily on adjacent sites, and the electrons are transferred from the reducing to the oxidising species through the solid catalyst. The rate therefore depends not only on the concentrations at the surface but also on the potential taken up by the catalyst, and this potential in turn is a function of the concentrations of the electroactive species present. Equations (28) and (29) fail to represent the kinetics in these cases because k_{het} is no longer independent of concentration. These kinetics must accordingly be treated by an electrochemical method of analysis and this is done in Sect. 4.1.

1.6 DIFFUSION-CONTROLLED CATALYSES

1.6.1 Tests for mass transport control

Mass transport is much more likely to be rate-controlling in the heterogeneous catalysis of solution reactions than in that of gas reactions. The reason lies in the magnitudes of the respective diffusion coefficients [48]: for molecules in normal gases at 1 bar and 300 K these are 10^{-5} to 10^{-4} m^{2}s^{-1} while, for typical solutes in aqueous solution, they are 10^{-10} to 10^{-9} m^2 s^{-1}. The rate-determining step in many solution catalyses has indeed been found to be external diffusion of reactant(s) to the outer surface of the catalyst and/or diffusion of product(s) away from it [3, 6]. Another possibility is internal diffusion within the pores of the catalytic solid, a step that often determines the rates of catalysed gas reactions [49–51]. It is clearly an essential part of a kinetic investigation to ascertain whether any of these steps control the rate of the overall catalytic process. Five main diagnostic criteria have been employed for this purpose:

(1) Variation of stirring. More effective agitation of the solution around the solid catalyst does not affect chemically- or surface-controlled processes but markedly increases external diffusion rates. Figure 4 illustrates how the

84

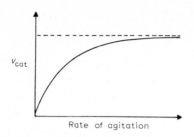

Fig. 4. Variation of catalytic rate with the rate of agitation of the solution. The dotted line marks the asymptotic limit of infinitely fast stirring.

overall reaction rate can respond to increased agitation of the solution, being diffusion limited at low flows and surface-controlled at sufficiently high ones [52]. This test should always be carried out, preferably with a rotating catalytic disc (see below). The rate of pore diffusion is little influenced by stirring but mass transport in short macropores should increase under turbulent flow conditions.

(2) Variation of catalyst area. The catalytic rate is proportional to the total surface area, A, external and internal, for reactions controlled by surface kinetics. In the case of internal or pore diffusion control, the rate is proportional to $A^{1/2}$ and is also a function of the catalyst shape and size [49, 53]. Under an external diffusion regime, the catalytic rate is proportional to the external surface area of the catalyst, A_{ex}.

(3) Magnitude of the catalytic rate. The rate of a reaction controlled by external diffusion can often be calculated when the hydrodynamic conditions are well established (see below). This calculated value will either be close to the observed rate or else much larger than it; if the former, diffusion control can be inferred while in the latter case the reaction may safely be taken as either pore- or surface-controlled.

Calculations of this kind invariably require a value for the diffusion coefficient, D. Certain source books are extremely useful for locating literature diffusion data [54, 55] but, should no experimental value be available, a fair estimate can be made by means of the Stokes–Einstein equation

$$D = \frac{kT}{6\pi r\eta} \tag{30}$$

where k is the Boltzmann constant, T the absolute temperature, r the effective radius of the solute species, and η the solvent viscosity. The equation is based on a model of a spherical solute diffusing through a continuous incompressible medium under conditions of "stick", i.e. when the solvent immediately adjacent to the sphere moves along with it. Under conditions of "slip", when the sphere drags no solvent with it, the 6π factor must be replaced by 4π [55, 56]. Related equations appropriate for solutes that are ellipsoids, flat discs, rod-shaped, or chain-like have been given elsewhere [56].

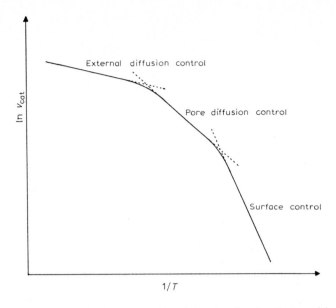

Fig. 5. Arrhenius plot for a catalysed reaction showing the transition between diffusion control at high temperatures and surface control at low temperatures.

(4) Variation of reactant concentrations. The observed reaction orders can provide pointers to the catalytic mechanism in cases where theoretical equations exist for both surface-controlled and diffusion-controlled situations (cf. Sect. 4).

(5) Variation of temperature. The activation energies of purely diffusion-controlled catalyses are low, generally 10–$16\,kJ\,mol^{-1}$, whereas those for surface-controlled reactions are usually much higher. It follows that catalytic reactions that are surface-controlled at low temperatures may well become mass transport limited at high temperatures [48], as depicted in Fig. 5. However, it will be shown in the following sections that, in certain mass transport regimes, the rate constant depends also on kinetic or thermodynamic factors whose enthalpy terms then form part of the activation energy.

1.6.2 Pore diffusion

Many catalysts are porous, a feature which greatly increases their surface area [48]. Pores above $50\,nm$ in width are termed macropores, those with widths below $2\,nm$ are called micropores, and those of intermediate size are mesopores. Not only the size but also the shape of pores can vary widely and common descriptions refer to open and closed cylinders, slits, cones, spheroidal cavities, and ink-bottle shapes. The type of pore can frequently be identified from the shape of the hysteresis loop in physical gas adsorption experiments [32, 49]. The absence of hysteresis indicates that the pores are closed perfect cylinders, or that the solid is microporous or, indeed, non-

References pp. 159–166

porous. Comparison between the nitrogen adsorption isotherms for a given porous solid and for an appropriate non-porous solid also provides information on the type of pore present as well as on the pore size distribution. The methods and limitations of calculating the distribution of pore sizes from adsorption and desorption isotherms has been ably discussed by Lecloux [32]. An alternative technique for obtaining pore size distributions employs pressure penetration by mercury, a non-wetting liquid. The smaller the pore, the greater the pressure required to force mercury inside. For many years the practical limit for this technique was reached at widths of approx. 15 nm [32] but recent high pressure equipment allows smaller mesopores to be examined.

The whole of the internal surface area of a porous catalyst will be available for the catalytic reaction if the rates of diffusion of reactant into the pores, and of product out of them, are fast compared with the rate of the surface reaction. In contrast, if the reactant diffuses slowly but reacts rapidly, conversion to product will occur near the pore entrances and the interior of the pores will play no role in the catalysis. Ion exchange resins are typical examples of catalysts for which such considerations are important (cf. Sect. 2.3). The detailed mathematics of this problem have been treated in several texts [49–51] and we shall now quote some of the main theoretical results derived for isothermal conditions. The parameters involved tend to be those employed by chemical engineers and differ somewhat from those used elsewhere in this chapter. In particular, the catalyst material (active + support) is present in the form of pellets of volume V_p and the catalytic rates v_v are given per unit volume of pellet ($mol\,s^{-1}\,m^{-3}$). The decrease in v_v brought about by pore diffusion is then expressed by an effectiveness factor, η, defined by

$$\eta = \frac{\text{observed or actual catalytic rate}}{\text{catalytic rate with no diffusion limitation}} \tag{31}$$

This means that the rate of reaction based on the pellet volume is given by

$$v_v = \eta k_v c_A^n \tag{32}$$

where n is the kinetic order of the surface reaction, c_A is the concentration of A at the outside surface of the catalyst pellet, and k_v is the corresponding rate constant (s^{-1} for $n = 1$). For a slab of catalyst with sealed edges and a simple pore structure, η is related by the equation

$$\eta = \frac{1}{\psi} \tanh \psi \tag{33}$$

to the Thiele modulus ψ which, for first-order reactions, is given by

$$\psi = L \sqrt{\frac{k_v}{D_{eA}}} \tag{34}$$

L is half the thickness of the catalyst slab and D_{eA} is the effective diffusivity of A defined by

$$D_{eA} = \frac{\varepsilon}{\tau} D_A \tag{35}$$

where ε is the internal void fraction of the catalyst (usually 0.3–0.8 for a catalyst pellet). If ρ_b is the bulk density of the pellet and ρ_{tr} the true density of the powder of which it is made, then [50]

$$\varepsilon = 1 - \frac{\rho_b}{\rho_{tr}} \tag{36}$$

The tortuosity factor, τ, allows for the zig-zag nature and constrictions of the diffusion path along the pores. Experimental values of τ are commonly 3–4 but values may range from 1.5 to over 10 [51].

When D_{eA} is large and k_v is small, ψ will be small enough for tanh $\psi \to \psi$ and $\eta \to 1$. The catalysis is then surface-controlled, as frequently happens with gas reactions proceeding in macropores. On the other hand, with D_{eA} small and k_v large, $\psi \gg 1$ so that tanh $\psi \to 1$ and $\eta \to 1/\psi$. This latter condition is more likely to be met with in catalysed solution reactions, particularly with mesoporous and microporous catalysts. To find out which situation applies to a given first-order reaction, one can combine eqns. (32) and (34) and rearrange

$$\frac{v_v L^2}{D_{eA} c_A} = \eta \psi^2 \tag{37}$$

If pore diffusion poses no hindrance $\psi \ll 1$, $\eta = 1$ and so $\eta \psi^2 \ll 1$ whereas in the case of pore diffusion limitation $\psi \gg 1$, $\eta = 1/\psi$ and $\eta \psi^2 \gg 1$. The type of rate control can therefore be deduced by determining whether the left-hand side of eqn. (37) is much smaller or much greater than unity. This so-called Weisz–Prater criterion [51] can be applied when experiments have been performed with only a single particle size. Alternatively, experiments can be carried out with two different sizes of catalyst pellet. Provided k_v and D_{eA} do not change (which is not true if the smaller pellet is made by cutting the larger one), then if pore diffusion is fast the effectiveness factor will be unity for both pellets and v_v will be the same for the two samples. However, strong pore diffusion control will lead to $\eta = 1/\psi \propto 1/L$ so that

$$\frac{(v_v)_1}{(v_v)_2} = \frac{L_2}{L_1} \tag{38}$$

The volume-based rates are then inversely proportional to the pellet sizes [51].

Aris [50, 51] found that a normalised modulus defined by

$$\psi = \frac{V_p}{A_{ex}} \sqrt{\frac{k_v}{D_{eA}}} \tag{39}$$

allowed η versus ψ curves for all shapes of pores to be superimposed under isothermal conditions. For a pellet in the form of a slab of cross-sectional area $A_{ex}/2$ and thickness $2L$, the ratio $V_p/A_{ex} = (A_{ex}/2)(2L)/A_{ex} = L$, as in eqn. (34). For a spherical pellet of radius R, the ratio equals $(4/3)\pi R^3/4\pi R^2 = R/3$. Equation (39) is still restricted to first-order processes: for an irreversible surface reaction of order n, the appropriate modulus is given by

$$\psi = \frac{V_p}{A_{ex}} \sqrt{\frac{(n + 1)k_v c_A^{n-1}}{2D_{eA}}} \quad (n > -1) \tag{40}$$

which reverts to the former value when $n = 1$. As before, the asymptotic effectiveness factor for extremely low diffusivity equals $1/\psi$. The catalytic rate in the event of strong pore diffusion resistance is therefore obtained by combining eqns. (32) and (40)

$$v_v = \left(\frac{1}{\psi}\right)k_v c_A^n = \frac{A_{ex}}{V_p} \sqrt{\left(\frac{2}{n + 1}\right) D_{eA} k_v c_A^{n+1}} \tag{41}$$

The observed order of reaction is thus $(n + 1)/2$ and only equals the true order n for first-order processes. If a reaction between A and B is pseudo-first order in A with B in excess, then $k_v = k_v' c_B$ and the apparent kinetic orders will be 1 with respect to A and 1/2 with respect to B. It is assumed here that B itself causes no diffusion limitation, i.e. that $D_{eB} \gg D_{eA}$.

Two other aspects are worth comment. Equation (41) was derived for conditions of a large Thiele modulus and so the reaction concerned will have almost gone to completion near the entrances of the pores. This is the reason why the catalytic rate $v_v V_p$ (mol s^{-1} per pellet) is proportional to the external surface area A_{ex}. However, it is probably more realistic to regard A_{ex}/V_p as an inverse distance parameter that depends upon the pellet's size and shape. If the volume-based rate constant k_v is then replaced by the areal rate constant k_{cat}, where $Ak_{cat} = V_p k_v$, the rate v_v becomes proportional to $A^{1/2}$ [49, 53]. Koros and Nowak [53] have pointed out that this result could allow a distinction to be drawn between reactions subject to pore diffusion control and those subject to surface control where the rate is proportional to A. The method they proposed consists of making pellets by mixing small catalyst particles in different proportions with particles of a suitable inert powder of similar diffusional characteristics (e.g. a sample of thoroughly poisoned catalyst, or of an inert support): in this way, the catalyst area can be varied while still keeping the pellet dimensions constant.

Inspection of eqn. (41) also shows that the only temperature-dependent parameters are the rate constant and the diffusion coefficient which appear together under the square root sign. The overall activation energy is thus given by

$$E^{\neq} = \frac{1}{2}E_k^{\neq} + \frac{1}{2}E_D^{\neq} \tag{42}$$

It is sometimes stated that $E^{\neq}$ is just equal to $E_k^{\neq}/2$, an approximation that may be acceptable for gaseous reactants. For solutes in solution, however, the activation energy of diffusion cannot be neglected.

Further progress depends upon a greater understanding of the properties of liquids in pores. The most promising approach is offered by computer simulation studies [57]. Grand canonical ensemble Monte Carlo calculations have already shown that the density profile of the liquid in a capillary is a strongly oscillating function of the distance normal to the pore walls [58]. To obtain information on transport properties such as diffusion coefficients, one must turn to molecular dynamics. Work in this field by Davis and his group [59] has recently demonstrated that the liquid's self-diffusion coefficient parallel to the pore walls (D_{11}) can be as much as a factor of two less in narrow channel pores than in the bulk fluid. The D_{11} value appears to correlate inversely with the average density of the liquid in the pore and not with the local density. In pores wider than 9.5–11.5 liquid diameters, the walls no longer affect the self-diffusion along the pore axis. These findings have been corroborated by molecular dynamics studies in cylindrical pores [60]. It is to be hoped that future calculations will attempt to predict the diffusion coefficients of solutes in narrow pores. Measurements in such systems are extremely difficult to carry out and recent experiments in an admittedly broad pore (a 2 mm diameter capillary) are therefore of particular interest. Liukkonen and co-workers [61] found that the diffusion coefficient of NaCl in a dilute aqueous solution was 75% greater at the walls of this capillary than in the bulk solution, a result in line with the phenomenon of "surface conductivity" [62]. Yet this finding clearly runs counter to the trend in the self-diffusion calculations in much narrower pores. It rather looks at this stage as if electrolytes near polar walls behave quite differently from non-electrolytes.

1.6.3 External diffusion

Every solid catalyst in solution is surrounded by a "stagnant" diffusion layer which reactants must cross in order to reach the surface. The resulting concentration profile is sketched in Fig. 6. The rate of the reactant's arrival at the solid/liquid interface is determined by its concentration gradient at that interface, $(dc/dx)_{x=0}$. The diffusion layer therefore has the same effect on the rate as does the simplified layer shown by the dotted lines [63]. The thickness of this so-called Nernst layer is designated δ. It follows from Fick's first law of diffusion that the number of moles of reactant A, n_A, that reach the surface in unit time is given by

$$-\frac{dn_A}{A\,dt} = \frac{D_A(c_A - c_{As})}{\delta_A} \tag{43}$$

where c_{As} is the concentration of A at the surface and A is the external area of the catalyst which was referred to as A_{ex} in Sect. 1.6.2. In the steady state, this diffusion rate will be equal to the rate at which A reacts at the surface.

References pp. 159–166

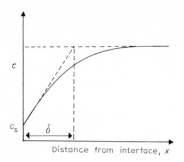

Fig. 6. Concentration profile of a reactant in the neighbourhood of a catalyst surface.

For the dispersal of the product, P, from the catalyst surface into the bulk solution, we can write similarly

$$\frac{\mathrm{d}n_P}{A\,\mathrm{d}t} = \frac{D_P(c_{P_s} - c_P)}{\delta_P} \tag{44}$$

The thicknesses of the Nernst layers are not necessarily the same for each solute, as will be made clear presently.

The value of the length δ clearly plays a pivotal role in solution catalysis. In a completely quiescent solution δ should, in principle, tend towards infinity but, in practice, tiny convection currents in the liquid brought about by thermal and mechanical fluctuations create finite layer thicknesses of around 0.1 mm [64] to 0.5 mm. Stirring the solution reduces the value of δ. With fast laminar flow at the surface, δ typically attains a value of 10 μm while turbulent flow can decrease the thickness to around 0.1 μm [64, 65]. The effective diffusion layer thickness around colloidal particles is much smaller still: to a first approximation, δ is then equal to the particle radius. For more exact values, the published hydrodynamic equations [66] should be consulted. It follows that, by altering the hydrodynamic regime and/or by changing the catalyst dimensions, we can at will control the rate of mass transport to the surface of the catalyst. In this way we can increase the overall rate of catalysis of many reactions and even change the rate-determining step (Fig. 4). The ultrasonic promotion of certain heterogeneously catalysed solution reactions [67] is partly due to microstreaming and cavitation turbulence produced by the ultrasound.

The transition between surface and diffusion control is easily demonstrated by considering a first-order surface reaction. In the steady state the rate at which reactant A diffuses to the surface equals the rate at which it is consumed by the chemical reaction, so that

$$v_{cat} = kc_{As} = \frac{D_A(c_A - c_{As})}{\delta_A} \tag{45}$$

Elimination of the unknown surface concentration c_{As} and rearrangement leads to the two equations

$$v_{\text{cat}} = \frac{kD_A c_A}{D_A + k\delta_A} \tag{46}$$

and

$$\frac{1}{v_{\text{cat}}} = \frac{1}{kc_A} + \frac{\delta_A}{D_A c_A} \tag{47}$$

When δ_A is large, $v_{\text{cat}} \rightarrow D_A c_A/\delta_A$ and the reaction is predominantly diffusion-controlled while, at small values of δ_A, $v_{\text{cat}} \rightarrow kc_A$ and the catalytic rate is controlled by the surface reaction. Thus sufficiently rapid stirring can often purge the system of mass transport effects and permit the kinetics at the surface to be studied. The situation has been graphically illustrated in Fig. 4.

Early workers in the field used paddle stirrers and similar devices to vary the Nernst layer thickness. Their results could be expressed by empirical relationships of the type

$$\frac{D}{\delta} \propto (\text{rotation speed})^\beta \tag{48}$$

with $\beta \leqslant 1$, the value of β and the proportionality constant being functions of the geometry of the stirrer and the design of the reaction system [8, 64]. A more sophisticated hydrodynamic arrangement was introduced in 1969 by Spiro and Griffin [68] when they employed a large rotating platinum disc as a catalytic tool. Small rotating disc electrodes of about 1 mm diameter had already proved their worth in electrochemical research [69] and the use of much larger discs of 4–6 cm diameter were to prove equally valuable in heterogeneous catalysis. As shown by Levich [64], the thickness of the diffusion layer is uniform over the whole surface of such a disc and is given by the equation

$$\delta_i = 0.643 D_i^{1/3} v^{1/6} f^{-1/2} = \frac{Le}{\sqrt{f}} \tag{49}$$

where D_i is the diffusion coefficient of the diffusing species i, v is the kinematic viscosity of the solution (dynamic viscosity divided by density), and f is the rotation frequency in Hertz. Small correction terms worth about 3% were evaluated later [63, 70] although they may be partly compensated by other factors such as edge effects [69]. Diffusion-controlled reactions are therefore easily distinguished from surface-controlled reactions: the rates of the former are proportional to the square root of the rotation speed while the rates of the latter are independent of speed. For intermediate control, the reciprocal rate varies linearly with $1/\sqrt{f}$ according to eqn. (47).

The Levich equation holds only under conditions of laminar or streamline flow, which means that the dimensionless Reynolds number

$$\text{Re} = \frac{2\pi f R^2}{v} \tag{50}$$

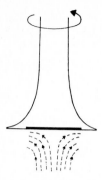

Fig. 7. Sketch of a large catalytic disc set in a trumpet-shaped former. The dotted curves indicate the lines of flow below the disc.

where R is the radius of the disc, must not exceed 2×10^5 [63]. For a 5 cm diameter disc immersed in an aqueous solution at 25°C, this imposes an upper speed limit of 3000 rev. min^{-1}. A further minor limitation arises from the fact that the theoretical model on which eqn. (49) is based assumes that the disc rotates in an infinite volume of fluid. In practice, this requirement is satisfied provided that the disc surface is at least 1 cm above the bottom of the (flat) vessel and that the diameter of the vessel is at least twice the maximum diameter of the disc and former [63, 71]. It is helpful to set the catalytic disc into an inert trumpet-shaped former as depicted in Fig. 7 so that the convective flow above the disc is pushed sideways and does not interfere with the crucial streamline flow patterns on the underside of the disc. These patterns are shown as dotted curves in the diagram. The model also requires the disc to be as flat as possible [63, 64] although even with a deliberately roughened surface the rates of diffusion-controlled reactions qualitatively retain their rotation dependence [72]. A further development would be the introduction of a ring electrode around the disc to monitor the formation of appropriate products or intermediates, as is done in electrode kinetics [69, 73].

Another hydrodynamic device that could be adapted for future catalytic research is the wall jet which was first described by Glauert [74] and developed mainly by Albery [75–77]. Here the mixed solution would be forced through a jet so as to impinge normally on to the centre of a stationary catalytic disc. As the vertical flow lines in Fig. 8 show, only fresh solution that has just come through the jet can reach the disc surface [75, 76]. The average thickness of the diffusion layer has been shown to be [77, 78]

$$\bar{\delta}_i = 2.28 D_i^{1/3} \nu^{5/12} R^{5/4} V_f^{-3/4} a^{1/2} \tag{51}$$

where the numerical coefficient incorporates a calibration constant [78] and where V_f is the volume flow rate of the solution issuing from the circular nozzle of the jet of diameter a. The disc surface is therefore not uniformly accessible to reagents under conditions of mass transport control while for

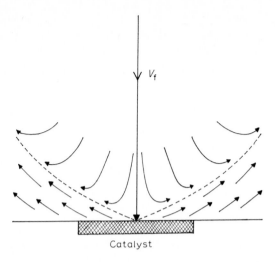

V_f

Catalyst

Fig. 8. Schematic flow lines for a wall jet catalytic disc.

a wholly chemically controlled reaction, where the flux is much smaller, the catalytic rate would be uniform over the surface. With a given wall jet cell in which the disc radius R is fixed, the type of control could be deduced from the dependence of the catalytic rate on the jet diameter and the volume flow rate. Several different flow rates could be tested successively with a given reaction mixture. It should be pointed out that each run with specified physical conditions would provide just one experimental point, the solution running off the disc during jet flow being analysed for reactant lost or product formed during contact with the catalyst. A correction would need to be applied for any parallel homogeneous reaction.

Let us now consider the effects of external diffusion control on reaction (II) between A and B when the surface kinetics are first order in each reactant. Because v_A mol of A react with v_B mol of B at the surface, the steady state fluxes of A and B towards the surface from the bulk solution will be in the ratio of v_A to v_B. By equations like (43), their concentration gradients and their concentration differences are seen to be of similar magnitude. If initially the solution contains much more B than A

$$c_B \gg c_A > (c_A - c_{As}) = O(c_B - c_{Bs}) \qquad (52)$$

Hence $c_{Bs} \approx c_B$. It follows that eqns. (45)–(47) can be applied to this reaction but k is now a pseudo-first-order rate constant equal to $k_{het}c_B$. With a rotating disc catalyst, a plot of c_A/v_{cat} will then vary linearly with $1/\sqrt{f}$ according to the equation

$$\frac{c_A}{v_{cat}} = \frac{1}{k_{het}c_B} + \frac{v_A Le}{D_A \sqrt{f}} \qquad (53)$$

where by eqn. (2) c_A/v_{cat} is equal to $-(v_A A/V)/(\mathrm{dln}c_A/\mathrm{d}t)$. If $k_{het}c_B \gg D_A/\delta_A$, the reaction will be completely diffusion-controlled and the plot will pass through the origin. Such a catalytic reaction will appear to the experimenter to be first order in A and zero order in B. This situation has been encountered in certain redox reactions catalysed by platinum and will be discussed further in Sect. 4.5.

In the second type of external diffusion control, the catalyst is so powerful that the surface reaction is virtually at equilibrium. Let the reaction be the general one in eqn. (II). Then, in the steady state, eqn. (45) must be replaced by [79]

$$v_A v_{cat} = \frac{D_A(c_A - c_{As})}{\delta_A} \tag{54}$$

so that

$$c_{As} = c_A - \frac{v_{cat} v_A \delta_A}{D_A} \tag{55}$$

Similarly for the product P, the diffusive flux in eqn. (44) equals $v_P v_{cat}$ whence

$$c_{Ps} = c_P + \frac{v_{cat} v_P \delta_P}{D_P} \tag{56}$$

Since the surface reaction is effectively at equilibrium

$$K_s = \frac{c_{Ps}{}^{v_P} c_{Qs}{}^{v_Q} \cdots}{c_{As}{}^{v_A} c_{Bs}{}^{v_B} \cdots} = \frac{\prod_{prod} [c_P + (v_{cat} v_P \delta_P/D_P)]^{v_P}}{\prod_{react} [c_A - (v_{cat} v_A \delta_A/D_A)]^{v_A}} \tag{57}$$

Solving for v_{cat} produces a general but complex rate equation. It is more instructive, therefore, to examine a few special cases. First, we shall look at the unimolecular transformation

$$A \xrightarrow{\text{diffn.}} A_{ads} \rightleftharpoons P_{ads} \xrightarrow{\text{diffn.}} P \tag{XII}$$

Here $v_A = 1$, $v_P = 1$, and

$$K_s = \frac{c_P + (v_{cat} \delta_P/D_P)}{c_A - (v_{cat} \delta_A/D_A)} \tag{58}$$

Rearrangement leads to the relation

$$v_{cat} = k_{obs} [c_A - (c_P/K_s)] \tag{59}$$

where

$$\frac{1}{k_{obs}} = \frac{\delta_A}{D_A} + \frac{\delta_P}{K_s D_P} \tag{60}$$

Equation (59) is the desired first-order rate equation expressed in terms of the bulk concentrations of reactant and product (it is, of course, only the concentrations in the bulk solution that can easily be monitored in the laboratory). The observed rate constant, k_{obs}, will be directly proportional to the square root of the rotation frequency if the catalyst is present in the form of a rotating disc since both δ_A and δ_P then vary inversely with $\sqrt{f}$. The analogues of eqns. (59) and (60) with pressures instead of concentrations, and with the D/δ terms replaced by mass transport rate constants, are well known to chemical engineers [80].

Let us now return to the general equation (57) but consider only the early stages of the reaction when the bulk concentrations $c_P, c_Q, \ldots$ are sufficiently small to be neglected. The initial reactant concentrations, in contrast, will be large enough to allow the denominator to be expanded binomially. Retaining only the first two terms and rearranging gives [79]

$$\frac{1}{v_{cat}} = \frac{1}{v'_{cat}} + \sum_{react} \frac{v_A{}^2 \delta_A}{D_A c_A (v_P + v_Q + \ldots)} \tag{61}$$

$$v'_{cat}{}^{(v_P + v_Q + \ldots)} = K_s \prod_{prod} (D_P/v_P \delta_P)^{v_P} \prod_{react} c_A{}^{v_A} = k_{obs}{}^{(v_P + v_Q + \ldots)} \prod_{react} c_A{}^{v_A} \tag{62}$$

In most real situations, the summation function in eqn. (61) will make only a minor contribution to $1/v_{cat}$ and we may accordingly adopt the approximation

$$v_{cat} \approx v'_{cat} \tag{63}$$

By taking the $(v_P + v_Q + \ldots)$th root of the right-hand side of eqn. (62), we can confirm that the reaction is wholly transport-controlled: the product of the (D_P/δ_P) terms is then a geometric mean (D/δ) raised to the first power. Were the catalyst introduced in the form of a rotating disc, v_{cat} would be proportional to $\sqrt{f}$. Less straightforward and more intriguing are three other properties of v_{cat}. First, the overall rate constant comprises not only several diffusion terms but also K_s. Both hydrodynamic and thermodynamic factors therefore govern the observed rate constant and its activation energy $E^{\neq}$

$$(v_P + v_Q + \ldots) E^{\neq} = \Delta H_s^{\circ} + \sum_{prod} v_P (E_D^{\neq})_P - \sum_{prod} v_P (E_\delta^{\neq})_P \tag{64}$$

The rate constant and activation energy can thus be predicted from auxiliary data provided the hydrodynamic flow conditions are properly defined, as they are for a rotating disc catalyst or a wall jet. An example that has been found to fit this situation well, the catalysis by a platinum rotating disc of the reaction between aqueous $Fe(CN)_6^{3-}$ and I^- ions [6], will be discussed in Sect. 4.4: for this system, the overall activation energy was actually *negative* because the exothermic enthalpy change for the reaction outweighed the positive activation energy contributions from the diffusion terms.

The kinetic orders in eqn. (62) are also unusual, being entirely predictable from the stoichiometric coefficients of the reaction. The resulting orders are mostly fractional. One could easily misunderstand this outcome as indicating Freundlich adsorption of the reactants, a warning against too facile an interpretation of the observed rate law. The third unexpected feature of v_{cat} is the kinetic response to the addition of one of the products (P, say) to the initial reaction mixture. An analysis of the effect requires the full eqn. (57) which, to a first approximation, leads to

$$v_{cat}''^{(v_Q + \ldots)} \approx K_s \prod_{\substack{\text{all prod} \\ \text{except P}}} (D_Q/v_Q\delta_Q)^{v_Q} c_P^{-v_P} \prod_{\text{react}} c_A^{v_A} \tag{65}$$

Not only does the concentration of the added product P now appear in the rate equation but the kinetic orders of the reactants themselves have altered. With only two products P and Q, for instance, the kinetic order of A changes from $v_A/(v_P + v_Q)$ to v_A/v_Q. This prediction, too, has been fulfilled in the example mentioned [6].

1.7 LINEAR FREE ENERGY RELATIONSHIPS

In homogeneous media, empirical linear correlations have been found to exist between the logarithms of the rate constants, k, of a given type of reaction and the logarithms of the equilibrium constants, K, of the same or a second reaction that is subject to the same variations of reactant structure [81]. Such relations apply to both organic [82, 83] and inorganic reactions [84]. Because $\log k$ is related to the Gibbs free energy of activation and $\log K$ to the standard Gibbs free energy change, these correlations are known as linear free energy relationships or LFER for short. The most widely used is the Hammett equation [81, 82, 85]

$$\log(k_{ij}/k_{0j}) = \rho_j\sigma_i \tag{66}$$

in which k_{ij} is the rate constant of a given reaction j when one of the reactants bears a substituent i and k_{0j} is the corresponding rate constant without any such substituent. The reaction constant ρ_j depends only upon the reaction in question and the experimental conditions of medium and temperature. The substituent parameter σ_i, on the other hand, is characteristic of the type and position of the substituent i and is independent of the reaction. It is given by

$$\sigma_i = \log(K_i/K_0) \tag{67}$$

where K_0 is the dissociation constant of benzoic acid and K_i that of the appropriately substituted benzoic acid. Similar linear relationships exist between $\log k_{ij}$ and $\log K_i'$ where K_i' is the equilibrium constant of the reaction itself or of some other related reaction or even a suitable structural parameter.

The Hammett equation describes reasonably well the inductive and mesomeric influences on the rate constant when there are substituent changes in *meta-* and *para*-substituted aromatic compounds. *Ortho*-substituted aromatics and substituted aliphatic compounds can be represented by the Taft equation [81, 83]

$$\log (k_{ij}/k_{0j}) \quad = \quad \rho_j^* \sigma_i^* + s_j E_{si} \tag{68}$$

where E_{si} is a steric substituent parameter [86] and σ_i^* a polar substituent parameter calculated from the experimental rates of acid and alkaline hydrolysis of the appropriate esters. This equation, too, has been widely applied in homogeneous systems.

In 1967, Kraus [87] and Mochida and Yoneda [88] independently showed that heterogeneously catalysed gas reactions could also be fitted by eqns. (66) and (68). Many more examples of heterogeneous LFER were listed and discussed in Kraus' second review in 1980 [89]. On reflection, it is rather remarkable that surface reactions, most of them hydrogenations of alkenes and aromatics at several hundred °C, should be correlated by σ and E_s parameters that depend on equilibrium properties of aqueous solutions at 25°C. Only a minority of the heterogeneous reactions involved liquids because of the dearth of data in this field, and these reactions, too, were mainly hydrogenations. An early and trend-setting example was provided by Cerveny and Ruzicka [90]. These workers measured the initial hydrogenation rates of 15 substituted alkenes $R_1R_2C = CR_3R_4$ in ethanol at 20°C using three different supported platinum catalysts in a reactor shaken sufficiently strongly (1000 min^{-1}) to lift the kinetics into the region of chemical control. The results on each catalyst could be fitted by the equation

$$\log v_0 \quad = \quad \rho_j^* \sigma^* + s_j E_s + \text{constant} \tag{69}$$

where the σ^* and E_s symbols stand for the summations of these parameters for the various R_i substituents in given alkenes. They concluded that these catalysed hydrogenations were mainly influenced by steric factors. Equation (69) introduces the important point that many LFER of heterogeneous reactions have been applied to the rates rather than to the rate constants. One does not therefore know a priori whether it is the rate constants k or the Langmuir adsorption coefficients K or both that are responsible for the Hammett or Taft correlation. Maurel and Tellier [91] were among the first to attempt a separate assessment of these factors. They measured the hydrogenation rates of liquid alkenes on a Pt/SiO$_2$ catalyst at 20°C both separately (when the rates were zero order in alkene and so led to values of k_{het}) and in competition (when the rates were proportional to $k_{het}K_{alk}$). Combination of the data yielded values of the relative adsorption coefficients of the various alkenes. The resulting $\log K_{rel}$ values varied linearly with σ^*. The correlations of K for other reactions have been summarised by Kraus [89]. It is hardly surprising that adsorption coefficients should be dependent on substituent effects since Pearson [92, 93] had pointed out earlier that atoms in bulk metals would act as soft acids and so preferentially adsorb soft rather than hard bases. Barclay [94] subsequently rationalized specific adsorption at metal electrodes in terms of the Pearson soft and hard acid–base (SHAB) principle. More detailed application of this principle to catalysis in solution will be reserved for subsequent sections. Returning to the main theme, we may conclude that there is now sufficient theoretical and experimental evidence [89] to say that adsorption coefficients as well as surface rate constants depend on electronic and steric factors and will follow appropriate LFER equations. These relationships can prove extremely useful for predictive purposes. It would nevertheless be desirable to test the applicability of the Hammett and Taft equations with a simpler catalysed reaction that does not involve a gaseous phase reactant. The heterogeneously catalysed solvolyses of substituted t-butyl halides have been suggested [79] as suitable candidates for such a project.

Linear free energy relationships do not apply to homogeneous reactions that are fast enough to be diffusion limited. According to the Smoluchowski equation [81, 95, 96], their rate constants are then proportional to the sum of the diffusion coefficients of the two reactants. Diffusion coefficients depend upon size and shape [55, 56] and only indirectly, and in a minor way, on electronic field effects so no correlation with σ or σ^* parameters would be expected. The same is true for one category of diffusion-limited heterogeneously catalysed reactions. This comprises the first-order reactions that follow eqn. (47) and pseudo-first-order reactions that obey eqn. (53): in both cases, the rate constants simply equal $D_A/v_A\delta_A$ provided this quantity is much smaller than the surface rate constant k. Plots of $\ln k_{obs}$ versus σ or σ^* will therefore be horizontal plateaux of zero slope as in the homogeneous situation, the main difference being that the heterogeneous rate constants will increase with increased stirring of the solution.

The second type of heterogeneously catalysed reaction subject to external diffusion control behaves quite differently. Where the catalysis is strong enough for the reaction to be almost at equilibrium on the surface, the rate constant will contain both diffusion and thermodynamic terms. Equation (60) for unimolecular reactions is one example and another is eqn. (62) which applies to the initial stages of a general reaction. In the latter case [79]

$$(v_P + v_Q + \ldots) \ln k_{obs} = \ln K_s + \sum_{prod} v_P \ln (D_P/v_P\delta_P) \tag{70}$$

where K_s is the equilibrium constant of the reaction under study. A plot of $\ln k_{obs}$ against $\ln K_s$ will therefore give a line of slope $1/(v_P + v_Q + \ldots)$. Not only is the slope finite, in contradistinction to the first category, but its value is solely determined by the stoichiometric coefficients of eqn. (II). If P and Q are the only products and $v_P = v_Q = 1$, the slope will be 0.5. Thus experimenters who study heterogeneous reactions of this type under constant stirring conditions will find finite correlations between the rate constants and σ or σ^* parameters. They may well conclude, incorrectly, that the catalysis is surface controlled and go on to draw wrong inferences about the reaction mechanism. It is therefore essential always to vary the hydrodynamic flow conditions to check whether external diffusion is playing a major role in the catalysis [79].

Cases of internal or pore diffusion control make up the third category. As shown in eqn. (41), the observed rate constant under such conditions is proportional to $\sqrt{kD}$ where k is the rate constant of the surface reaction.

Thus

$$\ln k_{\text{obs}} = \frac{1}{2} \ln k + \frac{1}{2} \ln D + \ln (\text{dim}) \tag{71}$$

where "dim" is a function of the internal and external dimensions of the catalyst. It is to be expected that $\ln k$ will correlate linearly with σ or with σ^* and E_s parameters and $\ln k_{\text{obs}}$ will accordingly fit the corresponding LFER but with half the slope. Once again the mere existence of an LFER has proved no guarantee of pure surface control. Further experiments to determine the rate-controlling step are always needed in order to interpret the resulting ρ, ρ^* and s parameters correctly.

1.8 GENUINE AND ILLUSORY HETEROGENEOUS CATALYSIS

True heterogeneous catalysis by a solid requires that it increases the rate of a specified reaction without undergoing any chemical change itself. This latter condition should always be checked. It is not sufficient merely to show that the solid increases the rate of loss of a reactant or the rate of formation of a product since either of these observations may be caused by other types of solid/solution interaction. In particular, such changes can also arise from chemical attack on the solid, or through homogeneous catalysis by species emanating from the solid, or from some new or side reaction. Theoretical and experimental approaches can be employed to test for these effects. Thermodynamic calculations and stability diagrams of the catalyst will exclude certain reactions and point to the possibility of others. Whether these are sufficiently fast to produce the observed rate increases can then be ascertained by appropriate kinetic tests. Details are given in the following sections together with some illustrative examples from the literature.

1.8.1 Thermodynamic stability of the solid

In a given medium, every solid possesses an intrinsic solubility which ranges from the infinitesimal to the significant. To quote a well-known example, solid silver chloride in water at 25°C is in equilibrium with $\sqrt{K_{\text{AgCl}}} = 1.3 \times 10^{-5}\,\text{mol dm}^{-3}$ of Ag^+ and Cl^- ions, where K_{AgCl} is the solubility product of the salt. This concentration of silver ions is high enough to catalyse homogeneously many of the reactions that solid AgCl catalyses heterogeneously [97, 98]. However, the concentration of dissolved silver(I) will change by several orders of magnitude if the solution contains chloride ions from another source (as a reactant or product or a co-ion of either, or as a supporting electrolyte). As c_{Cl^-} increases, the concentration of silver(I) at first decreases sharply by the common ion effect and then rises again at high chloride levels through the formation of complex ions like $AgCl_2^-$ and $AgCl_3^{2-}$. In 2.5 mol dm^{-3} KCl solution the solubility exceeds $1 \times 10^{-3}\,\text{mol dm}^{-3}$ [99]. The composition of the solution at any given chloride ion concentration can be calculated from tables of stability constants [100, 101] modified, where possible, by activity coefficient corrections [102, 103]. Complex formation by exogenous species will also result in dissolution of catalytic solids. Silver chloride, for instance, will dissolve in the presence of CN^- or $S_2O_3^{2-}$ according to equations such as

$$AgCl(c) + 2\,S_2O_3{}^{2-} \rightarrow Ag(S_2O_3)_2^{3-} + Cl^- \tag{XIII}$$

Provided some solid AgCl remains behind, the equilibrium constant of this reaction equals $K_{AgCl}K_{stab}$ where K_{stab} is the stability constant of the complex ion. It should be added that each new complex ion species introduced into the solution, whether $AgCl_2^-$, $AgCl_3^{2-}$, $Ag(CN)_2^-$ or $Ag(S_2O_3)_2^{3-}$, will exert its own specific homogeneous catalytic effect on the reaction in question.

Another type of solid/solution interaction is metathesis or ion exchange which actually alters the composition of the solid as well as that of the solution. Thus in the presence of bromide ions, AgCl will react to form the less soluble AgBr

$$AgCl(c) + Br^- \rightarrow AgBr(c) + Cl^- \tag{XIV}$$

These two solids will, of course, exhibit different catalytic effects. The equilibrium constant of reaction (XIV) is K_{AgCl}/K_{AgBr} which in water at 25°C equals 500. The tendency for reactions like (XIII) and (XIV) to occur can thus be calculated from known solubility products and stability constants. In the case of solids like oxides or carbonates, one must also take into account the relevant acid–base equilibrium constants because here dissolution can occur if the solution is made sufficiently acid or alkaline.

Catalyst stability is also at risk from oxidation (corrosion) or reduction of the solid or of one of its constituents. Whether a particular redox reaction is thermodynamically feasible can best be judged from stability diagrams. In the most popular ones the equilibrium or Nernst potential E is plotted against pH. The theory underlying these so-called Pourbaix diagrams has been explained in several elementary accounts (e.g. ref. 104) and also in the introduction to Pourbaix's massive compilation of data and diagrams for both metallic and non-metallic elements [105]. A list of Pourbaix diagrams published by subsequent workers has been provided by House [106]. Current calculations employ more modern thermodynamic data [107] and indeed Pourbaix diagrams are nowadays drawn by appropriate computer programmes [108].

The Pourbaix diagram for platinum [109] is shown in Fig. 9. The domain of stability of water itself lies between the lower broken diagonal line a (below which water can be reduced to hydrogen) and the upper broken diagonal line b (above which water may be oxidised to oxygen). Platinum metal is stable below the lowest full diagonal line which stands for the reaction

$$PtO(c) + 2\,H^+ + 2\,e^- \rightleftharpoons Pt(c) + H_2O \tag{XV}$$

and obeys the equation

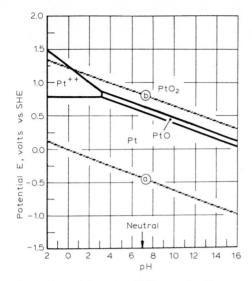

Fig. 9. Potential versus pH diagram for the system platinum/water at 25°C (after Lee [109]).

$$E = 0.980 - 0.0591 \, \text{pH} \tag{72}$$

The short horizontal line at low pH represents the reaction

$$\text{Pt}^{2+} + 2 \, \text{e}^- \rightleftharpoons \text{Pt(c)} \tag{XVI}$$

with the concentration of platinum ions taken as $1 \times 10^{-6} \, \text{mol dm}^{-3}$. It is conventionally assumed that a substance does not corrode if its solubility is smaller than this figure. Figure 9 shows that a sufficiently strong oxidising agent can attack even as noble a metal as platinum. An early illustration of this point comes from the work of Delépine in 1905 [110]. He discovered that the "catalysis" by platinum of the reduction of hot sulphuric acid by ammonium ions

$$4 \, \text{H}^+ + 3 \, \text{SO}_4^{2-} + 2 \, \text{NH}_4^+ \rightarrow 3 \, \text{SO}_2 + 6 \, \text{H}_2\text{O} + \text{N}_2 \tag{XVII}$$

really proceeded in two stages: in the first stage the hot acid attacked the metal to form Pt(IV) ions and SO_2 and, in the second stage, the Pt(IV) ions were reduced back to the metal with the concomitant oxidation of NH_4^+ ions to dinitrogen. A similar scenario was later established by Millbauer [111] for the apparent catalysis by platinum of the reduction of hot sulphuric acid by dihydrogen.

Attack on platinum is much more likely to occur when the solution contains ions capable of complexing platinum ions. Although relatively few couples possess a sufficiently high standard potential to oxidise platinum in $1 \, \text{mol dm}^{-3}$ HClO_4 where $E^{\ominus}(\text{Pt}^{2+}/\text{Pt}) = 1.19 \, \text{V}$, many more are thermodynamically capable of doing so in halide media since $E^{\ominus}(\text{PtCl}_4^{2-}/\text{Pt})$ and $E^{\ominus}(\text{PtCl}_6^{2-}/\text{Pt})$ are only about 0.75 V [107]. Thus Cohen and Taylor [112] found that dissolution of platinum was the real reason why the metal ap-

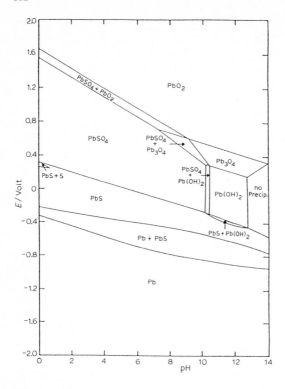

Fig. 10. Potential versus pH diagram for the system PbS/water at 25°C. (After Pritzker and Yoon [118].)

peared to catalyse the reduction in chloride media of Np(VI) to Np(V) and Np(IV). This example links the effect of complexing discussed at the beginning of Sect. 1.8.1 with that of redox attack. A need clearly exists for Pourbaix diagrams which incorporate both aspects and a welcome start has been made in this direction by the recent construction of stability diagrams for silver and gold that show the effects of complexing with cyanide [113] and chloride [114].

Pourbaix diagrams have been drawn not only for elements but also for certain compounds [115], especially ones of geological and industrial importance. Metal sulphides have attracted particular attention [115–118] and Fig. 10 shows a recently published diagram for lead sulphide. Its narrow and pH-dependent stability region, bordered by the stability domains of the sulphate and of the metal, is typical for many sulphides. Not surprisingly, metal sulphides may falsely appear to catalyse or inhibit some solution reactions. For example, the rate of the redox reaction

$$2 \, Fe(CN)_6^{3-} + 3 \, I^- \rightarrow 2 \, Fe(CN)_6^{4-} + I_3^- \qquad \text{(XVIII)}$$

followed by analysis of the iodine produced, was found to be greatly increased in the presence of MoS_2 and CuS, decreased in the presence of PbS,

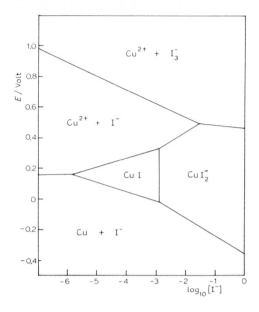

Fig. 11. Potential versus $\log_{10}[I^-]$ diagram for the system CuI/water at 25°C. (Based on the data of Groenewald [115].)

and apparently reduced to zero on the addition of Ag_2S, CdS, and HgS [119]. Calculations showed that both the oxidants $Fe(CN)_6^{3-}$ and I_3^- were thermodynamically capable of oxidising all the sulphides to sulphates and, in the case of MoS_2, to MoO_3 or MoO_4^{2-}. Test experiments with individual reactants and products confirmed that such attacks did take place, the rate of interaction varying from sulphide to sulphide. In some cases, the formation of sulphate is preceded by the formation of sulphur and thiosulphate [120] which resulted in an additional fall in the iodine concentration. Copper(II) sulphide is a special case for attack on it by ferricyanide released Cu^{2+} ions which produced extra iodine by the reaction

$$2\ Cu^{2+} + 5\ I^- \rightleftharpoons 2\ CuI(c) + I_3^- \tag{XIX}$$

None of the effects brought about by these metal sulphides could therefore be taken at face value. However, lead and other sulphides do act as genuine heterogeneous catalysts in the important process of froth flotation by which these minerals are separated from sand and other "gangue" materials [120–122].

A different type of thermodynamic diagram is appropriate for compounds whose stability is not pH-dependent. For metal halides, which are typical of this class, plots of E against the logarithm of the free halide concentration provide more relevant information. Figure 11 for CuI shows that the stability region of the solid is quite small although it continues to exist as an entity inside the CuI_2^- ion. Nevertheless, the addition of small amounts of CuI has been advocated [123] for speeding up the slow reaction

$$2 \text{ Fe(III)} + 3 \text{ I}^- \rightarrow 2 \text{ Fe(II)} + \text{I}_3^- \tag{XX}$$

sufficiently to render it useful for analytical purposes. Groenewald et al. [119] confirmed that the initial rate of iodine formation increased proportionately with the mass of CuI but they also discovered that the reaction no longer went to completion. The more CuI was added, the less iodine was seen. This removal of iodine by larger amounts of CuI was readily accounted for by the intervention of reaction (XIX). For analytical use, the problem can be overcome by restricting the mass of catalytic CuI and adding excess iodide ions to shift both reactions (XIX) and (XX) to the right.

1.8.2 Kinetic and other tests

The major kinds of untoward interaction between potential catalysts and reacting solutions can now be summarised. They fall conveniently into five categories.

(A) Homogeneous catalysis by ions or other species derived from the solid, either through its dissolution or as a consequence of chemical attack upon it.

(B) Chemical reaction between the solid and a reactant, forming soluble species.

(C) Chemical reaction between the solid and a product, forming soluble species.

(D) Chemical reaction between the solid and solution species to form a different insoluble solid either directly, or by metathesis, or by deposition of a layer of insoluble product on the surface of the original solid.

(E) Dissolution (physical or chemical) of the solid and its re-precipitation.

All these situations have been encountered in the literature and several specific examples have been cited in Sect. 1.8.1. One type of interaction for which no example has yet been given is the deposition on to the catalyst of a solid product. This is not an uncommon occurrence. Thus in the platinum-catalysed reduction of copper(II) ions [124]

$$\text{Cu}^{2+} + \text{H}_2 \rightarrow \text{Cu(c)} + 2 \text{ H}^+ \tag{XXI}$$

the metallic copper gradually coats the platinum surface. The rate of the reaction then decreases rapidly because the catalysis proceeds by an electrochemical mechanism [3] which involves the half-reaction

$$\text{H}^+ + \text{e}^- \rightleftharpoons \tfrac{1}{2} \text{H}_2 \tag{XXII}$$

This is fast on platinum but very slow on copper [125]. In other reactions the solid formed is a good catalyst and autocatalytic behaviour results. For instance, silver metal strongly catalyses the reduction of silver ions by hydroquinone (H_2Q) and similar reducing agents

$$2 \text{ Ag}^+ + \text{H}_2\text{Q} \rightarrow 2 \text{ Ag(c)} + 2 \text{ H}^+ + \text{Q} \tag{XXIII}$$

The silver product is deposited on to the particles of silver [126, 127] or other catalyst [128] that were present initially. Reactions like (XXIII) are of crucial importance in the development of photographic film [129] (cf. Sect. 4.3).

Only some of the solid–solution interactions (A)–(E) can possibly take place in any given experimental situation. Any interactions not ruled out on thermodynamic grounds should then be tested for. A suitable choice of the following experiments is recommended for this purpose [115, 130].

(1) Weighing the solid, inspecting it visually, and examining it by X-ray and/or surface spectroscopic techniques and/or by cyclic voltammetry both before and after the experiment.

(2) Exposing samples of the solid separately to each reactant and each product of the reaction and monitoring any interactions. The results of this test will be deceptive if in the reaction mixture the attack on the solid is a concerted one by several species which may include the co-ion of another reagent, e.g. Cl^-.

(3) Analysing the solution for new species derived from the solid such as metal ions, complex ions, or co-ions. Their concentrations may be too small for detection by conventional means and it may be helpful to employ newer techniques like atomic absorption, ion chromatography, or radioactivity measurements after prior neutron irradiation of the solid. If new species are found, an experiment could be carried out to see if their deliberate addition to the homogeneous reaction mixture reproduces the "catalytic" rate.

(4) Removing the supposed catalyst (by filtering or centrifuging) part way through the reaction to see if the rate returns to the value recorded in the absence of solid.

(5) Comparing the rate of a reaction mixture to which an aliquot of spent reaction mixture has been added with the rate of a mixture containing an equal aliquot of a spent and catalysed reaction (with the solid removed).

(6) Monitoring the concentrations of either two reactants, or of two products, or of a reactant and a product, to check on the stoichiometry of the reaction in the presence of the solid.

(7) Testing whether the catalytic rate is proportional to the mass of solid (provided the catalysis is not controlled by internal diffusion) and whether the rate increases as expected when more solid is added during a run.

(8) Seeing if the kinetics in the presence of the solid are normal or show unusual behaviour like autocatalysis or auto-inhibition.

(9) Observing the effect on the rate of intermittently shaking or stirring the reaction mixture containing the solid. The rate will be high during periods of agitation and low in each quiescent period if the catalysis is both heterogeneous and controlled by external diffusion. The test will also appear to be positive in the event of a diffusion-controlled attack on the solid by the reactant whose concentration is being monitored.

The tests that can detect the various types of solid–solution interaction are shown by ticks in Table 1. The question marks indicate conditions where a positive response might be obtained, depending upon circumstances. The table can be used either to select suitable tests when a given type of interac-

TABLE 1

Summary of the suitability of different tests for detecting the five main types of solid-solution interaction

Type of interaction	Test 1	Test 2	Test 3	Test 4	Test 5	Test 6	Test 7	Test 8	Test 9
A	√	√	√	√	√		√	?	√
B	√	√	√	?	?	√			√
C	√	√	√	?	?	√			
D	√	√	?	?	?		?	√	
E	√	√	√	?		?			

tion is suspected, or to help identify the kind of interaction following a particular test. Almost all the tests are sensitive to the presence of homogeneous catalytic agents. The question as to whether a catalyst acts homogeneously, colloidally (microheterogeneously), or (macro)heterogeneously has lately been of special concern to chemists using complex organometallic compounds which are designed as homogeneous catalysts. The problem has been highlighted in a recent paper by Lewis and Lewis [31] who reported that a hydrosilylation reaction, ostensibly catalysed by homogeneous platinum–alkene complexes, actually involved the formation of platinum colloid as the key step. Crabtree et al. [132, 133] have discussed the distinguishing methods employed in this field and pointed out their limitations. Dynamic light scattering, for instance, will detect colloidal particles but cannot tell whether they are catalytically active. Anton and Crabtree [133] end by recommending two tests for metal-based catalysts, mercury and dibenzo[a, e]cyclooctatetraene (dct). Liquid mercury poisoned all heterogeneous metal catalysts such as carbon-supported or colloidal palladium and colloidal rhodium but not homogeneous ones like $RhCl(PPh_3)_3$. However, it is important to note that the test reaction employed was the hydrogenation of hexene or cyclohexene. Many hydrogenations proceed by the electrochemical mechanism mentioned in connection with reaction (XXI) and so involve the hydrogen couple (XXII) whose exchange rate on mercury is extremely slow [125]. Mercury would therefore fail as a distinguishing test for any chemical reactions that are catalysed by mercury. The dct test could have wider applicability. Rigid and tub-shaped, this molecule binds strongly to metal complexes but not to planar metal surfaces. Addition of dct at a level of 1 molecule per metal atom thus inhibits homogeneous organometallic catalysts while having little effect on heterogeneous ones.

Finally, a word of warning about adsorption. In the presence of finely divided solids of large surface area, significant amounts of reactant and/or product will be adsorbed from the solution. Test (2), above, is the most direct way of measuring the effect. Care must then be taken to distinguish between adsorption and chemical attack: the latter will give rise to new chemical species. As an illustration, a sample of PbS introduced into a dilute solution of $Fe(CN)_6^{3-}$ not only removed this species but also converted it to $Fe(CN)_6^{4-}$ [119]. Adsorption will also affect several of the other tests and its existence should always be borne in mind. Adsorption can even negate the very appearance of catalysis! Thus, Kolthoff [134] showed that charcoal, which seems to inhibit the reaction

$$IO_3^- + 5 I^- + 6 H^+ \rightarrow 3 I_2 + 3 H_2O \qquad (XXIV)$$

does, in fact, catalyse it when the adsorption of H^+ and I^- ions and especially of the product iodine are taken into account.

1.8.3 Stoichiometry of the reaction

In the heterogeneous catalysis of gas reactions it is a commonplace that the products of a reaction depend upon the catalyst used. For instance, at 280°C ethanol is oxidised to CH_3CHO over copper (a dehydrogenation catalyst) but is dehydrated to $CH_3CH_2OCH_2CH_3$ over an alumina catalyst [135]. The corresponding situation in solution catalysis is less well documented but an example from the author's laboratory will illustrate the point. It concerns the aquation of $[Co(NH_3)_5Br]Br_2$

$$Co(NH_3)_5Br^{2+} + H_2O \rightarrow Co(NH_3)_5(H_2O)^{3+} + Br^- \qquad (XXV)$$

which was followed by measuring the decrease in the optical absorbance of the reactant. The rate was found to be considerably enhanced on adding AgBr, HgS [136, 137] or platinum [136, 138]. Subsequent experiments [139] revealed that, in the presence of these solids, another reaction took place, the reduction of the complex ion to Co(II). This new redox process accounted for a significant fraction of the rate enhancement in the cases of AgBr and HgS and for the whole of it in the case of platinum. Thermodynamic calculations gave support to these findings. It is therefore advisable to check the stoichiometry of catalysed reactions, either by specifically testing for any suspected new product or by applying test (6), above.

1.9 INADVERTENT HETEROGENEOUS CATALYSIS

The fact that solids often catalyse solution reactions can pose problems for both solution kineticists and electrochemists. In the past they have rarely considered such possibilities whereas kineticists studying homogeneous gas reactions routinely test for the absence of catalysis by the walls of the containing vessel. Glass does not affect many reactions in solution although it has been found to catalyse the isotopic exchange reactions between $Co(en)_3^{2+}$ and $Co(en)_3^{3+}$ [140], $Co(EDTA)^{2-}$ and $Co(EDTA)^-$ [141], and Np(IV) and Np(V) [142]. Dacre and co-workers [143] observed catalysis by the reaction flask in the exchange of iodine between 1-iodo-2,4-dinitrobenzene and low concentrations of iodide ions in methanol and in 1-butanol. However, the greatest danger of inadvertent heterogeneous catalysis in solution work undoubtedly arises from traditional electrodes. Platinum, in particular, catalyses many redox reactions (see Sect. 4.1). The kinetic consequences are illustrated by the findings of Ford-Smith et al. [144, 145]. They determined the rate of the aqueous redox reaction

$$Br_2 + Tl(I) \rightarrow 2\ Br^- + Tl(III) \tag{XXVI}$$

by spectrophotometry and also from E.M.F. measurements with a platinum electrode on the assumption that its potential was controlled entirely by the more electrochemically reversible Br_2/Br^- couple. The two sets of rate constants differed considerably because, in the E.M.F. method, the platinum will have adopted a mixture potential [146] with concomitant surface catalysis. This effect is much smaller if one of the couples is very irreversible electrochemically. The E.M.F. method has therefore been applied more successfully to the bromination of various organic compounds [147].

Platinum electrodes are, of course, used for many other physicochemical purposes such as the determination of standard electrode potentials, $E^\ominus$. Here couples with low values of $E^\ominus$ like V(III)/V(II), Ti(III)/Ti(II), Cr(III)/Cr(II), and Eu(III)/Eu(II) all react with hydrogen ions to produce hydrogen gas under the catalytic influence of the platinum metal [3]. In order to measure the potentials of these redox couples one must decrease the rate of the catalytic process. One way of doing this is to lower the hydrogen ion concentration by using a buffer solution of fairly high pH [148]. A more common remedy is to replace the noble metal electrode by one of mercury, tin, or lead at which the H^+/H_2 couple is electrochemically irreversible [149]. The same device can be effective in conductance measurements where platinized platinum electrodes are normally employed [150]. These may induce undesirable side reactions such as the decomposition of hydrogen peroxide whether present as solute [151] or as solvent [152, 153], a problem that was resolved by using tin electrodes instead. A different and intriguing catalytic interaction between conductance electrodes and a solution was reported by Auerbach and Zeglin [154]. They discovered that the decrease with time of the conductance of aqueous sodium formate solutions was caused by the catalytic oxidation of the solute by oxygen adsorbed on the platinized electrodes

$$HCO_2^- + \tfrac{1}{2}O_2 \rightarrow HCO_3^- \tag{XXVII}$$

The trouble was overcome by filling the empty conductance cell and its oxygen-laden electrodes with hydrogen gas until the catalytic process

$$H_2 + \tfrac{1}{2}O_2 \rightarrow H_2O \tag{XXVIII}$$

had gone to completion, a clear case of heterogeneous catalysis being used on the principle of "set a thief to catch a thief". An alternative and quite general method of avoiding heterogeneous catalytic processes in conductance measurements is to employ an electrodeless cell with an audiofrequency transformer bridge [150].

Other types of electrode can also produce unwanted catalytic effects. This was demonstrated by Archer and Spiro [136] when they deliberately followed the aquation of $Co(NH_3)_5Br^{2+}$ both spectrophotometrically and potentiometrically with AgBr/Ag electrodes to monitor the bromide ions formed. The two rates agreed in stirred solution but, in the absence of stirring, the potentiometric rate was nearly ten times greater. This arose from two types of interaction at the electrode: reduction of $Co(NH_3)_5Br^{2+}$ by the underlying metallic silver [155] and catalysis by AgBr of the aquation and reduction of the substrate [139]. It should be emphasized that such catalytic effects cannot be circumvented by the use of a smaller electrode since the potential responds to the concentrations at the surface of the electrode, whatever its size.

Titration methods of following reactions are not always immune either. A precipitate formed during the titration may introduce a completely unsuspected catalytic effect, as the following example illustrates. In a study of the rate of the Menschutkin reaction

$$Et_3N + EtI \rightarrow Et_4N^+ + I^- \tag{XXIX}$$

in methanol, Spiro and Barbosa [156] employed the Volhard method [157] of acidifying the solution, adding excess aqueous $AgNO_3$ to precipitate the iodide ions and back-titrating the excess silver ions with KSCN using ferric ions as indicator. They found, however, that the net $AgNO_3$ titre required for a given reaction mixture increased with the amount of excess $AgNO_3$ added. The explanation proved to be catalysis by the AgI and then by the AgSCN precipitates of the reaction [158]

$$EtI + Ag^+ + H_2O \rightarrow EtOH + AgI(c) + H^+ \tag{XXX}$$

Thus the peculiar titration results were due to the heterogeneous catalysis, not of the original reaction, but of a new reaction that had arisen from the analytical procedure. Precipitates formed by the reagents themselves can also prove catalytically active. For instance, the catalysis by rare earth metal salts of the hydrolysis of acid phosphonate esters was actually due to the development of metal hydroxide gels [159]. A quite different example concerns reactions of mercury(I) salts where the disproportionation

$$Hg_2^{2+} \rightleftharpoons Hg^{2+} + Hg(l) \tag{XXXI}$$

invariably occurs and where the metallic mercury can act as an unsuspected catalyst [160]. Solution kineticists would therefore be well advised always to carry out blank tests to check the possible catalytic effect of any solid introduced in no matter how innocent a guise.

2. Substitution reactions

This section, and the ones following, describe kinetic and mechanistic studies that have been carried out on various types of catalysed solution reactions. The numerous qualitative observations of catalytic phenomena that appear in the literature will usually be mentioned only where they are relevant to the mechanistic discussion. Further references to such reports can often be found in the introductory sections of papers dealing with more quantitative investigations.

2.1 SOLVOLYSIS OF TERTIARY ORGANIC HALIDES

It has been argued that insight into catalytic mechanisms can most readily be obtained by choosing reactions whose rate-determining step is unimolecular. The reactions best known to be unimolecular in homogeneous media are undoubtedly the solvolyses of tertiary butyl halides [161–163]. Their S_N1 mechanisms are typified by the solvolysis of t-butyl bromide in an ethanol + water medium

$$(CH_3)_3CBr \overset{rds}{\rightleftharpoons} (CH_3)_3C^+ + Br^-$$

$$(CH_3)_3C^+ \xrightarrow{H_2O, EtOH} (CH_3)_3COH + (CH_3)_3COC_2H_5 \qquad (XXXII)$$

$$+ (CH_3)_2C = CH_2 + H^+$$

Barbosa et al. [41] studied the heterogeneous catalysis of this reaction in the classical solvent of 80 vol.% (55.2 mol.%) EtOH + H_2O at 25°C, using a pH-stat technique to measure the rate. Silver, silver bromide, silver sulphide, mercury(I) bromide, and mercury(II) sulphide were all found to be good catalysts (cf. Fig. 12). With AgBr the catalytic rate was not affected by

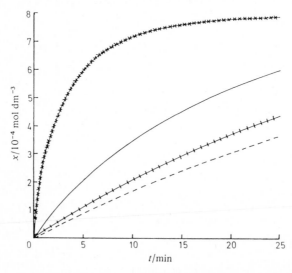

Fig. 12. Catalytic effects of silver bromide on the rate of solvolysis of t-BuBr (8.91×10^{-4} mol dm^{-3}) in 80 vol.% EtOH + H_2O (50 cm^3) at 25°C. - - - -, Homogeneous reaction; ———, addition of 0.5 g AgBr; ⊦⊦⊦⊦⊦⊦, addition of 0.5 g AgBr + 8.91×10^{-4} mol dm^{-3} KBr; ×⊷×⊷×⊷×, addition of 1.0 g AgBr from another source. (After Barbosa et al. [41].)

adding t-BuOH but it was considerably reduced in the presence of small quantities of bromide ion as Fig. 12 demonstrates. It follows that the $-$Br and not the t-Bu$-$ end of t-BuBr was adsorbed on the surface so that t-BuBr molecules and Br$^-$ ions competed for Ag$^+$ ion sites (cf. Sect. 1.4.4.). This is consistent with Pearson's SHAB principle [92, 93] (cf. Sect. 1.7) according to which bromide, a fairly soft base, would be expected to adhere well to soft acid sites such as Ag$^+$ and Hg$_2^{2+}$. The SHAB concept therefore explains why the silver and mercury salts acted catalytically and also why BaSO$_4$ and SiO$_2$ did not affect the solvolysis rate, and Al$_2$O$_3$ hardly at all, for the acid sites on these surfaces are hard. Carbons produced only minor effects and coarse platinum powder none.

Silver bromide was chosen as the catalyst for a quantitative study. In view of the competitive adsorption between reactant t-BuBr and product Br$^-$ ions, the results were interpreted by means of reaction scheme IV in Sect. 1.5.1 which in this case becomes

$$
\begin{array}{llll}
t\text{-BuBr} \xrightarrow{\ k_{hom}\ } & Br^- & & \left\{\begin{array}{l} t\text{-BuOH} \\[1ex] \\ t\text{-BuOEt} \\[1ex] \\ Me_2C{=}CH_2 \end{array}\right. \\[1ex]
\Big\updownarrow \qquad\qquad & \Big\updownarrow & +\,H^+\,+ & \\[1ex]
t\text{-BuBr(ads)} \xrightarrow{\ k_{het}\ } & Br^-\,(ads) & &
\end{array} \qquad\qquad \text{(XXXIII)}
$$

The experimental rates fitted eqn. (14) based on Langmuir adsorption, with adsorption coefficients $K(t\text{-BuBr}) = 400$ dm^3 mol^{-1} and $K(\text{Br}^-) = 3800$ dm^3 mol^{-1}. It seems very reasonable, both on steric and electrostatic grounds, that bromide ions should absorb more strongly on the silver ion sites. The rate constant k_{het} was found to be 10^2 s^{-1}, more than 10^5 times greater than k_{hom}. This large increase in rate was attributed partly to the more aqueous composition of the solvent layer at the AgBr surface and partly to the inductive electron shifts in the t-BuBr . . . Ag$^+$ system which facilitated cleavage of the C$-$Br bonds. The reaction of the carbocations Me$_3$C$^+$ released at the surface was also affected by the changed local solvent composition and structure. The water structure near the interface will have been enhanced both by the lower ethanol content and by preferential adsorption on the polar sites, so making the H$_2$O molecules less receptive acceptors for protons. Accordingly, no alkene at all was formed by the heterogeneous pathway compared with 11 mol.% by the homogeneous route. The heterogeneous process also gave 57 mol.% t-BuOH and 43 mol.% t-BuOEt whereas the homogeneous reaction led to 61 mol.% t-BuOH and only 28 mol.% t-BuOEt. This indicated that the interfacial alcohol molecules, though present in smaller concentration than in the bulk medium, were less strongly hydrogen-bonded and so competed more effectively than the highly structured water molecules for addition to the reactive Me$_3$C$^+$ ions.

Mortimer and Spiro [164] subsequently examined the catalysis by AgBr of the t-BuBr solvolysis in pure methanol at 25 and 40°C where the interpretation of the results was not complicated by changes in the interfacial solvent composition. The purity of the methanol was found to be very important: ionic impurities in particular decreased the catalytic effect, presumably by preferential adsorption on active sites. The rate equation (14) did not describe the total body of data as well as before, somewhat different values of the parameters being required for different experimental conditions. This could have been caused by slow desorption or by electrostatic repulsion between adsorbed bromide ions. The average adsorption coefficient for Br$^-$ was again much greater than for t-BuBr but the difference decreased as the temperature was raised. The surface rate constant k_{het} was ca. 10^6 times greater than k_{hom} at 25°C and the activation energy of the catalysed process was close to zero compared with 97 kJ mol^{-1} for the homogeneous solvolysis. As in the ethanol + water system, the catalysed process produced far less alkene (2 mol.%) than did the homogeneous reaction (17 mol.%). This can be understood if the methanol molecules solvating the silver ion sites became less basic and thus poorer acceptors of protons from the carbocations. In both solvents, therefore, the solvolysis at the silver bromide surface was much faster than in the bulk solution and it discriminated selectively against alkene formation.

Carbons, despite their high specific areas, had surprisingly little effect on the solvolysis of t-BuBr. A tertiary halide was therefore sought that would absorb sufficiently well on graphitic carbon surfaces to allow the heterogeneous rate to be measured. Mortimer and Spiro [36] found PhCH$_2$CMe$_2$Cl to be a suitable substrate. Its solvolysis kinetics were studied in 50 vol.% (23 mol.%) EtOH + H$_2$O at 40°C in the presence of two different carbons, an activated charcoal and Akzo Ketjenblack E.C. Both carbons inhibited the solvolysis. Competitive adsorption experiments showed that the phenyl and not the chloride end of the substrate molecule sat on the surface sites, for the degree of inhibition remained unchanged on adding NaCl or CH$_3$CMe$_2$OH but became much smaller when product PhCH$_2$CMe$_2$OH or naphthalene was added. The addition of benzene, which could easily penetrate into the charcoal structure, had no effect on the rate, which indicated that the substrate and the larger aromatic molecules were mainly restricted to surface positions. The reaction scheme can thus be written

$$
\begin{array}{ccc}
\text{PhCH}_2\text{CMe}_2\text{Cl} & \xrightarrow{\;k_{hom}\;} & \text{PhCH}_2\text{CMe}_2\text{OR} \\
\updownarrow & & \updownarrow \qquad + \text{H}^+ + \text{Cl}^- \\
\text{PhCH}_2\text{CMe}_2\text{Cl (ads)} & \xrightarrow{\;k_{het}\;} & \text{PhCH}_2\text{CMe}_2\text{OR (ads)}
\end{array}
\qquad\qquad \text{(XXXIV)}
$$

$$
R = H \text{ or } Et
$$

The rate is then given by eqn. (11) which can be recast in the form

$$
\frac{dx}{dt} = k_{hom}(a - x) + \frac{A}{V}(k_{het} - k_{hom})c_{A_{ads}} \qquad\qquad (73)
$$

It may be noted that its derivation [43] contained no assumptions about the rate or the extent of adsorption of either reactant or product. The homogeneous rate constant, k_{hom}, was measured in separate experiments and the numbers of moles of substrate adsorbed, $Ac_{A_{ads}}$, were determined by the method described in Sect. 1.4.3. The $c_{A_{ads}}$ values fitted a Langmuir isotherm with $K(PhCH_2CMe_2Cl) = 3400\,dm^3\,mol^{-1}$. The heterogeneous rate constant, k_{het}, evaluated from eqn. (73) was found to be one-quarter as large as k_{hom} and the activation energy of the surface reaction was ca. $135\,kJ\,mol^{-1}$ compared with $94\,kJ\,mol^{-1}$ for the homogeneous reaction. Although this system exhibited inhibition rather than catalysis, it provides a prime example of how heterogeneous rate data may be obtained without any assumption about the type of isotherm or the rates of sorption, provided only that an appreciable fraction of the substrate is adsorbed. In fact, a closer examination of the experimental data indicated that the reactant (or the carbocation) desorbed rather slowly from the carbon surface. A major reason for the lower rate of the surface reaction in this case was the preferential adsorption of ethanol by the carbon; it is known that the bulk solvolysis rate decreases as the mole fraction of ethanol rises. Moreover, for steric reasons the transition state on the surface was less likely than the transition state in the bulk solution to assume a cyclic phenonium form. Certain other carbons such as Grafoil [165], which possessed smaller surface areas and adsorbed the substrate less well, had no effect on the solvolysis whereas silver metal and silver chloride catalysed it.

2.2 HYDROLYSIS OF TERTIARY BUTYL ACETATE

The reaction

$$t\text{-BuOAc} + H_2O \rightarrow t\text{-BuOH} + HOAc \qquad (XXXV)$$

also proceeds via a carbocation intermediate but is extremely slow in homogeneous solution [166]. This fact, coupled with the ester's low solubility and its slow rate of dissolution, has made it necessary to study the hydrolysis at higher temperatures (typically, 60°C) where adsorption on surfaces is small. As expected, alumina which proffers hard acid sites exhibited slight catalysis [167] while surfaces which lacked this property either had no effect on the rate (e.g. silver) or slightly inhibited the reaction (silver chloride, gold) [168]. This unpromising situation appeared to be transformed when Despíc et al. [169] reported very strong catalysis of the reaction by electrically pulsed gold or silver electrodes. This phenomenon, which they termed non-faradaic electrocatalysis, came into operation only when the electrode was pulsed at 0.5 or 1 Hz over a suitable potential range in the double layer region. The effect was re-investigated by Freund et al. [168] who took care to remove dissolved dioxygen and to avoid loss of ester by volatilisation. They also set out to highlight the phenomenon by subjecting the system to alternating periods of pulsed potential and floating (open circuit) potential. Disappointingly, no catalysis was detected during any of the pulsing regimes. A similar set of experiments was then performed on the solvolysis of t-BuBr in 80 vol. % EtOH + H_2O (see Sect. 2.1) with an electrode of silver, a metal known to catalyse this reaction [41]. It was reasoned that this would reinforce any non-faradaic effect. Once again, however, the results were negative. The maximum possible catalytic rate that could be expected can be simply calculated from the frequency of pulsing and the assumption of complete monolayer coverage each time [170]: its magnitude is, on the one hand, sufficiently large to have been easily detected by Freund et al. [168] and, on the other, far too small to explain the large rates reported by Despic et al. [169]. The only fair conclusion that can presently be drawn about the phenomenon of non-faradaic electrocatalysis is best expressed by the traditional Scottish verdict of "not proven".

2.3 ESTER SOLVOLYSIS AND FORMATION

The rates and mechanisms of ester formation and solvolysis have been thoroughly investigated in homogeneous media [163, 171]. Many of these reactions can be heterogeneously catalysed [31]. The catalysis of esterifications and transesterifications has usually been carried out at high temperatures where the reactants are gaseous, the most popular catalysts being silica gel, alumina, silica-aluminas, and organic polymer cation exchange resins in the hydrogen form. In contrast, the effect of solids on ester hydrolyses has been mainly studied in aqueous solutions in the temperature range 25–45°C, with polymeric ion exchangers and mixed oxides as the predominant catalysts. All these researches have been ably reviewed by Beránek and Kraus in Vol. 20 of this series [31]. The present section will therefore deal only with certain recent studies and will focus on the catalytic effects of carbons and of weak acid ion exchange resins.

As has already been illustrated in Sect. 2.1, carbons with their graphitic surface structure often adsorb aromatic compounds well and thereby affect their reaction rates. In order to test what influence carbons would have on the solvolysis of an aromatic ester, Spiro and Mills [172] carried out exploratory experiments on the alkaline hydrolysis of benzyl acetate at 25°C

$$PhCH_2OOCCH_3 + OH^- \rightarrow PhCH_2OH + CH_3COO^- \qquad \text{(XXXVI)}$$

One gram of various solids was added to $200 \, cm^3$ of reaction mixtures that were $1 \times 10^{-3} \, mol \, dm^{-3}$ in each of the reactants. Insoluble inorganic salts and metals, including CaF_2, $BaSO_4$, HgS, Hg, Pt, and Si, caused no significant change in the rate. Organic solids were a different story. In the presence of graphitised Black Pearls carbon, the hydrolysis rate decreased to one-third of its homogeneous value whereas even 0.2 g of Carbolac 1 carbon black increased the rate 6-fold and 1 g of anthracene produced a doubling of the rate. These diverse responses may be related to the different structures and specific surface areas of the carbons [173], to the presence of quinonoid and other functional groups on the surfaces [174–176], and to the ability of Carbolac to adsorb hydroxide ions as well as aromatic molecules.

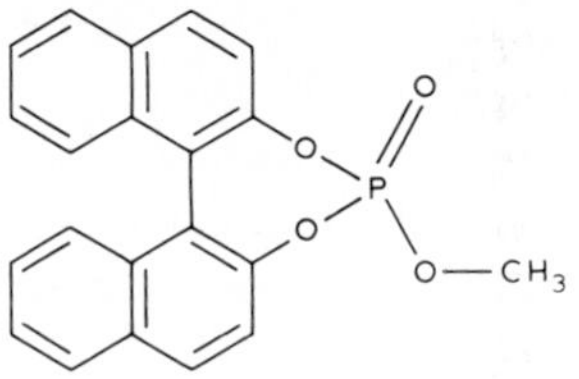

Fig. 13. Structure of the methyl phosphate ester of optically active 1,1′-bi-2-napthol studied by Hoyano and Pincock [177].

The solvolysis of a more sophisticated substrate, the methyl phosphate ester of optically active 1, 1′-bi-2-naphthol (Fig. 13), was studied by Hoyano and Pincock [177]. In the solvent 1,2-bis(2-methoxyethoxy)ethane (triglyme) at 190°C, the rate of loss of the ester's activity was not altered by adding Norite or Spheron 6 but it was much increased by these two carbons, as well as by Carbolac 1 and Sterling FT, when the ester was dissolved in the hydroxylic solvent 2-(2-ethoxyethoxy)ethanol at 130°C. One gram per dm^3 of any of these carbons increased the initial rate approximately 10-fold. The catalysed reaction was shown to be ester solvolysis and not racemisation by the fact that optically active 1,1′-bi-2-naphthol was isolated from a larger scale run. The specific rotation of this product was, however, so much smaller than that of the original ester that the same rate was obtained by polarimetric as by spectrophotometric measurements. The authors pointed out some puzzling features of this system: not only was the catalytic rate insensitive to the type of carbon employed but it also remained the same when half or twice the original amount of Norite was added. This suggested the possibility that the real catalyst was a homogeneous one, an impurity substance extracted from the carbons until it reached a constant saturated concentration. Careful experiments along the lines of test (4) in Sect. 1.8.2 excluded this interpretation. It rather looks, therefore, as if a joint homogeneous–heterogeneous effect might have been operative (cf. Sect. 2.6, below).

Before leaving the carbon scene, mention must be made of the strong and often specific catalytic effects exhibited by lamellar compounds of graphite. The structure of graphite itself is a layer one in which sheets of regular hexagonal nets of carbon atoms are held together by much weaker van der Waals forces. It is therefore possible to intercalate a variety of other species between these layers to form lamellar compounds [178]. Of particular interest for esterification purposes is C_{24}^+ HSO_4^-·2H$_2$SO$_4$, a blue crystalline compound prepared by electrolysing 98% sulphuric acid with a graphite anode [179] or simply by treating graphite with sulphuric acid containing a small amount of nitric acid [180]. The addition of this graphite bisulphate to equimolar amounts of a carboxylic acid and an alcohol in dry cyclohexane at room temperature was found to give high yields (often > 95%) of the ester in a few hours. The solid was especially efficient in forming formates and acetates and in esterifying primary alcohols [179]. Ethyl esters can also be synthesized with its aid from the carboxylic acid and $HC(OEt)_3$ [180]. The catalytic solid appears to act by taking up the carboxylic acid from the solution to form a very reactive species which then interacts with the alcohol. A secondary effect would seem to be an interaction between the graphite bisulphate and the water produced, so shifting the esterification equilibrium to completion. Kagan and his co-workers have taken out a patent on this method of esterification [181].

Organic polymeric cation exchange resins incorporating sulphonic acid groups have been employed for many years to catalyse esterifications and ester hydrolyses [31, 182]. It has been firmly established that the catalytic agent in these cases is the mobile counter ion in the swollen resin, H^+, acting in a completely analogous fashion to hydrogen ions in homogeneous acid catalysis. However, in homogeneous solution there are two main categories of acid catalysis. In one, the reaction is specifically or exclusively catalysed by hydrogen ions, an example being the hydrolysis of 1, 1-dimethoxyethane (DME) [183]

$$CH_3CH(OMe)_2 + H_2O \rightarrow CH_3CHO + 2\ MeOH \qquad (XXXVII)$$

Into the other category fall reactions subject to general acid catalysis, e.g. the hydrolysis of ethoxyethene (ethyl vinyl ether, EVE) [184]

$$CH_2{:}CHOEt + H_2O \rightarrow CH_3CHO + EtOH \qquad (XXXVIII)$$

Here not only hydrogen ions but all acid species, such as any weak acid HA, contribute to the catalysis according to the equation [185]

$$k_{obs} = k_o + k_{H^+} c_{H^+} + \Sigma\ k_{HA} c_{HA} \qquad (74)$$

where k_o is the "spontaneous" rate in water. According to Brønsted, the stronger the acid HA (i.e. the bigger its dissociation constant K_{HA}), the larger will be the corresponding rate constant k_{HA} [185]

$$k_{HA} = GK_{HA}{}^\alpha \quad (0 < \alpha < 1) \qquad (75)$$

The Brønsted relation (75), the earliest authenticated linear free energy relation, has been extensively tested in homogeneous solution. Until recently, no work had been done to find out whether general acid catalysis could also be exhibited by weak acid ion exchangers containing undissociated carboxylic acid groups. Gold and Liddiard [183, 184] therefore carried out kinetic experiments with a methacrylic acid–divinyl benzene copolymer (Zerolit SRC 41) as catalyst for the hydrolysis reactions of the esters DME and EVE, the former known to be subject to specific H^+ catalysis and the latter to general acid catalysis. Both reactions were followed by analysing the acetaldehyde produced. The concentration of undissociated methacrylic acid in the resin was varied by partial neutralisation.

In order to express their results quantitatively, Gold and Liddiard employed a two-phase model [186] and made use of a number of simplifying assumptions. The water-swollen resin was treated as an entirely permeable second phase which did not preferentially take up or reject the various small molecular solutes. The partition coefficients between solution and resin of the ester (λ) and of the other molecular solutes were therefore taken as near unity. The distribution of mobile ions between the phases was calculated by the Donnan membrane equilibrium equations. It was considered that the distribution equilibria for all the solutes would be maintained throughout, on the premise that the rates of transfer of solute species between the phases would be much faster than the rates of chemical reaction. Diffusion processes were also assumed to be too rapid to be rate-controlling, although there was some indication that this was not entirely true for the faster reaction (XXXVIII) which showed a slight increase in the catalysed rate per unit mass of resin as the resin particle size decreased. With this theoretical framework, the first-order rate constants, k_{obs}, could be expressed by the equations

$$k_{obs} = k_{aqu} + k_{resin} \tag{76}$$

$$k_{aqu} = (k_0 + k_{H^+} c_{H^+}) \left(\frac{V}{\lambda \bar{V} + V} \right) \tag{77}$$

$$k_{resin} = (\bar{k}_0 + \bar{k}_{H^+} \bar{c}_{H^+} + \bar{k}_{HA} \bar{c}_{HA}) \left(\frac{\lambda \bar{V}}{\lambda \bar{V} + V} \right) \tag{78}$$

In these equations a superscript bar denotes properties of the resin phase, absence of a bar refers to the aqueous solution phase, V is volume, and HA stands for the undissociated acid groups inside the resin.

The experiments showed that the resin was a more powerful catalyst for the hydrolysis of EVE than for that of DME. The reason lies in the fact that, for DME, $\bar{k}_{HA}$ was zero while for EVE it was $7.5 \times 10^{-4}\,dm^3\,mol^{-1}\,s^{-1}$ at 25°C when the resin was half neutralised; this meant that 97% of its catalytic effect was then caused by the undissociated acid groups. The value for $\bar{k}_{HA}$ expected for EVE from its hydrolysis kinetics in aqueous solution [187] and the Brønsted relation was $5 \times 10^{-4}\,dm^3\,mol^{-1}\,s^{-1}$ [186], in reasonable agreement with the value obtained. General acid catalysis within the resin was thus established. Further experiments revealed that $\bar{k}_{HA}$ decreased steadily with progressive neutralisation of the resin. This effect was attributed to a decrease in the acid strength of carboxylic acid groups when neighbouring carboxylate groups became ionised, just as with polyprotic acids where the second dissociation constant is always much smaller than the first. By introducing a transfer matrix method, the authors were able to evaluate the rate constant k_{uu} for catalysis by carboxylic acid groups flanked on either side by other undissociated acid groups; it was satisfying to note that these values of k_{uu} showed no systematic trend with the degree of resin neutralisation [184].

References pp. 159–166

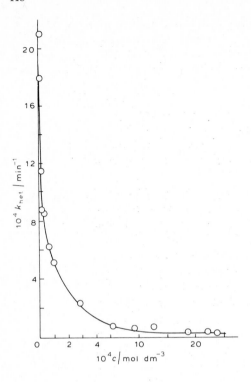

Fig. 14. Variation of the heterogeneous rate constant for the reaction of $Co(NH_3)_5Br^{2+}$ in the presence of HgS (1 g in 500 cm³ solution) at 25°C. (After Archer and Spiro [137].)

2.4 HYDROLYSIS AND FORMATION OF COBALT(III) COMPLEXES

Inorganic as well as organic substitution reactions can be heterogeneously catalysed. Not surprisingly, catalytic effects have been observed most often in connection with inert complexes like those of cobalt(III) [188]. Bjerrum in his classic book on metal ammine formation [189] noted several examples of the catalysis of cobaltammine reactions by charcoal, mercury, and platinum in macroscopic and in colloidal form, and he made good use of this effect in an easy and cheap synthesis of $Co(NH_3)_6^{3+}$ salts under ambient conditions in place of the high pressure preparation used previously [190]. About the same time, Bailar and Work [191] drew attention to the fact that reactions in which nitrogen atoms are coordinated to cobalt or chromium were particularly susceptible to catalysis by charcoal and that silica gel and Raney nickel were also good catalysts. A brief review has been given of these and related observations [136].

The first detailed kinetic study of these catalytic effects was carried out by Archer and Spiro [136–138] on the bromopenta-amminecobalt(III) ion which slowly aquates in homogeneous solution according to the equation

$$Co(NH_3)_5Br^{2+} + H_2O \rightarrow Co(NH_3)_5(H_2O)^{3+} + Br^- \qquad (XXV)$$

The rate of disappearance of the reactant was greatly increased on adding solid AgBr, HgS, Pt, Pd, and Au. Test (4) in Sect. 1.8.2 confirmed that the rate enhancement was heterogeneous in nature. The effect was repressed on adding additional halide ions and this fact, together with the failure of insoluble salts like $BaSO_4$ with hard acid sites to influence the rate, strongly suggested that the reactant was adsorbed by its halide end on soft acid sites on the surfaces. For each solid, the first-order heterogeneous rate constant rose dramatically as the initial concentration of the reactant decreased, as illustrated in Fig. 14. This behaviour was consistent with Langmuir adsorption of the reactant followed by rate-determining decomposition of the adsorbed species. It was also noted that in the presence of silver, mercury, or charcoal, the cobalt(III) complex ion rapidly reduced to cobalt(II). Not until a subsequent investigation [139] employing an even more sensitive analytical test for cobalt(II) [192] was it discovered that cobalt(II) had also been produced in the earlier experiments with AgBr, HgS, and platinum. Large fractions of the rate enhancements by these three solids had therefore been caused by reduction rather than aquation of the reactant. With platinum present, even the product $Co(NH_3)_5(H_2O)^{3+}$ formed by the parallel homogeneous route was slowly reduced to cobalt(II). [It might therefore be expected that, in the presence of AgI, the analogous iodocomplex $Co(NH_3)_5I^{2+}$ could be reduced to Co(II), a process partnered by the oxidation of released iodide ions to iodine, but Rustad [193] did not find this (Sect. 2.5).] As was mentioned in Sect. 1.8.3, these experiments illustrate that the stoichiometry of a catalytic reaction cannot be taken for granted. The same point applied also to two other kinetic studies in the literature, the carbon-catalysed hydrolysis of $Co(NO_2)_6^{3-}$ [194, 195] and of $Co(NH_3)_6^{3+}$ [195]. Both were later found to generate appreciable amounts of cobalt(II) [139]. There was evidence in all these cases of photochemical involvement.

Since the synthesis of $Co(NH_3)_6^{3+}$ is greatly facilitated by the use of carbon as a catalyst [190], Mureinik [196] undertook a careful kinetic study of the reaction

$$Co(NH_3)_5(OH)^{2+} + NH_4^+ \rightleftharpoons Co(NH_3)_6^{3+} + H_2O \qquad \text{(XXXIX)}$$

No reaction at all took place at 25°C in the absence of carbon so that the measured rates could be completely ascribed to the action of the catalyst, Decolorizing Charcoal C177. The concentrations of both cobalt complexes were spectrophotometrically monitored with time and it was noted that the sums of the concentrations of the two species were always 2–3% short of the initial concentrations. Since the intercepts of the first-order rate plots at zero time also gave concentrations 2–3% lower than the initial values, these apparent discrepancies clearly pointed to a small amount of fast adsorption. The rates were independent of the shaking speed which marked the catalysis as surface-controlled. The kinetics of this surface reaction were, however, extremely complicated. Mureinik systematically varied the concentrations of the relevant species: he found that the plot of the effective first-order rate constant, k_{eff}, against the concentration of NH_4^+ possessed a Langmuir-like shape (cf. Fig. 1) but with a large finite intercept, that k_{eff} depended on the NH_3 concentration raised to the power 1.7 at low concentrations before levelling off at high ones, that k_{eff} at first decreased with increasing concentration of $Co(NH_3)_5OH^{2+}$ as in Fig. 14 but then levelled out to a sizeable plateau value at higher concentrations of complex ion and, most surprising of all, that k_{eff} increased proportionately to the mass of catalyst raised to the power 1.3. The paper suggested that the "extra 0.3" power could have arisen from the introduction into the solution of a new soluble species originating from the carbon surface, a likely species being CO_3^{2-} formed by adsorbed carbon dioxide dissolving in the alkaline reaction medium. If this species played an active role in the mechanism and was adsorbed by a Freundlich isotherm with an exponent of 0.3 the curious mass dependence could be understood. Mureinik made a valiant attempt to interpret the kinetics with an 8-step reaction scheme involving as adsorbed species NH_4^+, NH_3, OH^-, the new desorbed species and its protonated form (e.g. CO_3^{2-} and HCO_3^-) as well as $Co(NH_3)_5OH \cdot NH_3^{2+}$, $Co(NH_3)_3(NH_2)_2^+$, and $Co(NH_3)_6 \cdot NH_3^{3+}$. All these cobalt species had been previously proposed in the literature to explain well-documented phenomena in homogeneous solution. Even then, the resulting rate equation was not able to account for all the experimental findings. There was also evidence of a parallel reaction route in which cobalt(II) participated. Although no cobalt(II) was detected during the course of the reaction at a limit of detection of ca. 10^{-6} mol dm^{-3} [192], addition of extra Co(II) produced a linear rise in k_{eff}. It seemed quite possible that, in this ammoniacal medium, the cobalt(III) complex had electron-exchanged to the more labile cobalt(II) complex which, after ligand rearrangement, was reoxidised to the cobalt(III) form along the lines suggested by Dwyer and Sargeson [197] for a different cobalt reaction. Finally, it was interesting to find that the activation energy of the standard catalysed reaction mixture was only 6.5 ± 3 kJ mol^{-1}. This represents an almost exact numerical balance between the positive enthalpy of activation of the chemical reaction and the negative enthalpy of the relevant adsorption process-(es).

To sum up, the two detailed rate studies of carbon-catalysed cobalt(III) substitutions have shown these reactions to be far from simple. Further research is required before various mechanistic questions can be fully answered.

2.5 JOINT HOMOGENEOUS AND HETEROGENEOUS CATALYSES

Catalysed substitution reactions of an unusual kind are collected together in this section. In each case, the catalysis of the reaction by a homogeneous entity is assisted by the surface of a solid. The resulting reinforcement of catalytic effects is frequently described as synergistic. The homogeneous and heterogeneous catalysts quite often possess a species in common, for example Ag^+ ions and solid AgI, and many of the homogeneously catalysed reactions exhibit autocatalysis as a result.

The archetypal reaction in this category, and the one most thoroughly investigated, is the hydrolysis of ethyl iodide

$$EtI + H_2O \rightarrow EtOH + I^- + H^+ \qquad \text{(XL)}$$

Its rate is greatly increased in the presence of silver ions [158]

$$EtI + Ag^+ + H_2O \rightarrow EtOH + AgI(c) + H^+ \qquad \text{(XLI)}$$

The silver iodide solid which is produced autocatalyses the reaction. Burke and Donnan [198], studying the EtI + AgNO$_3$ reaction in ethanol, were the first to observe the autocatalysis and they tried, unsuccessfully, to account for it by adding the various soluble reaction products (HNO$_3$, EtOEt, EtNO$_3$) to the initial mixture. So imbued were they with the idea that solution reactions can only be homogeneously catalysed that it did not occur to them to add silver iodide. It was left to Senter [199], working one year later with the MeI + AgNO$_3$ reaction in water and in alcohol, to identify AgI as the catalytic agent responsible. It is now widely recognized that alkyl halide + silver salt reactions in hydroxylic solvents are autocatalysed by the silver halides formed [98].

To find out whether AgI is a specific catalyst for reaction (XLI), Walton and Spiro [158] carried out experiments in which a variety of different solids was added to the reaction mixture. They followed the reaction by its pH change to avoid inadvertent catalysis by a metal sensing device. Glass, Perspex, PTFE, and BaSO$_4$ had no effect on the rate; silica, silicon, silver, palladium and platinum increased the rate appreciably, and charcoal and silver chloride, bromide, iodide, arsenate, and phosphate markedly catalysed the reaction. For the last two salts, due allowance was made for the extra silver ions introduced into the solution. These results, together with the kinetics discussed below, are consistent with a Langmuir–Hinshelwood mechanism in which EtI molecules and Ag$^+$ ions are adsorbed on neighbouring sites on the surfaces of these solids and then react. The close proximity of the adsorbed reactants and the weakening of the C–I bond contribute to the faster surface rate. The relative effectiveness of the solids can be qualitatively understood in terms of Pearson's SHAB theory (cf. Sect. 1.7). Iodide is a soft base that will coordinate best to a surface site that is a soft acid (e.g. Ag$^+$) and the silver ion is a soft acid and so will coordinate best with a soft base site like iodide. This explains why silver iodide, which provides both kinds of site, was such a powerful catalyst. In contrast BaSO$_4$, which provides only hard acid and hard base sites, did not catalyse at all. Metals were relatively poor catalysts: although Pearson [93] has concluded that bulk metals can act as either soft acids or soft bases, their high electrical conductivity may make it difficult for alternate metal atoms on the surface to function as acid and base, respectively. The greater catalytic activity of charcoal can be ascribed in part to its lower conductivity and the possibility of π-bonding and partly to its much greater specific area.

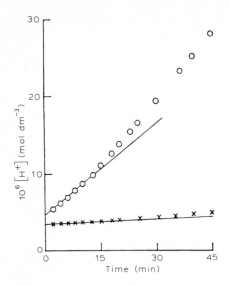

Fig. 15. Rates of hydrogen ion production in the reaction between EtI$(1 \times 10^{-3}\,\text{mol dm}^{-3})$ and AgNO$_3$$(1 \times 10^{-3}\,\text{mol dm}^{-3})$ at 5°C. $\times$, No added solid; $\circ$, addition of 1 g AgI in 500 cm^3 solution. The straight lines indicate the initial rates. (After Austin et al. [200].)

Walton and Spiro [158] also examined the effect of various solids on the related metal-ion catalysed reactions

$$n\,\text{EtI} + \text{M}^{n+} + n\,\text{H}_2\text{O} \rightarrow n\,\text{EtOH} + \text{MI}_n(\text{c}) + n\,\text{H}^+ \qquad \text{(XLII)}$$

where $\text{M}^{n+} = \text{Tl}^+$, Pb^{2+}, Hg^{2+}, and Hg_2^{2+}. All these reactions were strongly catalysed by freshly prepared silver iodide. Lesser degrees of catalysis were observed (for all except the Tl^+ reaction) on the addition of the corresponding metal iodide and of charcoal. A similar mechanism may be formulated for these cases.

The strong catalytic effects produced by silver iodide and by charcoal on reaction (XLI) were studied quantitatively by Austin et al. [200] with $0.02\,\text{mol dm}^{-3}$ KNO$_3$ as background electrolyte to stabilise the pH readings. Both catalytic reactions were independent of the stirring speed and were therefore surface-controlled. Most of the experiments were carried out at 5°C where the volatility of ethyl iodide was low and the homogeneous reaction made only a small contribution to the overall rate. Figure 15 shows how much more rapidly hydrogen ions were produced in the presence of added silver iodide. The autocatalytic nature of the reaction is also clearly evident at longer times. In the initial stages, the homogeneous reaction followed the simple rate law

$$\frac{d[\text{H}^+]}{dt} = k_{\text{hom}}[\text{EtI}]\,[\text{Ag}^+] \qquad (79)$$

where the square brackets represent concentration. When either silver iodide or activated charcoal were present, the initial heterogeneous rates could be well represented by equations of the form

$$\frac{d[H^+]}{dt} = k_{cat}[EtI]^\alpha[Ag^+]^\beta \qquad (0 < \alpha, \beta < 1) \tag{80}$$

or

$$\frac{d[H^+]}{dt} = k'_{cat}\left(\frac{[EtI]}{1 + K[EtI]}\right)[Ag^+]^\beta \tag{81}$$

These equations are consistent with a Langmuir–Hinshelwood mechanism involving the rate-determining step

$$EtI(ads) + Ag^+(ads) \rightarrow products \tag{XLIII}$$

with Ag^+ adsorbed according to a Freundlich isotherm and EtI adsorbed by either a Freundlich or a Langmuir isotherm (cf. Sects. 1.4.2 and 1.5.3). A comparison of the rate constants per unit area of catalyst showed that AgI was more effective than charcoal by a factor of at least 40. The activation energies of the two catalysed processes were 36 and 44 kJ mol^{-1}, respectively, consistent with surface control and much smaller than the activation energy of the homogeneous reaction (82 kJ mol^{-1}).

Although many other examples of joint silver ion–silver halide catalysis of alkyl halide solvolyses are now known [98], their kinetics have not been studied in detail. Inorganic analogues also exist as shown by Rustad's finding [193] that the reaction

$$Co(NH_3)_5I^{2+} + Ag^+ + H_2O \rightarrow Co(NH_3)_5(H_2O)^{3+} + AgI(c) \tag{XLIV}$$

was catalysed by silver iodide. Spectrophotometric evidence indicated complete conversion to the aquocomplex but cobalt(II) was not specifically tested for. The silver iodides were prepared by precipitation in solution and their surface areas were measured by eosine dye adsorption. In the reaction mixtures some 5% of the iodocomplex was rapidly adsorbed. The initial pseudo-first-order rate constants (with respect to iodocomplex) at 45°C decreased with increasing silver ion concentration (cf. Fig. 14), especially at low ionic strengths, although more complicated behaviour was also observed. The rates of catalysis correlated better with the moles of AgI per dm^3 than with the surface area. In order to explain his results, Rustad postulated that adsorbed silver ions tended to migrate into the bulk silver iodide while iodide ions and iodocomplex ions adsorbed competitively on the surface by Langmuir isotherms. It was further assumed that two reaction paths contributed to the overall catalytic rate. In one, the mobile silver ions reacted directly with iodocomplex ions adsorbed on "unassisted" surface sites; in the other, these silver ions reacted with iodocomplex ions sitting on "assisted" sites surrounded by two adsorbed iodide ions to lessen charge repulsion with the Ag^+ ions. The first pathway became relatively more important with ageing of the silver iodide and at higher ionic strengths.

A further case of cooperative homogeneous–heterogeneous catalysis was reported by Barbosa and Spiro [201] for the Menschutkin reaction

$$Et_3N + EtI \rightarrow Et_4N^+I^- \tag{XLV}$$

in benzene. The homogeneous reaction is extremely slow. The rate was slower still in the presence of silver iodide or Black Pearls carbon because of non-reactive adsorption. However, the addition of silver nitrate speeded up the rate by a factor of ca. 10^4, and silver nitrate plus either silver iodide or carbon led to a further increase in the rate. The product in these runs was $Et_4N^+NO_3^-$. A model in which the reactants and silver ions were suitably positioned on the surface was put forward to explain the results. More quantitative work is currently being carried out on these systems[202].

The final examples of synergistic catalysis to be considered in this section were recently discovered by Franklin and his group [203, 204]. The homogeneous catalysts were cationic surfactants, the heterogeneous ones platinum and other surfaces, and the reaction in question was the alkaline hydrolysis of ethyl benzoate. Such ester hydrolyses are known to be catalysed by cationic micelles although not by monomeric surfactant species [205]. The authors showed that the hydrolysis of their ester was barely catalysed by platinum alone, a point supported by some unpublished experiments of Ohag and Moseley [206] in which the alkaline solvolysis of ethyl benzoate in 85 wt. % EtOH + H$_2$O was not even catalysed by platinized platinum. Yet in the presence of platinum foil, cationic surfactants both below and above their critical micelle concentration were found to catalyse the ethyl benzoate hydrolysis. At constant surfactant concentration, the catalysis increased linearly with the area of metal present (nickel, platinum). When the area of the solid was held constant, an increase in the surfactant concentration produced a series of rises and falls in the rate. Such patterns appeared not only on metal surfaces but also in glass and in Teflon vessels. The authors postulated that on a polar surface the surfactant ions and their halide co-ions were initially arranged in a monolayer as depicted in Fig. 16. This hydrophobic film was assumed to solubilize the ester molecules and lead to catalysis, in the manner of an inverted micelle. The second layer would be

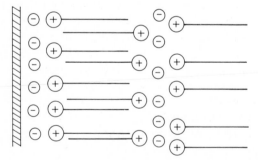

Fig. 16. Schematic diagram illustrating the formation of a hydrophobic surfactant film on a polar surface. (After Franklin et al. [204].)

hydrophilic and produce a lesser or even negative effect (a point not fully explained) while the third layer, once more hydrophobic, would again catalyse. On a non-polar surface like Teflon the reverse sequence would appear. It was also suggested that in some circumstances the first layer would be formed with the molecules lying flat on the surface together with the halide ions, and that such a film would act as a strong inhibitor. Several aspects of this intriguing system remain to be elucidated and further research would clearly be of interest.

3. Isomerisation reactions

3.1 RACEMISATION OF 1, 1'-BINAPHTHYLS

Pincock and his co-workers [39, 40, 207–209] have carried out a thorough kinetic study of the heterogeneous catalysis of the racemisation of 1, 1'-binaphthyls

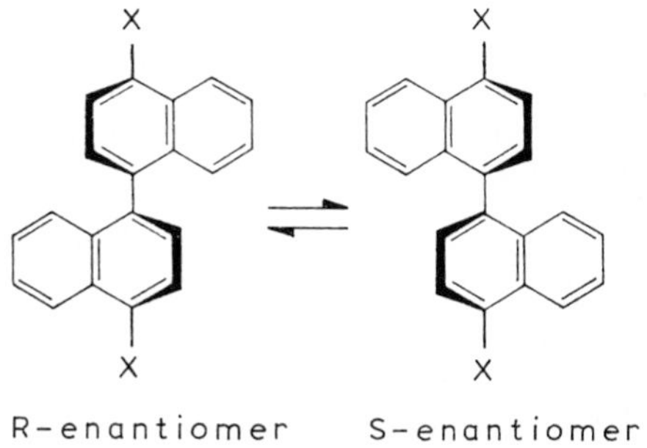

(III')

R-enantiomer   S-enantiomer

The interconversion of the two enantiomeric forms occurs by rotation around the central 1-1' bond joining the two naphthalene units. This process is sterically hindered. The homogeneous racemisation of either optical form of 1, 1'-binaphthyl (X = H) was therefore quite slow with a half-life of ca. 12 h in acetone at 20°C, but almost complete racemisation occurred within minutes on adding $1 \, \mathrm{g \, dm^{-3}}$ of several active carbons. Their effectiveness increased in the sequence graphite < Sterling FT < acetylene black < Spheron 6 < Norit SG1 < Carbolac 1, which qualitatively paralleled their specific surface areas. Figure 17 shows first-order kinetic plots for the uncatalysed run and several catalysed runs with different carbons. Rate constants k_{obs} were derived from the initial slopes and were found to be independent of the stirring speed. The catalysis was therefore not limited by external diffusion to the surface. Surface control was also consistent with the sensitivity of the catalytic rates to poisoning by polyaromatic impurities, as described in Sect. 1.4.4. Reproducible results were therefore only obtained by using well-purified binaphthyls.

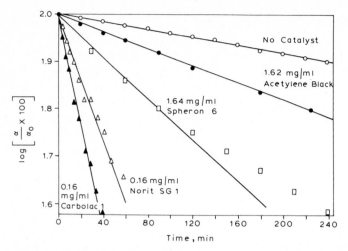

Fig. 17. First-order plots of the logarithm of the optical rotation α against time for the racemisation of 0.015 mol dm^{-3} 1,1'-binaphthyl in acetone at 20°C in the presence of various carbon catalysts. (After Pincock et al. [39].)

The following reaction scheme was proposed [39, 40]

$$R \underset{}{\overset{k_{unc}}{\rightleftharpoons}} S$$

$$R + C^* \underset{k_{-1}}{\overset{k_1}{\rightleftharpoons}} C^*B \underset{k_1}{\overset{k_{-1}}{\rightleftharpoons}} C^* + S \qquad \text{(XLVI)}$$

$$I + C^* \underset{k_{-1}}{\overset{k_1}{\rightleftharpoons}} C^*I$$

Here R and S are the two enantiomeric forms of binaphthyl, C* stands for the active sites on the carbon, C*B is the catalyst–binaphthyl complex, I is an inhibitor and C*I the inactive catalyst–inhibitor complex. On the assumption of steady state concentrations of C*B and C*I, Pincock et al. [39, 40] derived the equation

$$k_{obs} = k_{obs, unc} + \frac{k_1[C^*_{total}]}{1 + (k_1/2k_{-1})[\text{binaphthyl}] + k_1[I]/k_{-1}} \qquad (82)$$

It may be noted that the denominator of the heterogeneous term resembles the format expected from the Langmuir adsorption equation (8) even though Scheme (XLVI) has not stipulated equilibrium between bulk and adsorbed concentrations. As eqn. (82) predicts, k_{obs} was indeed found to increase linearly with the mass of carbon (Spheron 6) added. The rate constant also fell with rising binaphthyl concentration (cf. Fig. 14) but the equation fitted these results only qualitatively. However, the fall in rate with added inhibitor could be fitted better, a plot of $1/(k_{obs} - k_{obs,unc})$ rising linearly with increasing concentration of inhibitor. The relative inhibitory properties in acetone increased with aromatic size in the order benzene, nil; naphthalene,

0.05; anthracene, 1.3; pyrene, 10; perylene, 13. The figure for the sterically hindered compound 2,6-di-*tert*-butylnaphthalene was only 0.002.

The poisoning produced by these polyaromatics was attributed to their adsorbing competitively with the 1,1'-binaphthyl on the carbon surface. That the inhibition was greater the larger the planar aromatic molecule, and that it almost disappeared when that molecule was sterically hindered, pointed strongly to competition on the surface with a planar form of the binaphthyl. Such a planar form, adsorbed on the graphite-like sites of the active carbons, seemed an attractive possibility for the transition state of the catalysed racemisation. It was also consistent with the much smaller catalytic activities of platinum and nickel [208, 209]. However, this simple model had to be modified after further careful experiments by Hutchins and Pincock [207]. These workers racemised several 4,4'-disubstituted 1,1'-binaphthyls (X $= NH_2$, CH_3, Br, NO_2) in chloroform solutions at 24°C in the presence and absence of $1\,g\,dm^{-3}$ Norit SG1 carbon. Catalysis was observed in all cases. Although the effect of adding the carbon was less pronounced than for the unsubstituted binaphthyl (X $=$ H), it was clear that the binding of the binaphthyl substrate to the carbon surface was not very stereocritical. The same conclusion emerged from Hammett plots (cf. Sect. 1.7) of $\log k_{obs}$ versus σ (the substituent constants for *para* groups). The rate constants for the disubstituted binaphthyls catalysed by carbon actually fell on a better straight line than did those for the uncatalysed racemisation, with slopes ρ of -0.57 and -0.88, respectively. Electron-donating substituents therefore had less influence on the catalysed than on the uncatalysed reaction. A lower sensitivity to charge donation from substituent groups could be readily understood if the catalyst surface itself had donated electrons to the adsorbed substrate. This deduction was supported by evidence that the binaphthyl radical anion racemises extremely rapidly [210] and by calculations which showed that the two naphthalene moieties in the radical anion are inclined towards a coplanar geometry [211].

To round off their investigation, Hutchins and Pincock [207] prepared a series of modified carbons in order to discover which feature of the carbon surface was responsible for the catalytic action. Carbon blacks and activated charcoals contain not only graphite-like basal planes but also, attached mainly to edge sites, a variety of carboxylic, quinoid, lactone, and phenolic groups [174–176]. The number of these functional groups can be markedly increased by oxidation with nitric acid [212] and decreased by reduction with lithium aluminium hydride. Neither oxidative not reductive treatment was found to affect the racemisation rates on carbon. The catalysis therefore did not involve these functional groups nor did it take place on edge sites. Halogenation experiments were then carried out to vary the state of the polyaromatic basal planes as these sites readily adsorb Cl_2 or Br_2. The halogenation of Spheron 6 was found greatly to increase its catalytic activity, perhaps because the halogen improved the electron-donating properties of these surfaces. Deliberate attempts were also made to provide a negatively

128

charged and a positively charged graphite surface by preparing intercalation compounds of potassium–graphite [213], $C_{22}K$, and ferric chloride–graphite, C_6FeCl_3, respectively. In the event both showed reasonable catalytic activity, the rates with $C_{22}K$ being erratic and sensitive to impurities.

Taking all the evidence together, the catalytic mechanism can be well represented by the equation [214]

Optically active   Transition state   Racemic mixture
enantiomer

(XLVII)

The transition state is pictured as an essentially planar electron-accepting binaphthyl molecule loosely bound on electron-donor sites of the graphitic basal planes of the carbon. The enthalpy and entropy of activation for the catalysed reaction on Spheron 6 in acetone were 66 kJ mol^{-1} and -99 J K^{-1} mol^{-1}, respectively, compared with 92 kJ mol^{-1} and -25 J K^{-1} mol^{-1} for the homogeneous racemisation in a wide range of solvents [40]. The changes in both parameters are consistent with a catalytic route via an adsorbed transition state, with the restrictions in its degrees of freedom being reflected in the very negative $\Delta S_{het}^{o\neq}$ while $\Delta H_{het}^{o\neq}$ is smaller because it includes a negative enthalpy of adsorption term.

Hutchins and Pincock discovered that the racemisation of 1,1'-binaphthyl was also catalysed, though to a much smaller extent, by platinum [208] and by Raney nickel [209]. Nickel in heptane solutions at 25°C both catalysed and reduced the substrate. Addition of small amounts of sulphur or dodecanethiol poisoned the reduction reaction but larger amounts stopped the racemisation too. The authors believed that two functionally different types of site were involved, hydrogen atom donor sites being responsible for the reduction reaction and electron donor sites for the isomerisation. Runs with partially poisoned Raney nickel gave first-order plots, though the slopes were not reproducible. With platinum as catalyst in ethanol solutions at 25°C, the kinetic plots were also first order and difficult to reproduce. By careful experimentation it was possible to show that the first-order rate constants decreased as expected with increasing binaphthyl concentration (cf. Fig. 14). Surprisingly, however, the catalytic rates appeared to be zero order in the concentration of platinum. This was not due to catalysis by a homogeneous species with which the solution was saturated because the reaction could be stopped by filtering off the platinum (Test 4 in Sect. 1.8.2). The heterogeneous nature of the catalysis was also consistent with inhibitory effects produced by the addition of cyclohexene or cyclohexane. When air was injected into the reaction mixture, there was only a pause in the progress of the reaction after which it proceeded as before; the duration of the pause was roughly proportional to the volume of air injected. This observation showed that on platinum the racemisation and hydrogenation (of oxygen) reactions occurred on the same site, in contradistinction to Raney nickel. On both these metals the catalysis probably took place on adsorption sites that were electron-donating but less stereoselective than on carbon. Since this racemisation has now been shown to be catalysed by several surfaces which have in common the possibility of acting as reversible electron donors, Pincock [214] has recently suggested that the isomerisation of 1,1'-binaphthyl could serve as a convenient test reaction for active electron donor sites on the surfaces of other solids such as oxides.

3.2 MUTAROTATION OF GLUCOSE

A reaction exhibiting general acid catalysis, the ester hydrolysis (XXXVIII), has been discussed in Sect. 2.3. The present section deals with a classic reaction which is subject to both general acid and base catalysis in homogeneous media, the mutarotation of D-glucose.

α-D-Glucose $\rightleftharpoons$ (open chain) $\rightleftharpoons$ β-D-Glucose (XLVIII)

The interconversion of these ring diastereoisomers is believed to take place through the open-chain hydroxyaldehyde form and thus involves the breaking and reforming of a semi-acetal link [185]. The homogeneous kinetics of reaction (XLVIII) and related reactions in the presence of a large number of acids and bases have been extensively documented [215]. The possibility that a mutarotation could be similarly catalysed by acid and base groups on the surfaces of solids was first demonstrated by Tanabe et al. [216] with α-D-tetramethylglucose in benzene. Only recently, however, has the subject been tackled quantitatively by Dunstan and Pincock [217, 218] in what must be regarded as a pioneering study.

The solid chosen for their work on reaction (XLVIII) was alumina whose surface possesses several types of Brønsted and Lewis acidic and basic functional groups ($-Al^+-$, $-OH^{\delta+}$, $-O^{\delta-}H$, $-O^-$, and defect sites) [175, 219] which could be potential catalysts for the mutarotation of glucose. Woelm alumina did show pronounced catalytic activity: some catalysis was also displayed by Black Pearls carbon but none by graphite or silica [214]. All the experiments were carried out in the aprotic solvent dimethyl sulphoxide at 25°C. The progress of the mutarotation from the pure α (or β) form to the equilibrium mixture ($\alpha/\beta = 0.6$) was determined by optical rotation (θ). First-order plots of $\ln(\theta_t - \theta_\infty)$ against time, t, in the presence of alumina were characterised by a fairly rapid initial decrease in rotation due to adsorption of glucose followed by a line curving gently because of a slow deactivation of the catalyst. The rate constants were accordingly obtained from the slopes at the point at which adsorption equilibrium had just been attained. The catalysis was marked as surface-controlled after experiments with different degrees of stirring and different particle sizes had excluded both external and internal diffusion as rate-limiting. The heterogeneous mutarotation was thus represented by the equation

$$G_\alpha + C \underset{k_{-1}}{\overset{k_1}{\rightleftharpoons}} G_\alpha C \underset{k_4}{\overset{k_3}{\rightleftharpoons}} G_\beta C \underset{k_1}{\overset{k_{-1}}{\rightleftharpoons}} C + G_\beta \qquad \text{(XLIX)}$$

where G denotes glucose and C the catalyst. The rate constants for adsorption (k_1) and desorption (k_{-1}) of α- and β-glucose were taken to be the same, and adsorption and desorption were assumed to be fast relative to the interconversion reaction itself. The resulting rate equation, derived by treating the powdered dispersed solid catalyst in the same way as a homogeneous catalytic system, was given by

$$k_{obs} = \frac{(k_3 + k_4)\,[C]}{(1/K) + [G_0]} \qquad (83)$$

where $K = k_1/k_{-1}$ and where [C] is the concentration of catalyst and [G_0] that of the total glucose. It is interesting that an equation of the same format emerges from the two-phase model in Sect. 1.5.2 if the homogeneous rate can be neglected, if the fraction of glucose adsorbed is small, and if the adsorbed concentration can be expressed by a Langmuir isotherm. On this interpretation K becomes the Langmuir adsorption coefficient and [C] must be replaced by Ac_{mono}/V.

Dunstan and Pincock carried out a carefully planned series of experiments to ascertain the number and nature of the active adsorption sites. They began by determining the adsorption isotherms on the alumina using equilibrated glucose solutions. An analysis of the results revealed the existence of three kinds of site: 0.7×10^{-4} mol of irreversible adsorption site, 1.0×10^{-4} mol of strong adsorption site, and 1.3×10^{-4} mol of weak adsorption site, all per gram of alumina. When the amounts of glucose adsorbed on these various sites were compared with the observed mutarotation rates over a range of glucose concentrations, the rates were found to correlate well only with the adsorption on the weak sites. These were therefore the active ones in the catalysis. Combination of adsorption and kinetic data through eqn. (83) yielded a value for ($k_3 + k_4$) of 5×10^{-3} s^{-1}. This may be compared with a rate constant for the homogeneous reaction in water of $4 \times 10^{-4} \text{ s}^{-1}$ [215].

Competitive adsorption experiments showed, somewhat surprisingly, that added water did not inhibit the catalytic mutarotation. Neither did methanol nor π-donors like benzene and naphthalene. However, the polyhydroxy-compounds methyl α-D-glucoside, iso-inositol (a conformational model for the cyclic form of glucose) and DL-glyceraldehyde (a model for the open-chain form of glucose) competed well with glucose for both overall adsorption sites and the catalytically active ones. The only inhibitor that discriminated between these two types of site was n-hexanal, which produced a greater percentage decrease in catalytic rate than in glucose adsorption. These experiments indicated that the catalytically active sites specifically adsorb polyhydroxy compounds and interact expecially well with aldehyde groups.

The research described so far refers to neutral aluminium oxide which had been used without modification. In their second paper [218], Dunstan and Pincock reported experiments with alumina that had been progressively heated. The results are summarised in Table 2. It can be seen that the more severe the thermal treatment, the smaller the BET (N_2) surface area although the particle size distribution was not much changed. The amount of glucose adsorbed decreased little up to the "800°C" treatment but then fell sharply with the two most strongly heated materials. The most interesting changes were shown by the kinetic data. The rate constant per unit area decreased steadily at first as the alumina was heated, falling to one third of

TABLE 2

Effect of thermal treatment on the catalytic activity of alumina
After Dunstan and Pincock [218].

Dehydration conditions	Wt. loss (%)	Surface area $(\text{m}^2\text{g}^{-1})$	Rate constant[a] per unit area $(10^{-6} \text{ s}^{-1} \text{ m}^{-2})$	Glucose adsorbed (%)
None		140	1.4	14
24°C, 4 days 0.01 Torr over P_2O_5	2.8	144	1.0	14
150 ± 5°C, 0.01 Torr, 2 days	5.7	130	0.7	14
600 ± 50°C, under dry N_2, 4 h	6.2	116	0.43	13
800 ± 50°C, under dry N_2, 4 h	6.2	100	0.7	10
1100 ± 50°C, under dry N_2, 3 h	7.2	14	3.9	3.5
1250 ± 50°C, under dry N_2, 6 h	7.9	6.2	36	3

[a]The first-order rate constants were obtained with a 0.05 mol dm^{-3} glucose solution in DMSO at 25°C containing 1.6 g alumina in 60 cm^3.

its original value with the "600°C" solid. Thereafter it rose and on the "1250°C" alumina reached a value over 25 times larger than on the untreated material. Moreover, the "1250°C" alumina produced strictly linear first-order plots with no deactivation of catalyst whereas curved plots were exhibited by all the other aluminas. This quite different behaviour was clearly related to the fact that, above 1100°C, alumina alters its crystal structure from the γ to the α form.

The thermal treatment also changed the nature of the catalytically active sites. According to the quoted alumina literature, gentle heating removes some adsorbed water while other water molecules form Brønsted acid $-OH^{\delta+}$ and base $-O^{\delta-}$ H groups on the surface. On further heating, these groups are eliminated to form Lewis base and acid sites, respectively. Above 300°C, defect sites are formed consisting of clusters of vacancies (Lewis acids) and neighbouring oxide ions (Lewis bases). The decrease in areal mutarotation rates on heating the oxide up to 600°C therefore indicated that the catalytic sites on "low temperature" aluminas were of the Brønsted type. The higher activity on the most strongly heated samples pointed to the involvement of a Lewis type of catalytic site. Dunstan and Pincock probed their identity further by selective masking. Treatment with dry CO_2 which reacted with strongly basic functional groups decreased the catalytic activities of the orginal, "800°C" and "1250°C" aluminas by 10, 27 and 85%, respectively. The active sites of the "1250°C" alumina were therefore predominantly basic in character and were likely to be oxide ions, $-O^-$. These results also suggested that on the original alumina the catalytic sites were mainly acid in character, a conclusion reinforced by experiments in which pyridine and the stronger base n-butylamine had been added to the alumina. The decrease and subsequent increase in alumina activity on heating was therefore due to a complete change in the nature of the catalytically active sites, from weak Brønsted acid to strong Lewis base.

3.3 RACEMISATION OF COBALT(III) COMPLEXES

Many optically active inorganic complexes are quite stable in homogeneous aqueous solution but racemise on the addition of charcoal and other solids [196, 220, 221]. The first systematic kinetic studies of this phenomenon were carried out by Mureinik and Spiro [35, 222, 223] and Totterdell and Spiro [46, 224] using the typical substrate $(+)_{589}$-$Co(en)_3^{3+}$. In an exploratory survey with over 50 different solids [222] they found that very few were able to racemise the perchlorate salt but several (especially Ag_2S, Sb_2S_3, HgS, Hg_2I_2, SiO_2-Al_2O_3, Ag, Hg, and certain carbons) caused racemisation of the iodide salt. In some instances chemical interaction had obviously occurred between the solid and the constituents in the solution; silver and mercury, for example, were seen to form the corresponding iodides. In almost every case the racemisation was accompanied by measurable adsorption of the substrate and by some reduction to Co(II). All three processes were therefore studied side by side in a more detailed investigation. A carbon black, Black Pearls 2 ungraphitised, was chosen as the catalyst. Its attached carboxylate and phenolic groups and adsorbed carbon dioxide [174–176, 225] rendered the solutions slightly acid but washing the carbon beforehand had virtually no effect on the extents of adsorption or the catalytic rates. All the experiments were carried out in dark reaction vessels at 40°C.

The surface area of Black Pearls 2 carbon was sufficiently large $(850 \, \mathrm{m^2 \, g^{-1}})$ for the rates and extents of adsorption to be easily determined from the initial decreases in either optical rotation (see Fig. 2) or optical absorbance. The equilibrium amounts of $Co(en)_3^{3+}$ and of I^- adsorbed per gram of carbon fitted Freundlich isotherms. The rates of adsorption exhibited first-order behaviour and led to half-lives of adsorption of 2 min for $Co(en)_3^{3+}$ and of 3 min for I^- [223]. These were much faster than the rate of racemisation or the rate of the slow accompanying carbon-catalysed redox reaction.

$$Co(en)_3^{3+} + 6 \, H^+ + I^- \rightarrow Co^{2+} + 3 \, H_2en^{2+} + \tfrac{1}{2} I_2 \qquad (L)$$

The species H_2en^{2+} was the predominant form of ethylenediamine in these slightly acid solutions. Under a dinitrogen atmosphere the moles of cobalt (II) produced corresponded closely to the moles of iodide lost. More iodide than this disappeared in air and more still under a dioxygen atmosphere, due to its carbon-catalysed oxidation. The iodine produced by both reactions was strongly adsorbed by the carbon. In the absence of iodide ions no reduction of $Co(en)_3^{3+}$ occurred at all, a strong indication that the carbon itself was not acting as a reductant. The cobalt(II) ions were partly adsorbed on the carbon surface in competition with, but much less strongly than, the $Co(en)_3^{3+}$ ions. Plots of the Co(II) concentration against time were linear and passed through the origin. Their slopes gave the rate, v_{red}, of the redox reaction (L) and could be expressed by the equation

$$v_{red} = k_{red} \, [Co(en)_3^{3+}]_{ads} \, [I^-]_{ads} - k'_{red} \qquad (84)$$

The partial adsorption of Co(II) ions or their slow re-oxidation on the surface probably accounts for the parameter k'_{red} and also for the fact that v_{red} passed through a maximum as the mass of carbon was increased. A similar variation with mass of catalyst was observed by Morawetz et al. [226] for electron-transfer reactions catalysed by polyelectrolytes, but their explanation should not apply to the present case where the mechanism is likely to involve electron transfer through the carbon (cf. Sect. 4).

In the absence of a catalyst, optically active tris(ethylenediamine)cobalt (III) iodide is completely inert in slightly acidic aqueous solution. No racemisation whatever was detected even after 1 month at 40°C. On the addition of Black Pearls 2 carbon to solutions of $Co(en)_3I_3$ the optical activity decreased with time as shown in Fig. 2. An initial rapid fall due to adsorption was always followed by a slower first-order loss of activity as a result of the catalysed racemisation. The rates of the surface racemisation, v_{sur}, were calculated from the slopes by means of eqn. (19) [35, 46]. Two alternative rate equations fitted these values well

$$v_{sur} = k_1 [Co(en)_3^{3+}]_{sur}^2 \, [I^-]_{sur} \qquad (85)$$

$$v_{sur} = k_2 [Co(en)_3^{3+}]_{sur} \, [I^-]_{bulk} \qquad (86)$$

where $k_1 = 3.5 \times 10^{-4} \, dm^6 \, mol^{-2} \, s^{-1}$ and $k_2 = 2.5 \times 10^{-3} \, dm^3 \, mol^{-1} \, s^{-1}$ at 40°C. Equation (85) implies that the isomerisation proceeds by the interaction on the carbon surface between two adsorbed complex ions and one adsorbed iodide ion. Simple ligand exchange may be ruled out since Sen and Fernelius [221] have shown that over charcoal the rate of exchange of labelled ethylenediamine with $Co(en)_3^{3+}$ (as the iodide) is slower than the rate of racemisation. However, intramolecular links of the kind $Co-NH_2\cdots NH_2-Co$ may be postulated to facilitate rearrangement or dissociation of other en groups. The iodide may play a bridging role to stabilise a bond-ruptured intermediate [35]. Equation (86), on the other hand, implies a Rideal–Eley type of interaction between an adsorbed complex ion and an iodide-ion in the bulk solution. This can be a realistic interpretation only if the iodide ion concentration in the outer Helmholtz plane, near the adsorbed complex ions, is proportional to the bulk iodide concentration. Such a proportionality would be invalidated by variations in the surface potential of the carbon under different reaction conditions [46]. However, it is not clear how much the carbon potential would vary in practice since it was largely controlled by the I_2/I^- couple. Certainly simple association between bulk I^- ions and bulk $Co(en)_3^{3+}$ is insufficient for reaction because ion-pairs are known to exist in aqueous solution [227] where the racemisation rate is zero. The crucial catalytic role played by the carbon could be further ex-

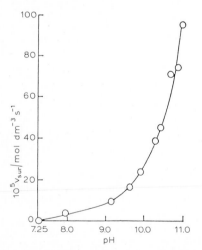

Fig. 18. Variation with pH of the catalysed rate of racemisation of 0.0015 mol dm^{-3} $(+)_{589}$-Co-(en)$_3$(ClO$_4$)$_3$ in NaOH solutions containing 25 mg Black Pearls 2 carbon in 25 cm^3 at 25°C. (After Totterdell and Spiro [224].)

plored by identifying the active sites through chemical modification of the surface as described in Sect. 3.1.

One of the aims of the research was to find out whether the racemisation proceeded via a labile Co(II) intermediate as suggested by Dwyer and Sargeson [197]. Mureinik and Spiro [35] showed that addition of Co(II) ions, or Co(II) ions together with extra H_2en^{2+} ions, did not increase the racemisation rate. In fact, the system itself was steadily accumulating Co(II) ions through reaction (L) yet no autocatalytic behaviour was observed (see Fig. 2). The values of v_{red} and v_{sur} also responded differently when the gas atmosphere was changed from dinitrogen to air to dioxygen, and on the addition of perchloric acid. The catalysed racemisation of the complex ion and its much slower reduction therefore proceeded along parallel but substantially independent paths in these acidic media.

In alkaline solutions the catalysed racemisation of $(+)_{589}$-$Co(en)_3^{3+}$ was found to be considerably faster. Totterdell and Spiro [224] were able to carry out these experiments with the perchlorate salt which had not been racemised at all in acidic solutions and, for convenience of study, the mass of Black Pearls 2 was reduced tenfold and the temperature lowered to 25°C. The sharp rise of the catalysed rate with increasing pH is depicted in Fig. 18. The homogeneous isomerisation was again too slow to be measured and the heterogeneous rates could be represented by the equation

$$v_{sur} = k_3 [Co(en)_3^{3+}]_{sur} [OH^-]_{sur} \tag{87}$$

with $k_3 = 9 \times 10^{-4}$ dm^3 mol^{-1} s^{-1} at 25°C. The complex ions and the hydroxide ions both adsorbed strongly on the carbon black and each assisted the other. The amount of hydroxide adsorbed fitted a Freundlich isotherm. However, its rate of adsorption on the acidic carbon black surface was slow, a point already noted in the literature [175, 225]. The complex ions adsorbed rapidly and attained monolayer coverage even at quite low cobalt concentrations. In solutions of higher NaOH concentration a colour change developed due to partial hydrolysis to cis-$Co(en)_2(OH)_2^+$. This was ruled out as a racemisation precursor because its rate of formation was found to be an order of magnitude slower than the rate of racemisation. The authors believed that the racemisation mechanism in NaOH solutions followed the general pattern for homogeneous base hydrolysis of octahedral cobalt(III) amine complexes, $S_N1(CB)$ [228]. The adsorbed OH^- first removes a proton from the adsorbed $Co(en)_3^{3+}$ and in the conjugate base so produced, a Co–N bond is broken to form the adsorbed five-coordinated intermediate $(en)_2Co(NH_2CH_2CH_2NH)^{2+}$. Rearrangement then takes place within this intermediate, perhaps by an intramolecular proton jump. An important role is clearly played by interaction with the carbon surface: the dissociation of Co–N ligands on charcoal has been noted for other reactions [229]. There was again no evidence for a redox pathway via Co(II) species. No Co(II) was detected by analysis and the racemisation rate did not increase when the reaction mixture was flushed out with dinitrogen to avoid its aerial oxidation.

The charcoal-catalysed racemisation of another cobalt complex was studied by Hammershøi and Larsen [230]. These authors had noted that the racemisation of $(+)_{589}$-$Co(en)_3^{3+}$ in acid solution required the presence of both a carbon and the soft counter ion I^-, and they reasoned that a cobalt(III) complex with a soft ligator like sulphur already present in the inner coordination sphere should also be racemised by charcoal. Their chosen complex, the unsymmetrical facial isomer of $(-)_{589}$-bis[di(2-aminoethyl)sulphide] cobalt(III), was indeed racemised in acid solution in the presence of Norit W charcoal. Some 10% of the complex ions were found to adsorb on the charcoal during the runs. The racemisation rates at 60°C, obtained by circular dichroism measurements, were first order as expected from eqn. (19). They showed an inverse variation with $[H^+]^{1.4}$ which was interpreted as a sign that the hydrogen ions competed efficiently with $Co(daes)_2^{3+}$ ions for active charcoal sites. It could also reflect participation by adsorbed hydroxide ions in the catalytic mechanism as had been found for $Co(en)_3^{3+}$ racemisation in alkaline solutions. The mass of charcoal was changed in only one run in which 1 g instead of 2 g was added per dm^3, with the curious result that the rate then dropped to 3% of its former value. Repeated attempts failed to detect any cobalt(II) in the reaction mixture although thermodynamic calculations indicated that some should have been present. The authors concluded that the racemisation did not proceed through any Co(II) intermediate. They proposed that the reaction entailed Co–S bond rupture brought about by distortion of the ligand through hydrophobic interaction with the charcoal surface.

It is noteworthy that in none of these researches, by two different groups using different cobalt complexes and acidic as well as alkaline media, was any Co(II) found to accompany the racemisations. These results seem to be in direct conflict with Dwyer and Sargeson's report [197] that cobalt(III) complexes in contact with activated carbon invariably produce small amounts of cobalt(II) complexes, and with their proposal that the racemisation of $(+)_{589}$-$Co(en)_3^{3+}$ proceeds through a labile $Co(en)_3^{2+}$ ion. In an effort to resolve this situation, Totterdell and Spiro [224] studied the racemisation of $(+)_{589}$-$Co(en)_3^{3+}$ in ethylenediamine solutions. No change in rotation was observed in homogeneous solution containing 0.01 mol dm^{-3} en at 25°C even after 1 week but quite rapid racemisation took place in the presence of Black Pearls 2 carbon. However, most of this catalysed reaction was attributable to the hydroxide ions formed by the hydrolysis of en to Hen^+ and H_2en^{2+}. Only trace amounts of Co(II) were detected in the reaction mixtures. Deliberate addition of cobalt(II) did not affect the rate nor was any Co(II) detected at the ends of these runs. Thus any cobalt(II) present in the en solutions appeared to be rapidly oxidised to $Co(en)_3^{3+}$ by dissolved or adsorbed dioxygen. A new experiment was then performed with solutions flushed out with dinitrogen before the addition of a larger amount of cobalt(II). Again no Co(II) could be detected in the final filtrate but the racemisation rate increased dramatically. Here the cobalt(II)) had evidently been present for long enough to enhance the reaction rate. Strong catalysis was also observed with a cathodically pretreated platinum disc in dinitrogen-flushed solutions of $(+)_{589}$-Co(en)$_3^{3+}$, Co(II), and en. These experiments confirm Dwyer and Sargeson's suggestion of a new catalytic pathway that becomes operative in the presence of Co(en)$_3^{2+}$. Its essential step is the electron exchange reaction

$$(+)_{589}\text{-Co(en)}_3^{3+} + (-)_{589}\text{-Co(en)}_3^{2+} \rightarrow (+)_{589}\text{-Co(en)}_3^{2+} + (-)_{589}\text{-Co(en)}_3^{3+} \quad \text{(LI)}$$

with subsequent rapid equilibration between the labile optically active forms of Co(en)$_3^{2+}$. The catalysis of reaction (LI) by carbon and by platinum almost certainly proceeds by an electrochemical mechanism of electron transfer through the solid catalyst (see Sect. 4). Now electrochemical evidence [231] has demonstrated that the half-reaction on platinum

$$\text{Co(en)}_3^{3+} + \text{e}^- \rightleftharpoons \text{Co(en)}_3^{2+} \quad \text{(LII)}$$

is fast only if both species possess identical chemical stoichiometry. Ions such as Co(en)$_2^{2+}$ or Co(en)$^{2+}$ are electrochemically inactive. This nicely explains why the addition of cobalt(II) had no effect on the catalysed racemisations in acid or alkaline media but increased the rate only in solutions containing a sufficiently high concentration of ethylenediamine to convert a sizable percentage of cobalt(II) ions to Co(en)$_3^{2+}$.

4. Oxidation–reduction reactions

4.1 PRINCIPLES OF ELECTRON TRANSFER CATALYSIS

The general redox reaction

$$v_{\text{ox2}}\text{Ox}_2 + v_{\text{red1}}\text{Red}_1 \rightarrow v_{\text{red2}}\text{Red}_2 + v_{\text{ox1}}\text{Ox}_1 \quad \text{(LIII)}$$

expresses the interaction of the two couples

$$v_{\text{ox1}}\text{Ox}_1 + \text{e}^- \rightleftharpoons v_{\text{red1}}\text{Red}_1 \quad \text{(LIV)}$$

$$v_{\text{ox2}}\text{Ox}_2 + \text{e}^- \rightleftharpoons v_{\text{red2}}\text{Red}_2 \quad \text{(LV)}$$

Such redox reactions are frequently catalysed by platinum [3], other noble metals [232], silver [126–128], and carbons [233] which are all electron-conducting solids. This fact points to a simple catalytic mechanism whereby the electron is transferred from Red$_1$ to Ox$_2$ through the solid phase, as depicted in Fig. 19. In contrast to other bimolecular catalytic mechanisms (Sect. 1.5.3), the two reactants do not need to occupy neighbouring sites. Since the catalytic rate depends upon the coupled transfers of an electron from Red$_1$ to the solid and from the solid to Ox$_2$, the kinetics are best treated in electrochemical terms.

The previous sentence can now be restated by saying that, at the catalyst/solution interface, the net anodic current due to couple (LIV) must equal the

138

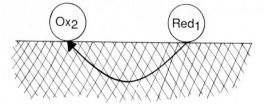

Fig. 19. Catalytic mechanism of electron transfer through the solid.

net cathodic current produced by couple (LV). This equality of currents is achieved by the catalyst adopting an appropriate mixed or mixture potential E_{mix} [3, 146] which lies between the equilibrium (Nernst) potentials of the two couples. The catalyst can then act simultaneously as an anode for couple (LIV) and as a cathode for couple (LV). The quantitative treatment is greatly simplified by the additivity principle of Wagner and Traud [234]. This postulates that in the reaction mixture the two couples behave independently, each exhibiting its own individual current. Figure 20 illustrates the situation. Curve 1 is the current–potential plot of couple (LIV), curve 2 that of couple (LV), and E_{mix} marks the potential at which the two contributing currents algebraically add up to zero. The modulus of each of these two currents, I_{mix}, or of the corresponding current densities i_{mix}, is by Faraday's law directly proportional to the catalytic rate

$$v_{cat} = \frac{I_{mix}}{FA} = \frac{i_{mix}}{F} \tag{88}$$

where F is the Faraday constant. It follows that kinetic equations for v_{cat}

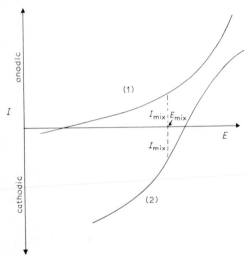

Fig. 20. Schematic electrochemical diagram showing the effect of two redox couples present together but not in equilibrium with each other. (After Spiro [146].)

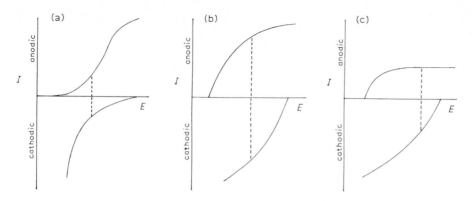

Fig. 21. Schematic current–potential curves for various categories of mixed couples. (a) Two irreversible couples; (b) two reversible couples; (c) two couples whose E_{mix} value lies within the plateau region of one of them.

can, in principle, be derived from the well-known equations for current–potential curves [235, 236]. Several examples are given in the next paragraph.

The shape of the current–potential curve of a given couple like (LV) depends on the rate of exchange of the couple at the electrode surface at equilibrium (the exchange current density), the concentrations of the electroactive species, and on the hydrodynamic flow conditions. A couple with a slow rate of exchange is termed electrochemically irreversible whereas one with a fast rate is called electrochemically reversible. Curve 1 in Fig. 20 represents an irreversible couple. As the potential departs from the equilibrium (Nernst) value, the current first rises linearly with a small slope, then at greater potentials increases exponentially (the Tafel region) and finally reaches a diffusion-limited plateau value independent of further changes in potential. Curve 2 in Fig. 20 is a typical one for a reversible couple: near the equilibrium position the current rises steeply with increasing potential before curving over to the plateau value, and it is diffusion-controlled throughout. Some special cases of mixed couples will now be considered.

(a) Both couples are irreversible and E_{mix} lies in the Tafel regions of both couples [Fig. 21(a)]. It can then be shown that [237]

$$v_{cat} = k_{el1}^{r_2} k_{el2}^{r_1} \exp\left[\alpha_2 z_2 r_1 F(E_2^0 - E_1^0)/RT\right] c_{ox2}^{r_1 n_2} c_{red1}^{r_2 n_1} \tag{89}$$

where

$$r_1 = 1 - r_2 = \frac{(1 - \alpha_1) z_1}{(1 - \alpha_1) z_1 + \alpha_2 z_2} \tag{90}$$

and where k_{el} is the rate constant of the electrochemical exchange reaction of the subscripted couple, α its cathodic transfer coefficient, z its charge-transfer valence, and E^0 its formal electrode potential in the reaction medium; R is the gas constant and T the absolute temperature. For a given chemical system at a given temperature, the factors preceding the concentration terms in eqn. (89) are constant and together form the catalytic rate constant. Usually n_1, the electrochemical reaction order of Red_1 in the reverse of reaction (LIV) and n_2, the electrochemical reaction order of Ox_2 in reaction (LV), will both be unity. The catalytic rate is then proportional to the concentrations of the reactants raised to (different) fractional powers. The superficial impression given by such a rate equation is that reaction (LIII) proceeds by a Langmuir–Hinshelwood mechanism with Red_1 and Ox_2 adsorbed side by side by Freundlich isotherms (Sect. 1.5.3). In fact, eqn. (89) was derived from a quite different model in which Red_1 and Ox_2 do not need to occupy adjacent positions and in which no assumption was made about their adsorption isotherms. Experiments that can differentiate between these models are outlined below.

(b) The mixture potential again lies within the Tafel regions of the two couples but they are now less irreversible and allowance must be made for the contribution of diffusion. Then [238]

$$\frac{1}{v_{cat}} = \frac{1}{v_{cat_\infty}} + \dot{F}\left(\frac{r_1 n_2}{L_{ox2}} + \frac{r_2 n_1}{L_{red1}}\right) \tag{91}$$

where L_j is the limiting diffusion current density of species j. From Fick's first law and Faraday's law

$$L_j = \frac{FD_j c_j}{v_j \delta_j} = k_{trj} c_j \tag{92}$$

If the reaction has been studied at a rotating disc catalyst spinning at a speed f (Sect. 1.6.3), the diffusion layer thickness δ_j is given by the Levich equation (49) so that

$$\frac{1}{v_{cat}} = \frac{1}{v_{cat_\infty}} + \left(\frac{r_1 n_2}{\chi_{ox2}} + \frac{r_2 n_1}{\chi_{red1}}\right) \frac{1}{\sqrt{f}} \tag{93}$$

where

$$\chi_j = \frac{1.554 D_j^{2/3} v^{-1/6} c_j}{v_j} \tag{94}$$

The kinematic viscosity of the solution, v, should not be confused with the stoichiometric coefficient of j, v_j [cf. eqn. (LIII)]. A plot of $1/v_{cat}$ against $1/\sqrt{f}$ will therefore be a straight line. Its intercept will yield a value for v_{cat_∞}, the surface-controlled catalytic rate, which is given by eqn. (89).

(c) Both couples are reversible with E_{mix} situated in the steeply rising portions of the two current–potential curves [Fig. 21(b)]. The reaction is then so strongly catalysed that it is virtually at equilibrium on the surface and the overall process is totally transport-controlled [238]. An analysis of this situation leads to the following equation for the initial catalytic rate:

$$\frac{1}{v_{cat}} = \frac{F}{W} + \left(\frac{v_{ox2}}{L_{ox2}} + \frac{v_{red1}}{L_{red1}}\right) \frac{F}{(v_{ox1} + v_{red2})} \tag{95}$$

where the dominant term W is given by

$$W^{(v_{ox1} + v_{red2})} = k_{tr,ox1}^{v_{ox1}} k_{tr,red2}^{v_{red2}} \exp\left[F(E_2^0 - E_1^0)\right] c_{ox2}^{v_{ox2}} c_{red1}^{v_{red1}} \tag{96}$$

Equations (95) and (96) are actually special cases of eqns. (61) and (62) in Sect. 1.6.3. At a rotating disc catalyst, W and the two L values are directly proportional to the square root of the disc rotation speed so that v_{cat} will vary directly with $\sqrt{f}$. In contrast, the potential E_{cat} taken up by the catalytic disc, which is equal to E_{mix} on the electrochemical model, will be completely independent of rotation speed. Addition of either of the products Red_2 and Ox_1 to the reaction mixture brings about a dramatic change in the rate law [238]. A catalytic system that behaves in this way is reviewed in Sect. 4.4.

(d) The mixture potential lies within the diffusion-limited plateau region of one couple [say (LIV)], the limiting current for the other couple being much larger [Fig. 21(c)]. The degree of electrochemical reversibility of the couples is now immaterial. It follows from eqns. (92) and (88) that [237]

$$v_{cat} = \frac{D_{red1} c_{red1}}{v_{red1} \delta_{red1}} = \frac{k_{tr,red1} c_{red1}}{F} \tag{97}$$

If a rotating disc catalyst is employed

$$v_{cat} = \frac{D_{red1} \; c_{red1} \; \sqrt{f}}{v_{red1} \; Le_{red1}} = \chi_{red1} \sqrt{f} \tag{98}$$

where the parameters Le and χ are expressed in terms of measurable properties in eqns. (49) and (94). The catalytic rate is therefore proportional to the square root of the rotation speed; it is, moreover, first order in Red_1 and completely independent of the concentration of Ox_2. An orthodox kinetic interpretation of such a rate law would have been monolayer coverage of the surface by Ox_2 but of course no such assumption was made in the electrochemical derivation. The situation depicted in Fig. 21(c) is bound to occur whenever the concentration of Red_1 falls below a certain critical value. An entirely analogous situation arises when the concentration of Ox_2 becomes sufficiently small. A chemical example is given in Sect. 4.5.

As has been seen, some of the rate equations based on the electrochemical model resemble traditional catalytic formulae. One or more of the following tests should therefore be applied to check that the catalysis did indeed proceed by electron transfer through the solid.

(1) Comparison of the predicted and observed kinetics not only with respect to reaction orders but also with respect to hydrodynamic flow (rotation speed of catalytic disc). Moreover, judicious changes in the experimental conditions should alter the rate equations in a predictable way. In case (c), for example, addition of one of the products should alter the reaction orders of the reactants. In most cases, lowering the concentration of one of the reactants should bring about compliance with case (d).

References pp. 159–166

(2) Comparison of the observed value of v_{cat} with that calculated from the I_{mix} value obtained from independent current–potential measurements of the two contributing couples. The potential taken up by the catalyst, E_{cat}, should also agree with the electrochemically determined E_{mix}. However, systems are known in which the Wagner and Traud additivity rule breaks down [239, 240] yet the catalytic mechanism still proceeds by electron transfer through the solid. One such case is discussed in Sect. 4.4.

(3) Carrying out a specially designed experiment in which the two reactants are in contact with the catalyst but are separated from each other [241–243]. If reaction (LIII) still takes place under these circumstances, it must do so via electron transfer through the solid. Examples of such experiments are described in Sects. 4.2 and 4.3.

(4) Imposing upon the catalyst a potential equal to the equilibrium potential of one of the couples [say (LIV)]. This will stop the overall reaction (LIII) if the latter proceeds by an electrochemical mechanism but should influence the rate much less if the catalytic mechanism is a non-electrochemical one. This test, initially suggested by Wagner and Traud [234], was later refined by Wagner [244] and applied by Takehara [245] for distinguishing between mechanisms in the noble-metal catalysed hydrogenations of organic compounds.

(5) Measurement of the steady-state potential of the catalyst and of its polarization characteristics in the reaction mixture may indicate whether the catalytic reaction proceeds predominantly by an electrochemical or a non-electrochemical route [234, 244, 245]. However, the interpretation of the results is likely to be more complex than in Tests (2), (3), or (4) and the original papers should be consulted for the details.

The following sections deal with specific oxidation–reduction reactions whose catalysis has been shown to proceed by the electrochemical mechanism. Catalysed redox reactions with gaseous reactants or products, many of which follow a similar mechanistic path, will only be mentioned in passing as was indicated in the Introduction (Sect. 1.1). However, in view of the fact that numerous publications have appeared on redox reactions involving gas/solid and gas/liquid as well as liquid/solid interfaces, it may be of interest to cite a few key review references. Birkett et al. [246] have recently discussed the mechanisms of liquid-phase metal-catalysed hydrogenations of organic compounds; an earlier classic book on this subject, dealing mainly with Soviet work, is that of Sokol'skii [247]. A survey of mainly inorganic platinum-catalysed reactions of dihydrogen and dioxygen has been given by Spiro and Ravnö [3] who also considered platinum-catalysed decomposition reactions that form gaseous products. Redox reactions producing either H_2 or O_2 and catalysed by noble metal and semiconductor colloids have been treated by Grätzel [248] as part of a review on possible schemes for photocatalytic water splitting.

4.2 ELECTRON EXCHANGE REACTIONS

When the two couples in the electron-transfer process are chemically identical, the reaction becomes one of electron exchange. Its rate can be determined by isotopically labelling a small fraction of one of the components as indicated in the equation

$$Ox + Red^* \rightleftharpoons Red + Ox^* \tag{VII}$$

The exchange reactions of several couples have been found to be catalysed by metallic platinum; just those reactions, in fact, where the couple Ox/Red is electrochemically reversible [3]. This was to be expected if the catalytic mechanism is one of electron transfer through the metal. Only very few quantitative studies of the catalytic kinetics have been carried out and these are described below.

As was explained in Sect. 1.5.2, any given isotopic exchange experiment will be kinetically of first order. For systems where the rates of adsorption and desorption are rapid, it follows from the two-phase model of catalysis and eqn. (23) that the observed first-order rate constant is given by

$$k_{obs} = \left(\frac{v_{hom} V_{sol} + v_{sur} V_{sur}}{V_{sol} + V_{sur}} \right) \left(\frac{1}{c_{ox,sys}} + \frac{1}{c_{red,sys}} \right) \tag{99}$$

where the symbol sys signifies the concentration of the substance in the whole system. The rate of exchange in the bulk solution phase is v_{hom} and that in the surface layer v_{sur} (mol dm^{-3} s^{-1}). The latter can be converted to the normal areal catalytic rate v_{cat} (mol m^{-2} s^{-1}) by eqn. (20), $v_{cat} = v_{sur} V_{sur}/A$. If the volume of the surface layer V_{sur} is, as usual, much smaller than the volume of the bulk solution V_{sol}, then eqns. (99) and (20) give

$$k_{obs} = v_{hom} + \frac{v_{cat} A}{V_{sol}} \left(\frac{1}{c_{ox,sys}} + \frac{1}{c_{red,sys}} \right) \tag{100}$$

The same equation had previously been derived by Fronaeus and Östman [249]. These workers made two further important contributions to the theory. Taking it for granted that the catalysed exchange proceeded by an electron transfer mechanism, they related v_{cat} to the concentrations of Ox and Red by the theory of electrode kinetics. For simple redox couples whose electrochemical reaction orders are unity, this leads to

$$v_{cat} = k_{el} c_{ox}^{1-\alpha} c_{red}^{\alpha} \tag{101}$$

where α is the cathodic transfer coefficient of the couple. Exactly the same equation is obtained from eqn. (89) for the situation where couples 1 and 2 are chemically identical. Appropriate equations for more complex systems have been given elsewhere [235, 249]. Fronaeus and Östman also recognized that fast catalysed reactions may be partly or wholly diffusion-controlled and derived the equation

$$\frac{1}{v_{cat}} = \frac{1}{v_{cat_\infty}} + \frac{\delta_{ox}}{D_{ox}c_{ox}} + \frac{\delta_{red}}{D_{red}c_{red}} \tag{102}$$

in which v_{cat_∞}, the catalytic rate for infinitely fast stirring, is equal to the quantity in eqn. (101). In the original version the diffusion layer thicknesses δ_{ox} and δ_{red} were assumed to be the same but it is now known that they depend somewhat upon the diffusing species. If the catalyst is present in the form of a rotating disc, the δ parameters can be expressed in terms of the rotation frequency and other measurable properties through eqn. (49). The resulting equation for v_{cat} bears a strong resemblance to eqn. (93).

Fronaeus and Östman [249] studied the isotopic exchange of the Ce(IV)/Ce(III) couple in 3 mol dm^{-3} HClO$_4$ at 0°C. The reaction mixture contained ^{141}Ce(III) tracer and the progress of the exchange was followed by periodically extracting Ce(IV) into ether and determining its gamma radioactivity. For every set of initial concentrations they performed two sets of measurements: one in the absence of platinum (to determine v_{hom}) and the other in the presence of platinum wire of surface area 40 cm^2 (to determine v_{cat}). In the latter runs the solution was shaken to keep the diffusion layer thickness constant. Good first-order plots were obtained in all cases. The catalytic rates determined from eqn. (100) were then analysed in terms of eqns. (102) and (101) and variants thereof. The authors concluded that the catalytic exchange rate was controlled jointly by the electron transfer process at the platinum surface and by the diffusion of Ce(IV) whereas the diffusion of Ce(III) was too rapid to affect the rate significantly. Moreover, their results indicated that the uptake of electrons from the platinum took place solely via a dinuclear hydrolysis product of cerium(IV), even though its concentration in the strongly acidic medium was only between 0.5 and 2.5% of the total cerium(IV) concentration. A corresponding mixed cerium(III)–cerium(IV) species was thought to operate in the reverse direction. Subsequent electrode impedance experiments were not able to confirm or rebut the second-order cerium(IV) kinetics [250]. The postulate of exclusive participation by cerium(IV) dimers in the heterogeneous exchange process was later strongly criticized by Greef and Aulich [251]. These workers found that the electrode kinetics of the Ce(IV)/Ce(III) couple in perchloric acid at rotating platinum electrodes were very dependent upon the state of the electrode surface, especially upon the degree of coverage by adsorbed oxygen and by adsorbed Ce(III) ions. These and other factors were put forward as more likely explanations for the unusual reaction orders reported for the heterogeneous Ce(IV) + Ce(III) exchange reaction.

The other major published study of electron transfer catalysis is that of Jonasson and Stranks [243]. They chose the Tl(III)/Tl(I) system in which there was no evidence of polynuclear species, and studied the exchange kinetics in 1.1 mol dm^{-3} HClO$_4$ at 35°C and other temperatures using β-emitting ^{204}Tl as a tracer. The exchange was catalysed so strongly by various platinum black surfaces that the much slower homogeneous reaction could be neglected in comparison. Adsorption equilibrium was complete within a few seconds when the system was stirred by dinitrogen bubbling although desorption was slower. Thallium(III) and thallium(I) adsorbed by independent Langmuir isotherms. In mixtures of the two species the adsorption was additive suggesting non-competitive adsorption on different types of site. Tl(III), which adsorbed more slowly, was thought to sit on $Pt(OH)_2$ sites while Tl(I) occupied vacant Pt sites. Since Tl^{3+} is regarded by Pearson as a softer acid than Tl^+ in spite of its greater positive charge [93], a case could actually be made out for the reverse arrangement.

The rate plots exhibited good first-order behaviour. Identical half-lives of exchange were obtained when either Tl(I) or Tl(III) were labelled initially. The rates of exchange were also constant over a wide range of dinitrogen bubbling rates; this fact, together with the evidence for faster adsorption and desorption, proved that the exchange reaction at the surface was rate-controlling. The rates fitted the equation

$$v_{cat} = k_{cat} c_{Iads} c_{IIIads} \tag{103}$$

where subscripts I and III stand for the corresponding thallium species. This was shown both by inserting the adsorbed concentrations from the experimental isotherms and also by substituting into eqn. (103) the appropriate Langmuir expressions for non-competitive adsorption. The resulting equation, whose form was that of eqn. (29), represented the variation in rate over the whole range of initial Tl(I) and Tl(III) concentrations. The activation parameters for k_{cat} obtained from experiments over the temperature range 0–60°C, were $\Delta H^{\neq} = 46\,\mathrm{kJ\,mol^{-1}}$ and $\Delta S^{\neq} = 43\,\mathrm{J\,K^{-1}\,mol^{-1}}$. The magnitude of the enthalpy of activation is consistent with surface control but not with diffusion control.

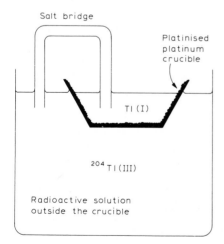

Fig. 22. The experimental arrangement whereby the isotopic exchange reaction Tl(I) + ^{204}Tl(III) → Tl(III) + ^{204}Tl(I) was catalysed by a platinised platinum crucible separating the two reactants. (After Jonasson [252].)

One possible interpretation of these kinetics is that adsorbed Tl(I) ions simply transfer two electrons directly to neighbouring adsorbed Tl(III) ions. The authors ruled out this adjacent-adsorption mechanism by three somewhat doubtful arguments. The first was based on the fact that the enthalpy of activation for the heterogeneous reaction was $21 \, kJ \, mol^{-1}$ smaller than for the homogeneous reaction while the entropy of activation was $138 \, J \, K^{-1} \, mol^{-1}$ larger. Both changes, however, would be consistent with a Langmuir–Hinshelwood mechanism. The authors also estimated that the effective increase in the concentrations of the reactants in the surface layer would have been insufficient to account for the catalysis, but their calculations employed the geometric and not the BET surface area of the platinum black and a layer thickness corresponding to the diffusion layer instead of the Helmholtz layer. Their third line of reasoning was the most interesting one. They carried out the experiment depicted in Fig. 22 in which a Tl(I) solution was placed inside a platinized platinum crucible and a Tl(III) solution outside it. The solutions were not stirred because of technical difficulties. There was $1.1 \, mol \, dm^{-3} \, HClO_4$ in both solutions and also in the short "salt" bridge connecting them. Oxidation of the inner solution and reduction of the outer solution then took place, with a half-life of ca. 9 h. This result completely excludes a mechanism of direct exchange of electrons between Tl(I) and Tl(III) ions adsorbed on adjacent sites and can only be explained by a transfer of electrons through the 5 mm thick wall of the platinum crucible. On the other hand, the experimental arrangement is essentially a short-circuited concentration cell and the observed reactions can be regarded as a direct consequence of the potential difference across the joined electrodes.

Although the above criteria do not rule out an adjacent-adsorption mechanism, the alternative mechanism of electron transfer through the solid more easily explains why the exchange reaction was not catalysed by the insulators SiO_2 [243] and BN [253] but was catalysed by a number of electron-conducting solids including graphite [253] and metallic platinum as well as by the semiconductors TiO_2 and brown thallium(III) oxide, $Tl_2O_3 \cdot H_2O$ [243, 254]. The catalytic effect of the latter oxide on the Tl(I) + Tl(III) exchange reaction has been studied by Hasany and Stranks [254, 255]. They found that the catalytic rate was proportional to the adsorbed concentration of Tl(I) (which could be fitted by either Langmuir or Freundlich isotherms) and independent of the bulk Tl(III) concentration. The amount of Tl(III) adsorbed was also independent of its bulk concentration, indicating monolayer coverage. Adsorption of the two reactants appeared to be additive. The results were therefore again consistent with eqn. (103). It is curious that with neither platinum black nor thallium(III) oxide did the catalytic rate obey an equation containing fractional powers of the concentrations that added up to unity, as expected from eqn. (101). The slight dependence of catalytic rate on the ratio c_{III}/c_I which was noted for both catalysts does not lead to the same relationship. However, it may be pointed out that most workers who have investigated the electrode kinetics of the Tl(III)/Tl(I) system at platinum electrodes have found the kinetic parameters to be very sensitive to the state of the electrode surface and to the adsorption of other species such as anions [256]. Variations in the surface conditions of the catalysts, which were not electrochemically pretreated, may therefore have distorted the concentration dependence.

Hasany [254] and Stranks have carried out similar tracer experiments on two other platinum-catalysed electron-exchange reactions, the cationic system $Co(en)_3^{2+} + Co(en)_3^{3+}$ and the anionic system Co(II)EDTA + Co(III) EDTA. Unfortunately this interesting research has remained unpublished apart from a short and not easily accessible note about the former system [257]. A brief summary of the results will therefore be given. In the cationic system in $0.2 \, mol \, dm^{-3}$ ethylenediamine solution the catalysis by platinum black was found to be partly diffusion-controlled. Less active "grey" platinum was accordingly used instead with a stirring speed above 2000 rev. min^{-1} where further increases in speed had no effect. The $Co(en)_3^{3+}$ ions adsorbed more strongly than $Co(en)_3^{2+}$, and the adsorbed concentrations of both ions could be reasonably well represented over a 40-fold concentration range by either Langmuir or Freundlich isotherms. In admixture the reactants adsorbed competitively on the same sites. The rate law for the catalysed exchange

$$v_{cat} = k_{cat} c_{IIads} c_{IIIads} \tag{104}$$

where subscripts II and III denote the oxidation states of the cobalt complexes, was consistent with rapid adsorption pre-equilibrium followed by rate-determining electron transfer between the adsorbed reactants. The rate constant depended upon the ratio of the reactant concentrations according to the equation

$$\ln k_{cat} = \ln k_{cat}^0 + \alpha \ln (c_{II}/c_{III}) \tag{105}$$

with $\alpha = 0.27$. If the adsorbed concentrations in eqn. (104) are expressed by Freundlich isotherms and the resulting equation combined with eqn. (105), we obtain, for low concentrations of the reactants

$$v_{cat} = k'_{cat} c_{II}^{0.92} c_{III}^{0.05} \tag{106}$$

The sum of the exponents is close to unity. This equation now resembles eqn. (101) and so supports the idea of an electrochemical exchange mechanism at the platinum surface. However, quite different exponents (0.24 and 0.76, respectively) are predicted by electrode kinetic data [231]. As the concentration of $Co(en)_3^{3+}$ rose further, the catalytic rate passed through a maximum because of the effect of competitive reactant adsorption.

With the Co(II)EDTA and Co(III)EDTA system [254] the experiments were carried out at pH 2 where the main cobalt(III) species was $Co(EDTA)^-$ (99%) while the main cobalt(II) species were $Co(HEDTA)(H_2O)^-$ (86%) and $Co(EDTA)^{2-}$ (11%). Platinum black surfaces were employed as catalysts but some problems were experienced with gradual loss of activity. The catalysis was again surface-controlled since the rates were independent of stirring speed as well as being much faster than the rates of adsorption and desorption. Co(II)EDTA adsorbed more strongly than Co(III)EDTA. The adsorption of both complexes could be adequately described by either Langmuir or Freundlich isotherms, and the two reactants adsorbed additively. No detailed study was made of the variation of the catalysed exchange rate with reactant concentration but the authors assumed that eqn. (104) would again apply. If so, the sum of the Freundlich exponents [0.23 and 0.47 for the cobalt(II) and cobalt(III) complexes, respectively] falls short of the figure of unity expected from eqn. (101). However, as was made clear in Sect. 3.3, evidence from electrode kinetics shows that electron transfer through the platinum will be rapid only between species with the same chemical stoichiometry, in other words between $Co(III)EDTA^-$ and $Co(II)EDTA^{2-}$ ions. The latter constitute only 11% of the cobalt(II) species and may well present adsorption characteristics different from those of the majority $Co(HEDTA)$ $(H_2O)^-$ ions. It is therefore possible that the catalysed exchange rate of the Co(II)EDTA + Co(III)EDTA system behaves similarly to that of the $Co(en)_3^{2+}$ + $Co(en)_3^{3+}$ system. The activation parameters of the two reactions were certainly the same within experimental error, with mean values of $\Delta H^{\neq} = 36\,kJ\,mol^{-1}$ and $\Delta S^{\neq} = 75\,J\,K^{-1}\,mol^{-1}$. This indicated to the authors that the energy barrier for the electron exchange at the platinum surface was largely independent of the nature and charge of the reactant species [254].

4.3 PHOTOGRAPHIC DEVELOPMENT

When a photographic film containing silver halide is exposed to light, a few silver ions are reduced to form tiny specks of metallic silver. The resulting latent image must be intensified 10^8–10^9 times to produce a visible photographic image [129]. This is done by reducing more silver ions with an appropriate chemical reducing agent called a developer

$$Ag^+ + Red \rightarrow Ag(c) + Ox \qquad\qquad (LVI)$$

The process is termed chemical development if the silver ions come from the silver halide crystal and physical development if they come from a solution phase [126, 129]. Reaction (LVI) is strongly catalysed by the photolytically formed silver sites and so these grow autocatalytically until the image becomes visible. Were it not for this catalytic effect the reaction would proceed slowly and uniformly over the whole film and just produce a fogged plate. Many inorganic and organic substances can act as developers but the compounds that have achieved greatest commercial importance are hydroquinone, p-aminophenol, p-phenylenediamine and their derivatives [258]. Colour photography is made possible with developers whose oxidation products form dyes with added coupling substances [259].

Most workers in the field agree that the catalysis of reaction (LVI) operates by an electrochemical mechanism. The overall reaction for chemical development can then be split up into anodic and cathodic contributions

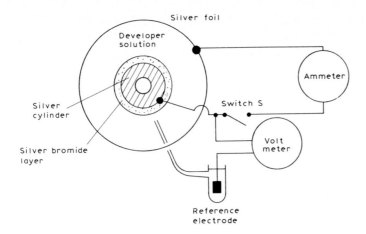

Fig. 23. Experimental arrangement used by Jaenicke and Sutter [241, 242] to test the electro-chemical mechanism of photographic development.

$$\text{Red} \rightarrow \text{Ox} + \text{e}^- \tag{LVII}$$

$$\text{AgBr(c)} + \text{e}^- \rightarrow \text{Ag(c)} + \text{Br}^- \tag{LVIII}$$

This was physically demonstrated by Jaenicke and Sutter [241, 242] with an experiment in which these two processes were spatially separated. As shown in Fig. 23, a central silver cylinder covered with a dense silver bromide layer formed the cathodic component of the system while the anodic component was a surrounding silver foil. Both components dipped into a flowing dioxy-gen-free solution of developer. If the electrochemical mechanism applied, the oxidation of the developer [reaction (LVII)] would take place only on the silver foil and the reduction of the silver bromide [reaction (LVIII)] only at the central AgBr/Ag interface. The potential difference between the two pieces of silver would simultaneously drive a current through the ammeter. This was indeed observed. The number of faradays of electricity passed during a 3 min run was found to account for 70% (with hydroquinone) to 88% (with p-phenylenediamine) of the number of moles of AgBr which had been reduced. The remaining 12–30%, which had appeared to arise from direct chemical reaction between the developer and the silver bromide layer, was shown to be caused by threads of silver that had grown through the AgBr layer during the reaction and so provided local anodic surfaces. The contri-bution of the direct reaction could also be determined simply by leaving the switch S open. Of the various reducing agents tested, only sodium stannite reacted directly with the silver bromide. The results obtained are thus consistent with an electrochemical mechanism for the catalysis of the de-velopment reaction (LVI) by silver particles. Test (3) in Sect. 4.1 has thereby been fulfilled.

Because of the importance of reaction (LVI) its kinetics have been studied by several groups of workers and the subject has been reviewed in some detail both by James in his treatise [126] and by Jaenicke [129]. Suffice it to say here that the kinetics of the catalysed development reaction have often been found to obey rate laws of the form

$$v_{cat} = k' c_{red}^{\alpha} c_{H^+}^{\beta} \quad (0 < \alpha, \beta \leqslant 1) \tag{107}$$

for chemical development and

$$v_{cat} = k'' c_{Ag^+}^{\gamma} c_{red}^{\delta} \quad (0 < \gamma, \delta \leqslant 1) \tag{108}$$

for physical development. These rate equations resemble theoretical relationships derived for special cases of the electrochemical mechanism (see Sect. 4.1). Another special case described by Bagdasaryan [260] for physical development assumed that the silver couple was electrochemically reversible while the developer couple was irreversible with E_{mix} lying in its Tafel region. The current–potential plots of the two couples thus resembled curves 2 and 1 in Fig. 20, respectively. This enabled him to derive the equation

$$v_{cat} = k c_{Ag^+}^{\gamma} c_{red} \quad (0 < \gamma < 1) \tag{109}$$

where $(1 - \gamma)$ was the cathodic transfer coefficient of the Ox/Red couple. In later years more complex kinetic formulae have been found necessary to fit the experimental data. However, workers in the photographic literature have usually been able to interpret their kinetic findings satisfactorily in terms of additive current–potential curves when adsorption, diffusion, and the surface reaction were all taken into account as well as the increasing size of the silver catalyst particles. Jaenicke [129] has shown that electrochemical considerations also go a long way towards explaining the phenomenon of "superadditivity" where two developing agents act synergistically [261].

4.4 THE FERRICYANIDE + IODIDE REACTION

The reaction between hexacyanoferrate(III) (commonly called ferricyanide and abbreviated below as Feic) and iodide ions

$$Fe(CN)_6^{3-} + \tfrac{3}{2} I^- \rightarrow Fe(CN)_6^{4-} + \tfrac{1}{2} I_3^- \tag{LIX}$$

provides an excellent example of interaction between two electrochemically reversible couples [case (c) in Sect. 4.1]. The reaction is fairly slow in homogeneous aqueous solution but its catalysis by platinum was noted as long ago as 1908 [262]. Quantitative studies of this catalysis were carried out at large anodically pretreated platinum discs by Spiro and Griffin [68] at 0°C and more especially by Freund and Spiro [6] at 5°C and other temperatures in a $1 \, mol \, dm^{-3} \, KNO_3$ medium. Figure 24 shows that the initial catalytic rates were directly proportional to the square root of the disc rotation speed, f, while the values of E_{cat}, the potential adopted by the platinum catalyst, were independent of f. Both these findings are exactly as predicted by the electrochemical model for case (c) where the surface reaction is so fast that

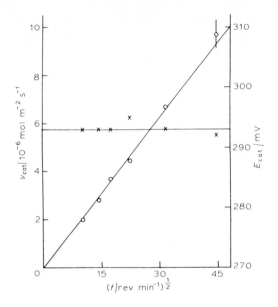

Fig. 24. Variation of the initial catalytic rate (O) and the catalyst potential ($\times$) with the square root of the platinum disc rotation speed in a reaction mixture containing 1×10^{-3} mol dm^{-3} K$_3$Fe(CN)$_6$, 0.05 mol dm^{-3} KI, and 1 mol dm^{-3} KNO$_3$ at 5°C. (After Freund and Spiro [6].)

the overall process becomes totally transport-controlled. The derived equation (96) also requires the kinetic orders of the reactants in the catalysed reaction to be

for Feic

$$\frac{v_{ox2}}{v_{ox1} + v_{red2}} = \frac{1}{1 + \frac{1}{2}} = \frac{2}{3}$$

for I$^-$

$$\frac{v_{red1}}{v_{ox1} + v_{red2}} = \frac{\frac{3}{2}}{1 + \frac{1}{2}} = 1$$

The experimental orders were 0.66 and 1.02, respectively, in excellent agreement with the electrochemical model. Another mathematical consequence of this model is a marked change in the kinetic orders when one of the products of the reaction is added to the initial mixture [238]. Thus if hexacyanoferrate(II) (commonly called ferrocyanide and abbreviated Feoc) is present at the start of the reaction, the theory forecasts that the order with respect to Feic and I$^-$ will increase to 2 and 3, respectively, while the order with respect to Feoc will be -2. In the event, a plot of lnv_{cat} versus ln[Feoc] was a curve that coincided with a line of slope -2 at the higher concentrations of ferrocyanide. At a constant initial Feoc concentration of 4×10^{-4} mol dm^{-3}, the orders with respect to Feic and I^- were found to be ca. 1.7_5 and 2.8, much higher than in the absence of added product. It is likely that even better agreement with the theoretical requirements would have been obtained with a larger initial Feoc concentration. Experiments with added tri-iodide were more difficult to carry out because reaction (LIX) was followed by measuring the optical absorbance of this species; even so, plots of $\ln v_{cat}$ against $\ln[I_3^-]$ were consistent with the predicted slope of $-1/2$.

The electrochemical model also made predictions about the potential taken up by the catalyst. For 5°C, to a first approximation, the theory led to

$$\frac{\partial E_{cat}}{\partial \ln[\text{Feic}]} = 8.0\,\text{mV}$$

$$\frac{\partial E_{cat}}{\partial \ln[I^-]} = -24.0\,\text{mV}$$

The experimental variations were 8.2_5 and $-23.2\,\text{mV}$, respectively. When the full theory [238] was used, the predicted and experimental potentials agreed on average to $\pm\ 1\,\text{mV}$ over a range of over $60\,\text{mV}$ in E_{cat}. Similar agreement had been found in an earlier purely E.M.F. study at 25°C [263].

The theory was equally successful in accounting for an unusual property of the catalytic rate – its negative temperature coefficient. A rise in temperature of 25°C almost halved the rate, the activation energy being $-16.9\,\text{kJ mol}^{-1}$. This can readily be understood from eqn. (96). The product of the two mass-transport rate constants k_{tr} gave a positive contribution to the activation energy of $+\ 15.2\,\text{kJ mol}^{-1}$ while the exponential term, a thermodynamic factor, led to a negative enthalpy contribution of $-32.3\,\text{kJ mol}^{-1}$ [6]. Their sum of $-17.1\,\text{kJ mol}^{-1}$ is in almost perfect agreement with the experimental value. This result, taken together with the results for the variations in rotation speed and in concentrations, firmly establish that the Feic $+ I^-$ reaction had rapidly come to equilibrium at the platinum surface and that the catalytic rate was totally diffusion-controlled. In contrast, the activation energy of the homogeneous reaction was $38.4\,\text{kJ mol}^{-1}$.

A final and even more direct vindication of the electrochemical model came from comparing the catalytic rate and the catalyst potential with independent electrochemical measurements, as set out in test (2) of Sect. 4.1. The cathodic current–potential curve of the Feic/Feoc couple and the anodic current–potential curve of the I_3^-/I^- couple were separately determined with the same anodically preconditioned platinum disc and are drawn as full lines in Fig. 25 (where all currents are taken as positive). Their point of intersection can be seen to lie close to the circled letter A which marks the data for the corresponding catalytic experiment. In numerical terms, the electrochemical intersection point is at $E_{mix} = 294\,\text{mV}$ (SCE) and $I_{mix} = 455 \pm 14\,\mu\text{A}$ while the catalytic runs yielded $E_{cat} = 295\,\text{mV}$ (SCE) and, by eqn. (88), $I_{cat} = FAv_{cat} = 479 \pm 6\,\mu\text{A}$. Similar good agreement was found at other rotation speeds [6] and at another temperature [68], all with anodically pretreated discs.

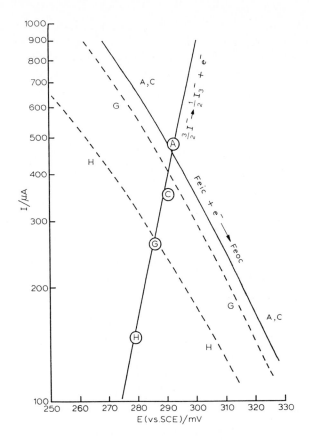

Fig. 25. The anodic current–potential curve for $0.05 \, \text{mol dm}^{-3}$ KI $+ \, 1 \times 10^{-6} \, \text{mol dm}^{-3}$ KI$_3$ and the cathodic current–potential curves for $1 \times 10^{-3} \, \text{mol dm}^{-3}$ K$_3$Fe(CN)$_6$ $+ \, 2 \times 10^{-6} \, \text{mol dm}^{-3}$ K$_4$Fe(CN)$_6$, all in $1 \, \text{mol dm}^{-3}$ KNO$_3$ at 5°C at a platinum electrode rotating at 500 rev. min^{-1}. All currents are taken as positive. The letters on the broken curves refer to the electrode preconditioning described in the text. Ringed letters mark the corresponding catalytic rates (converted to currents) and catalyst potentials in catalysed runs between $0.05 \, \text{mol dm}^{-3}$ KI and $1 \times 10^{-3} \, \text{mol dm}^{-3}$ K$_3$Fe(CN)$_6$ in $1 \, \text{mol dm}^{-3}$ KNO$_3$ at 5°C and 500 rev. min^{-1}. (After Spiro and Freund [239].)

A different story emerges from experiments by Spiro and Freund [239] with platinum discs subjected to other pretreatments. As Fig. 25 shows, cathodic preconditioning (marked by the letter C) did not alter the current–potential curve of either the I$_3^-$/I$^-$ or the Feic/Feoc couple but significantly lowered the catalytic rate (ringed letter C) and also made it less reproducible. When the pretreatment of the disc consisted of cathodic electrochemical conditioning followed by a 10 min immersion in a $0.05 \, \text{mol dm}^{-3}$ KI solution and rinsing (letter G), the catalytic rate was lower still and even the curve of the Feic/Feoc couple was shifted downwards. A pretreatment that included immersion of the disc in a reaction mixture containing both I$^-$ and I$_3^-$ produced more pronounced lowerings as shown by the Feic/Feoc curve and the catalytic point both marked H. The anodic current–potential curve of the I_3^-/I^- couple remained unchanged throughout. These phenomena can be understood in terms of iodide ion adsorption. Although iodide does not adsorb on anodically preconditioned platinum, it is known to exhibit specific adsorption on the cathodically pretreated metal [264, 265]. This adsorption has no effect on the diffusion-controlled oxidation rate of I^- and thus on the I_3^-/I^- curve but it does inhibit the reduction of Feic as shown by the decrease in the cathodic current of the Feic/Feoc couple in curves G and H. In the reaction mixture there are always iodide ions present and so the initial catalytic rate is less on cathodically pretreated surfaces, and less still if that surface had been previously exposed to iodide ions. The catalytic points therefore no longer coincide with the intersections of the current–potential curves of the contributing couples. To put it another way, the two couples present together no longer behave independently at a cathodically pretreated platinum surface. This is one of the few authenticated examples of the breakdown of the Wagner and Traud additivity principle. However, it is striking that all the catalytic points lie on the I_3^-/I^- current–potential curve which was unaffected by the various conditioning methods. This demonstrates that the catalysis of the Feic + I^- reaction still proceeded by electron transfer through the platinum metal, even though the ferricyanide reduction was hindered by adsorbed iodide and iodine species. It is relevant to add that the reaction has been found to be catalysed by other electron conductors like the noble metals iridium, palladium, rhodium, and ruthenium [232] as well as by charcoal [266], graphite and phthalocyanines [233] but not by the electrical insulators glass, silica, and barium sulphate [233].

4.5 THE IRON(III) + IODIDE REACTION

The rate of the reaction

$$Fe(III) + I^- \rightarrow Fe(II) + \tfrac{1}{2} I_2 \tag{LX}$$

is also not affected by the addition of silica and barium sulphate [233] but does increase in the presence of graphite and charcoal [233], and especially platinum [266] and other platinum metals [232]. These qualitative facts again suggest a catalytic mechanism of electron transfer through the solid. This inference was tested and shown to be valid in a recent quantitative investigation by Spiro and Creeth [240]. Using a large rotating platinum disc as electrode, they determined the current–potential curves of the Fe(III)/Fe(II) couple and of the I_2/I^- couple and then evaluated the mixture potential, E_{mix}, at which the currents, I_{mix}, were numerically equal. With the same disc, now acting as a catalyst, they measured the rate of the corresponding Fe(III) + I^- reaction. The catalytic rate, converted to a current by Faraday's law [eqn. (88)], agreed with I_{mix} to better than 5% while the potential adopted by the catalyst, E_{cat}, agreed with E_{mix} to better than 3 mV. Similar agreement was obtained over a wide range of initial Fe(III) and I^- concentrations. These results confirm the electrochemical mechanism of catalysis by test (2)

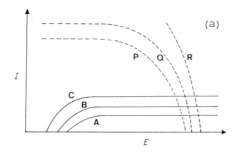

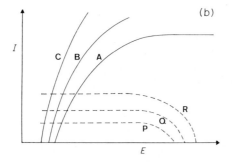

Fig. 26. Schematic current–potential curves for iodide oxidation (full lines, iodide concentrations increasing from A to B to C) and for iron(III) reduction [broken lines, iron(III) concentrations increasing from P to Q to R]. All currents are taken as positive. (a) Low iodide concentrations and high iron(III) concentrations. (b) High iodide concentrations and low iron(III) concentrations.

of Sect. 4.1 as well as the validity of the Wagner and Traud additivity hypothesis.

A comment needs to be made about the state of the platinum surface during these experiments. The solutions that were used contained 0.05 mol dm^{-3} HClO$_4$ to curtail iron(III) hydrolysis and most of the work was carried out at 5°C to decrease the contribution of the homogeneous reaction. Under these conditions the surface of the platinum at the catalytic potential was always in the reduced state, irrespective of its preconditioning treatment. As mentioned in Sect. 4.4, such a surface specifically adsorbs iodide ions [239, 264, 265]. In order to allow for their effect, the current–potential curves for the Fe(III)/Fe(II) couple had been determined in the presence of a small concentration (5×10^{-6} mol dm^{-3}) of potassium iodide. It is interesting that this treatment was sufficient to maintain the validity of the additivity hypothesis, in contrast to the results obtained for the Fe(CN)$_6^{3-}$ + I$^-$ system on cathodically pretreated platinum surfaces (Fig. 25).

The Fe(III) + I$^-$ reaction furnished the first example of a heterogeneously catalysed solution reaction whose kinetics change dramatically with a change in the ratio of the reactant concentrations [case (d) in Sect. 4.1]. When [Fe(III)]/[I$^-$] was low, the catalysed reaction was first order in iron(III) and zero order in iodide; as the ratio $[Fe(III)]/[I^-]$ increased, the order in iron(III) decreased and that in iodide increased until, at high $[Fe(III)]/[I^-]$ ratios, the catalysed reaction became zero order in iron(III) and first order in iodide. Each of these kinetic regimes could be studied separately in the laboratory because the standard potentials of the $Fe(III)/Fe(II)$ and I_2/I^- couples were sufficiently far apart. In all cases the catalysed reaction was totally diffusion-controlled, the values of v_{cat} being directly proportional to the square root of the disc rotation speed as predicted in eqn. (98). Moreover, at any given rotation speed the observed catalytic rate at a low concentration of either reagent agreed within experimental error with the value expected from eqn. (98) using the diffusion coefficient obtained from limiting current measurements.

The activation energy at low concentrations of iodide was $17 \pm 2\,kJ$ mol^{-1}, in good accord with the value of $17.2\,kJ\ mol^{-1}$ expected from a mass transport-controlled reaction at a rotating disc [eqn. (98)] with the diffusion coefficient given by the Stokes–Einstein equation (30). The activation energy of the homogeneous $Fe(III) + I^-$ reaction was $102\,kJ\ mol^{-1}$.

The results can be easily understood in terms of the current–potential curves in Fig. 26 which amplify the skeleton outline in Fig. 21(c). The situation at low concentrations of iodide is depicted in Fig. 26(a) where the limiting plateau currents increase from A to B to C in direct proportion to the iodide concentrations. Consider the points where these curves cross curve P: here the current due to oxidation of iodide is equal in magnitude to the current due to the reduction of iron(III). When both I^- and $Fe(III)$ are present together at a platinum surface these currents equal I_{mix} by the additivity hypothesis and thus, by Faraday's law, yield values of v_{cat} [eqn. (88)]. Since the limiting currents are proportional to the iodide concentrations, the catalytic rates are first order in iodide. Furthermore, any given iodide curve (say A) intersects the iron(III) current–potential curves P, Q, and R at the same current so that I_{mix} is independent of the iron(III) concentration and v_{cat} is zero order in iron(III). On the other hand, the mixture potentials at the intersection points can be seen to depend on both curves. In conformity with Fig. 26(a), the values of E_{cat} were found to decrease with increasing iodide concentration (at constant $[Fe(III)]$) and increased with increasing iron(III) concentration (at constant $[I^-]$). The results for reaction mixtures containing low concentrations of iron(III) and high concentrations of iodide can be similarly understood from the current–potential curves in Fig. 26(b). The electrochemical interpretation has therefore provided a satisfying explanation of the unusual kinetic results that were encountered in this heterogeneously catalysed system.

5. Concluding comments

This chapter should have helped to dispel the notion, unfortunately still prevalent, that only reactions involving gases can be heterogeneously catalysed. As the literature review in Sects. 2–4 has demonstrated, many kinds of solution reaction including substitution, isomerisation and redox, both inorganic and organic, are catalysed by the surfaces of suitable solids. A more detailed inspection shows that in some categories the type of reaction studied so far has been fairly restricted: solvolyses predominate among the substitution processes as do racemisations among the isomerisation reactions. Both hydrolysis and racemisation reactions were selected so frequently because they are kinetically of first order, which simplifies the theoretical analysis of the results. More adventurous choices would now be feasible with the benefit of our increased understanding of the subject. There can be little doubt that the catalysis of a wider range of solution reactions will provide fruitful projects for future research.

The favourite analytical method for following catalysed reactions has certainly been spectrophotometry. This technique has been upgraded in recent times by the availability of diode array instruments which will now make it easier to observe the changes in concentration of more than one reactant and/or product species, a considerable advantage for heterogeneously catalysed processes [Test (6), Sect. 1.8.2]. Second in popularity have been titration methods, either of samples or in situ as in pH-statting. Metal sensors in the form of E.M.F. or conductance electrodes have been and should be deliberately avoided in case they themselves catalyse the reaction, although some types of ion-selective electrode may prove sufficiently inert. Special analytical techniques have been required for certain types of reaction: optical rotary dispersion or circular dichroism for racemisations, and separation followed by radiochemical assay for isotopic exchange reactions. In view of the range of analytical methods employed, it is surprising that workers in this field have been reluctant to examine the surfaces of their catalysts by modern spectroscopic techniques. Electrochemists have now developed a range of in situ methods [267] that might well be applied to heterogeneous catalytic studies in solution. Among these could be UV–visible examination of thin layers of catalyst (a few nm thick) deposited on glass to form an optically transparent catalyst, reflection techniques, vibrational spectroscopy including FTIR and surface enhanced Raman scattering (SERS) on roughened catalyst surfaces as well as ESR for reactions involving radicals and paramagnetic species. Such techniques could prove valuable tools for increasing our understanding of the catalytic mechanisms.

The kinetics found for the reactions at the solution/solid interface show some marked similarities with those at gas/solid [9, 49], gas/liquid, and liquid/liquid interfaces [268]. Whenever one of the phases is a liquid rather than a gas, mass transport is apt to become rate-controlling because of the smaller diffusion coefficients of species in liquids. Many of the catalysed redox reactions in Sect. 4 were indeed partly or wholly diffusion-controlled. These systems could be converted to surface-controlled ones simply by reducing the size of the catalysing material; by using colloidal catalysts, for example [269]. In surface-controlled solution catalyses the rate-limiting steps, whether unimolecular or bimolecular, involve the reactions of adsorbed species. In a few systems where the catalysts possessed large areas (carbons, platinised platinum), the concentrations of the adsorbed species and their rate of reaction could be measured directly. In most systems, however, the kinetics of the surface reaction have had to be inferred from the rate law combined with appropriate adsorption isotherms. The Langmuir model has proved particularly useful here. Very often the well-known rate laws developed for the heterogeneous catalysis of gas reactions have proved applicable to solution catalyses after being modified to take account of parallel homogeneous pathways. Phenomena such as surface saturation and poisoning by competitive adsorption of another reactant or a product have been encountered at solution/solid as well as at gas/solid interfaces. However, rather more cases of non-competitive adsorption have been met with in solution catalyses, especially with insoluble salt catalysts like AgI and partially oxidised metals which provided two kinds of surface site. The study of heterogeneously catalysed redox reactions has also revealed a new surface mechanism. Here the two reactants need no longer sit side by side on the surface as in the Langmuir–Hinshelwood mechanism since their interaction takes place by the transfer of an electron through the bulk catalyst. Several examples of this type have been described in Sect. 4. In these cases it is now possible to predict the extent of catalysis and even the type of kinetics from purely electrochemical and hydrodynamic data. It is therefore fair to claim that certain kinds of heterogeneous catalysis in solution are now better understood than the much more extensively studied catalyses of gas reactions.

The most efficacious catalysts to date have been the noble metals, carbons, and some insoluble oxides and salts. As was emphasized in Sect. 1.8, tests should always be carried out to confirm that heterogeneous catalysis is the true reason for any observed rate increase. One of these tests requires the catalytic rate to rise proportionately with the mass or area of the catalyst. While most reaction systems have satisfied this criterion, the rates of several carbon-catalysed reactions in the literature have been reported as increasing either much more or much less than expected when larger amounts of the solid were added. Pore diffusion could account for only some of these results. Since carbons are cheap catalysts with large surface areas, their aberrant behaviour in this respect would be worth serious investigation.

Which catalyst should be chosen for a given reaction will depend upon chemical, steric, and mechanistic factors. The application of Pearson's soft and hard acid-base (SHAB) principle has often proved a valuable qualitative guide as to suitable surface sites for a particular reactant. In fact, certain solids actually owe their catalytic power to attached Brønsted or Lewis acid and base groups as exemplified by weak acid ion exchange resins (Sect. 2.3), alumina (Sect. 3.2), and sometimes charcoals. Steric aspects can be conveniently gauged beforehand by molecular models (see for example ref. 36) while more detailed information, for instance on lateral interactions on the surface, will ultimately be obtainable from computer simulation (see for example ref. 270). As regards mechanistic considerations, the prime example is provided by electron transfer reactions which have been shown to be catalysed by electron-conducting solids such as metals, graphitic carbons, and selected semiconductors. In future, however, catalysts will be not so much chosen as custom-built. This stage of development has already progressed quite far in the field of electrode kinetics with the production of a range of chemically modified electrodes [271] and a similar evolution is taking place in heterogeneous catalysis in solution. The simplest way of modifying the surface has been the specific adsorption of a suitable species such as a surfactant or iodide ion on platinum. Synergistic homogeneous and heterogeneous catalysis tends to operate in this fashion. The next step has involved chemical modification of the surface by means such as electrochemical preconditioning (of metals) or appropriate thermal pretreatment (as with alumina). Several workers in this and related catalytic areas have embarked on more extensive chemical changes of the solid: the doping of semiconductors, the deposition of metals on their surfaces, underpotential deposition of one metal on another, alloy formation, the incorporation of particles of the solid into microemulsions, and the synthesis of intercalation compounds of graphite and of pillared zeolites. Efforts in this direction will undoubtedly continue at an accelerating pace. Nevertheless, considerable basic research is still needed before we can achieve the eventual goal of mission-oriented catalyst preparation.

References

1   E. Millon, Ann. Chim. Phys., 13 (1845) 29.
2   G. Lemoine, C.R. Acad. Sci., 173 (1921) 7.
3   M. Spiro and A.B. Ravnö, J. Chem. Soc., (1965) 78.
4   M. Spiro, in J.N. Bradley, R.D. Gillard and R.F. Hudson (Eds.), Essays in Chemistry, Vol. 5, Academic Press, London, 1973, p.63.
5   K.J. Mysels, Pure Appl. Chem., 46 (1976) 73.
6   P.L. Freund and M. Spiro, J. Chem. Soc. Faraday Trans. 1, 79 (1983) 491.
7   R.S. Bradley, Trans. Faraday Soc., 34 (1938) 278.
8   L.L. Bircumshaw and A.C. Riddiford, Q. Rev., 6 (1952) 157.
9   G. Webb, in C.H. Bamford and C.F.H. Tipper (Eds.), Comprehensive Chemical Kinetics, Vol. 20, Elsevier, Amsterdam, 1978, Chap. 1.
10   J.J. Kipling, Adsorption from Solutions of Non-Electrolytes, Academic Press, London, 1965, Chaps. 7, 13, 15.
11   D.H. Everett, Pure Appl. Chem., 58 (1986) 967.
12   B. Dacre and P.A. Wheeler, J. Chem. Soc. Faraday Trans. 1, 77 (1981) 1285. B. Dacre, B. Savory and P.A. Wheeler, J. Chem. Soc. Faraday Trans. 1, 77 (1981) 1297.
13   J. Sobkowski and M. Szklarczyk, Electrochim. Acta, 25 (1980) 383.
14   M. Szklarczyk and J. Sobkowski, Electrochim. Acta, 26 (1981) 345.

160

15 K. Hachiya, M. Ashida, M. Sasaki, H. Kan, T. Inoue and T. Yasunaga, J. Phys. Chem., 83 (1979) 1866. K. Hachiya, M. Sasaki, T. Ikeda, N. Mikami and T. Yasunaga, J. Phys. Chem., 88 (1984) 27.

16 K. Hachiya, M. Ashida, M. Sasaki, M. Karasuda and T. Yasunaga, J. Phys. Chem., 84 (1980) 2292.

17 R.D. Astumian, M. Sasaki, T. Yasunaga and Z.A. Schelly, J. Phys. Chem., 85 (1981) 3832.

18 T. Ikeda, M. Sasaki, K. Hachiya, R.D. Astumian and T. Yasunaga, J. Phys. Chem., 86 (1982) 3861.

19 V.K.F. Chia, M.P. Soriaga and A.T. Hubbard, J. Phys. Chem., 91 (1987) 78.

20 R.J. Mureinik and M. Spiro, J. Chem. Soc. Dalton Trans., (1974) 2493. R.J. Mortimer and M. Spiro, J. Chem. Soc. Perkin Trans. 2, (1980) 1228.

21 W.J. Albery, P.N. Bartlett and A.J. McMahon, J. Electroanal. Chem., 182 (1985) 7.

22 H.-C. Chang and M. Aluko, Chem. Eng. Sci., 36 (1981) 1611.

23 M.G. Goodman, M.B. Cutlip, C.N. Kenney, W. Morton and D. Mukesh, Surf. Sci., 120 (1982) L453. D. Mukesh, M. Goodman, C.N. Kenney and W. Morton, Specialist Periodical Report: Catalysis, Vol. 6, Royal Society of Chemistry, London, 1983, p.1.

24 C.H. Giles, T.H. MacEwan, S.N. Nakhwa and D. Smith, J. Chem. Soc., (1960) 3973.

25 C.H. Giles and S.N. Nakhwa, J. Appl. Chem., 12 (1962) 266.

26 C.H. Giles, D. Smith and A. Huitson, J. Colloid Interface Sci., 47 (1974) 755.

27 C.H. Giles, A.P. D'Silva and I.A. Easton, J. Colloid Interface Sci., 47 (1974) 766.

28 G.D. Parfitt and C.H. Rochester, in G.D. Parfitt and C.H. Rochester (Eds.). Adsorption from Solution at the Solid/Liquid Interface, Academic Press, London, 1983, Chap. 1.

29 D.O. Hayward and B.M.W. Trapnell, Chemisorption, Butterworths, London, 2nd edn., 1964, Chap. V.

30 J.M. Thomas and W.J. Thomas, Introduction to the Principles of Heterogeneous Catalysis, Academic Press, London, 1967, Chaps. 2–4.

31 L. Beránek and M. Kraus, in C.H. Bamford and C.F.H. Tipper (Eds.), Comprehensive Chemical Kinetics, Vol. 20, Elsevier, Amsterdam, 1978, Chap. 3.

32 A.J. Lecloux, in J.R. Anderson and M. Boudart (Eds.), Catalysis: Science and Technology, Vol. 2, Springer Verlag, Berlin, 1981, Chap. 4.

33 H.J. van den Hul and J. Lyklema, J. Am. Chem. Soc., 90 (1968) 3010.

34 J.J. Kipling, Adsorption from Solutions of Non-Electrolytes, Academic Press, London, 1965, Chaps. 2, 17.

35 R.J. Mureinik and M. Spiro, J. Chem. Soc. Dalton Trans., (1974) 2493.

36 R.J. Mortimer and M. Spiro, J. Chem. Soc. Perkin Trans. 2, (1980) 1228.

37 J.J. Kipling, Adsorption from Solutions of Non-Electrolytes, Academic Press, London, 1965, Chap. 10.

38 J. Sobkowski and A. Wieckowski, J. Electroanal. Chem., 41 (1973) 373.

39 R.E. Pincock, W.M. Johnson, K.R. Wilson and J. Haywood-Farmer, J. Am. Chem. Soc., 95 (1973) 6477.

40 R.E. Pincock, W.M. Johnson and J. Haywood-Farmer, Can. J. Chem., 54 (1976) 548.

41 E.F.G. Barbosa, R.J. Mortimer and M. Spiro, J. Chem. Soc. Faraday Trans. 1, 77 (1981) 111.

42 J.M. Thomas and W.J. Thomas, Introduction to the Principles of Heterogeneous Catalysis, Academic Press, London, 1967, Chaps. 7, 9.

43 M. Spiro, J. Chem. Soc. Faraday Trans. 1, 73 (1977) 1825.

44 H.A.C. McKay, Nature (London), 142 (1938) 997.

45 H.A.C. McKay, J. Am. Chem. Soc., 65 (1943) 70.

46 P.D. Totterdell and M. Spiro, J. Chem. Soc. Faraday Trans. 1, 72 (1976) 1477.

47 P.D. Totterdell and M. Spiro, J. Chem. Soc. Faraday Trans. 1, 72 (1976) 1485.

48 G.C. Bond, Heterogeneous Catalysis: Principles and Applications, Clarendon Press, Oxford, 2nd edn., 1987, Chaps. 2, 5.

49 J.M. Thomas and W.J. Thomas, Introduction to the Principles of Heterogeneous Catalysis, Academic Press, London, 1967, Chaps. 4, 9.

50   R. Aris, The Mathematical Theory of Diffusion and Reaction in Permeable Catalysts, Clarendon Press, Oxford, 1975, Vol. I.

51   G.F. Froment and K.B. Bischoff, Chemical Reactor Analysis and Design, Wiley, New York, 1979, Chap. 3.

52   M. Boudart and G. Djéga-Mariadassou, Kinetics of Heterogeneous Catalytic Reactions, Princeton University Press, Princeton, 1984, Chap. 6.

53   R.M. Koros and E.J. Nowak, Chem. Eng. Sci., 22 (1967) 470.

54   Landolt-Börnstein Zahlenwerte und Funktionen, Vol. II, Part 5a, Springer Verlag, Berlin, 6th edn., 1969.

55   H.J.V. Tyrrell and K.R. Harris, Diffusion in Liquids, Butterworths, London, 1984, pp. 446–448; Chap. 6.

56   M. Spiro, in B.W. Rossiter and J.F. Hamilton (Eds.), Physical Methods of Chemistry, Vol. II, Wiley–Interscience, New York, 2nd edn., 1986, Chap. 8, pp. 686–690.

57   D. Nicholson and N.G. Parsonage, Computer Simulation and the Statistical Mechanics of Adsorption, Academic Press, London, 1982, Chaps. 4, 6.

58   I.K. Snook and W. van Megen, J. Chem. Phys., 72 (1980) 2907.

59   J.J. Magda, M. Tirrell and H.T. Davis, J. Chem. Phys., 83 (1985) 1888.

60   S.-H. Suh and J.M.D. MacElroy, Mol. Phys., 58 (1986) 445.

61   A. Gosman, S. Liukkonen and D. Passiniemi, J. Phys. Chem., 90 (1986) 6051.

62   D.J. Shaw, Introduction to Colloid and Surface Chemistry, Butterworths, London, 3rd edn., 1980, Chap. 7.

63   A.C. Riddiford, Adv. Electrochem. Electrochem. Eng., 4 (1966) 47.

64   V.G. Levich, Physicochemical Hydrodynamics, Prentice-Hall, Englewood Cliffs, NJ, 1962, Sect. 7; Sect. 11; Sects. 4, 25, 33.

65   M. Spiro and C.M. Page, J. Sci. Food Agric., 35 (1984) 925.

66   M. Spiro and P.L. Freund, J. Chem. Soc. Faraday Trans. 1, 79 (1983) 1649.

67   D. Bremner, Chem. Br., 22 (1986) 633.

68   M. Spiro and P.W. Griffin, J. Chem. Soc. Chem. Commun., (1969) 262.

69   F. Opekar and P. Beran, J. Electroanal. Chem., 69 (1976) 1.

70   J.S. Newman, Electrochemical Systems, Prentice-Hall, Englewood Cliffs, NJ, 1973.

71   R.R. Johnston, Ph.D. Thesis, University of London, 1965, Chap. II.

72   M. Spiro and D.S. Jago, J. Chem. Soc. Faraday Trans. 1, 78 (1982) 295.

73   W.J. Albery and M.L. Hitchman, Ring–Disk Electrodes, Clarendon Press, Oxford, 1971.

74   M.B. Glauert, J. Fluid Mech., 1 (1956) 625.

75   W.J. Albery and C.M.A. Brett, J. Electroanal. Chem., 148 (1983) 201.

76   W.J. Albery and C.M.A. Brett, J. Electroanal. Chem., 148 (1983) 211.

77   W.J. Albery, J. Electroanal. Chem., 191 (1985) 1.

78   J. Yamada and H. Matsuda, J. Electroanal. Chem., 44 (1973) 189.

79   M. Spiro, Faraday Discuss. Chem. Soc., 77 (1984) 275.

80   J.M. Smith, Chemical Engineering Kinetics, McGraw-Hill, New York, 1956, Chap. 9.

81   I.D. Clark and R.P. Wayne, in C.H. Bamford and C.F.H. Tipper (Eds.), Comprehensive Chemical Kinetics, Vol. 2, Elsevier, Amsterdam, 1969, Chap. 4.

82   O. Exner, in N.B. Chapman and J. Shorter (Eds.), Advances in Linear Free Energy Relationships, Plenum Press, London, 1972, Chap. 1.

83   J. Shorter, in N.B. Chapman and J. Shorter (Eds.), Advances in Linear Free Energy Relationships, Plenum Press, London, 1972, Chap. 2.

84   J.R. Chipperfield, in N.B. Chapman and J. Shorter (Eds.), Advances in Linear Free Energy Relationships, Plenum Press, London, 1972, Chap. 7.

85   L.P. Hammett, Physical Organic Chemistry, McGraw-Hill Kogakusha, Tokyo, 2nd edn., 1970, Chap. 11.

86   R. Gallo, Prog. Phys. Org. Chem., 14 (1983) 115.

87   M. Kraus, Adv. Catal., 17 (1967) 75.

88   I. Mochida and Y. Yoneda, J. Catal., 7 (1967) 393.

89   M. Kraus, Adv. Catal., 29 (1980) 151.

90   L. Cerveny and V. Ruzicka, Collect. Czech. Chem. Commun., 34 (1969) 1570.

91   R. Maurel and J. Tellier, Bull. Soc. Chim. Fr., (1968) 4191, 4650.

92   R.G. Pearson, J. Am. Chem. Soc., 85 (1963) 3533; Chem. Br., 2 (1967) 103; J. Chem. Educ., 64 (1987) 561.

93   R.G. Pearson, Science, 151 (1966) 172.

94   D.J. Barclay, J. Electroanal. Chem., 19 (1968) 318.

95   U.M. Gösele, Prog. React. Kinet., 13 (2) (1984) 63.

96   J. Keizer, Chem. Rev., 87 (1987) 167.

97   G.C. Lalor and D.S. Rustad, J. Inorg. Nucl. Chem., 31 (1969) 3219.

98   D.N. Kevill, in S. Patai and Z. Rappoport (Eds.), The Chemistry of Functional Groups. Supplement D: the Chemistry of Halides, Pseudo-Halides and Azides, Part 2, Wiley, New York, 1983, Chap. 20.

99   A. Seidell, Solubilities of Inorganic and Metal Organic Compounds, Vol. I, Van Nostrand, New York, 3rd edn., 1940.

100   L.G. Sillén and A.E. Martell, Stability Constants of Metal–Ion Complexes, Spec. Publ. No. 17, The Chemical Society, London, 1964; Spec. Publ. No. 25, The Chemical Society, London, 1971.

101   A.E. Martell and R.M. Smith, Critical Stability Constants, Plenum Press, New York, Vol. 1, 1974; Vol. 2, 1975; Vol. 3, 1977; Vol. 4, 1976; Vol. 5, 1982.

102   R.A. Robinson and R.H. Stokes, Electrolyte Solutions, Butterworths, London, 2nd edn., 1959.

103   Z.F. Zemaitis, Jr., D.M. Clark, M. Rafal and N.C. Scrivner, Handbook of Aqueous Electrolyte Thermodynamics, American Institute of Chemical Engineers, New York, 1986.

104   P. Delahay, M. Pourbaix and P. Van Rysselberghe, J. Chem. Educ., 27 (1950) 683.

105   M. Pourbaix, Atlas of Electrochemical Equilibria in Aqueous Solutions, Pergamon, Oxford, 1966.

106   C.I. House, in G.A. Davies (Ed.), Separation Processes in Hydrometallurgy, Ellis Horwood, Chichester, 1987, Chap. 1.

107   A.J. Bard, R. Parsons and J. Jordan (Eds.), Standard Potentials in Aqueous Solution, Dekker, New York, 1985.

108   P.B. Linkson, B.D. Phillips and C.D. Rowles, Miner. Sci. Eng., 11 (1979) 65.

109   J.B. Lee, Corrosion (N.A.C.E.), 37 (1981) 467.

110   M. Delépine, C. R. Acad. Sci., 141 (1905) 886, 1013.

111   J. Millbauer, Z. Phys. Chem., 77 (1911) 380.

112   D. Cohen and B. Taylor, J. Inorg. Nucl. Chem., 22 (1961) 151.

113   T. Xue and K. Osseo-Asare, Metall. Trans. B, 16 (1985) 455.

114   M.A. Diaz, Ph.D. Thesis, University of London, 1986, pp. 112–119.

115   T. Groenewald, D.I.C. Thesis, Imperial College, London, 1966, Chaps. 2, 3.

116   A.R. Burkin, The Chemistry of Hydrometallurgical Processes, Spon, London, 1966, Chaps. 3, 5.

117   E. Peters, Metall. Trans. B, 7 (1976) 505.

118   M.D. Pritzker and R.H. Yoon, Int. J. Miner. Process., 12 (1984) 95.

119   T. Groenewald, J.M. Austin and M. Spiro, J. Chem. Soc. Dalton Trans., (1980) 860.

120   P. Eadington and A.P. Prosser, Trans. Inst. Min. Met., 78 (1969) C74.

121   R. Tolun and J.A. Kitchener, Trans. Inst. Mining Met., 73 (1964) 313.

122   R. Woods, J. Phys. Chem., 75 (1971) 354.

123   A.I. Vogel, A Textbook of Quantitative Inorganic Analysis, Longmans, London, 2nd edn., 1951, pp. 356–357.

124   P. Boutry, O. Bloch and J.-C. Balaceanu, Bull. Soc. Chim. Fr., (1962) 988.

125   A.J. Appleby, H. Kita, M. Chemla and G. Bronoël, in A.J. Bard (Ed.), Encyclopedia of Electrochemistry of the Elements, Vol. IX, Part A, Dekker, New York, 1982, Chap. IX, a-3.

126   T.H. James, in T.H. James (Ed.), The Theory of the Photographic Process, Macmillan, New York, 4th edn., 1977, Chap. 13.

127   U. Nickel, N. Rühl and B.M. Zhou, Z. Phys. Chem. N.F.. 148 (1986) 33.

128   G.P. Faerman and E.D. Voeikova, Usp. Nauchn. Fotogr., 3 (1955) 174; 4 (1955) 150.

129   W. Jaenicke, Adv. Electrochem. Electrochem. Eng., 10 (1977) 91.

130   A.B. Ravnö and M. Spiro, J. Chem. Soc., (1965) 97.

131   L.N. Lewis and N. Lewis, J. Am. Chem. Soc., 108 (1986) 7228.

132   R.H. Crabtree, M.F. Mellea, J.M. Mihelcic and J.M. Quirk, J. Am. Chem. Soc., 104 (1982) 107.

133   D.R. Anton and R.H. Crabtree, Organometallics, 2 (1983) 855.

134   I.M. Kolthoff, Rec. Trav. Chim., 48 (1929) 298.

135   G.C. Bond, Principles of Catalysis, Royal Institute of Chemistry, London, 2nd edn., 1968, Chap. 1.

136   M.D. Archer and M. Spiro, J. Chem. Soc. A, (1970) 68.

137   M.D. Archer and M. Spiro, J. Chem. Soc. A, (1970) 73.

138   M.D. Archer and M. Spiro, J. Chem. Soc. A, (1970) 78.

139   R.J. Mureinik, A.M. Feltham and M. Spiro, J. Chem. Soc. Dalton Trans., (1972) 1981.

140   W.B. Lewis, C.D. Coryell and J.W. Irvine Jr., J. Chem. Soc., (1949) S386.

141   A.W. Adamson and K.S. Vorres, J. Inorg. Nucl. Chem., 3 (1956) 206.

142   J.C. Sullivan, D. Cohen and J.C. Hindman, J. Am. Chem. Soc., 76 (1954) 4275.

143   C.H. Bovington, D.F. Maundrell and B. Dacre, J. Chem. Soc. B, (1971) 767. D.F. Maundrell, C.H. Bovington and B. Dacre, J. Chem. Soc. B, (1972) 1284.

144   L.G. Carpenter, M.H. Ford-Smith, R.P. Bell and R.W. Dodson, Discuss. Faraday Soc., 29 (1960) 92.

145   M.H. Ford-Smith, N. Sutin and R.W. Dodson, Discuss. Faraday Soc., 29 (1960) 134.

146   M. Spiro, Chem. Soc. Rev., 15 (1986) 141.

147   R.P. Bell and E.N. Ramsden, J. Chem. Soc., (1958) 161.

148   H.N. McCoy, J. Am. Chem. Soc., 58 (1936) 1577.

149   For example, G. Jones and J.H. Colvin, J. Am. Chem. Soc., 66 (1944) 1573.

150   M. Spiro, in B.W. Rossiter and J.F. Hamilton (Eds.), Physical Methods of Chemistry, Vol. II, Wiley–Interscience, New York, 2nd edn., 1986, Chap. 8, pp. 721–727.

151   H.T. Calvert, Z. Phys. Chem., 38 (1901) 513.

152   E.S. Shanley, E.M. Roth, G.M. Nichols and M. Kilpatrick, J. Am. Chem. Soc., 78 (1956) 5190.

153   D.K. Thomas and O. Maass, Can. J. Chem., 36 (1958) 449.

154   F. Auerbach and H. Zeglin, Z. Phys. Chem., 103 (1923) 178.

155   M.D. Archer and M. Spiro, J. Chem. Soc. A, (1970) 82.

156   M. Spiro and E.F.G. Barbosa, React. Kinet. Catal. Lett., 3 (1975) 311.

157   I.M. Kolthoff and E.B. Sandell, Textbook of Quantitative Inorganic Analysis, Macmillan, New York, 3rd edn., 1952, pp. 545–546.

158   P.S. Walton and M. Spiro, J. Chem. Soc., (1969) 42.

159   F.M. Blewett and P. Watts, J. Chem. Soc. B, (1971) 881.

160   M. Spiro and J.M. Garnica Meza, unpublished work.

161   C.K. Ingold, Structure and Mechanism in Organic Chemistry, Bell, London, 1953, Chaps. VII, VIII.

162   P.B.D. de la Mare and B.E. Swedlund, in S. Patai (Ed.), The Chemistry of the Carbon–Halogen Bond, Part 1, Wiley, London. 1973, Chap. 7.

163   J. March, Advanced Organic Chemistry: Reactions, Mechanisms, and Structure, McGraw-Hill, New York, 3rd edn., 1985, Chap. 10.

164   R.J. Mortimer and M. Spiro, J. Chem. Soc. Perkin Trans. 2, (1982) 1031.

165   M. Bretz, J.G. Dash, D.C. Hickernell, E.O. McLean and O.E. Vilches, Phys. Rev. A, 8 (1973) 1589.

166   K.R. Adam, I. Lauder and V.R. Stimson, Aust. J. Chem., 15 (1962) 467.

167   M. Spiro and D.S. Jago, unpublished work.

168   P.L. Freund, M.C.P. Lima and M. Spiro, J. Electroanal. Chem., 133 (1982) 241.

169   A.R. Despić, D.M. Drazić, M.L. Mihailović, L.L. Lorenc, R. Adzić and M. Ivić, J. Electroanal. Chem., 100 (1979) 913.

170   P.L. Freund, Ph.D. Thesis, University of London, 1982, Chap. XI.

171   A.J. Kirby, in C.H. Bamford and C.F.H. Tipper (Eds.), Comprehensive Chemical Kinetics, Vol. 10, Elsevier, Amsterdam, 1972, Chap. 2.

172   M. Spiro and J.W. Mills, unpublished work.

173   R.T. Lowson, Ph.D. Thesis, University of London, 1968, Chaps. 4, 5.

174   R.E. Panzer and P.J. Elving, Electrochim. Acta, 20 (1975) 635.

175   H.-P. Boehm, E. Diehl, W. Heck and R. Sappock, Angew. Chem. Int. Edn., 3 (1964) 669; H.-P. Boehm and H. Knözinger, in J.R. Anderson and M. Boudart (Eds.), Catalysis: Science and Technology, Vol. 2, Springer Verlag, Berlin, 1981, Chap. 4.

176   J.B. Donnet, Carbon, 6 (1968) 161. J.-B. Donnet and A. Voet, Carbon Black, Dekker, New York, 1976, p. 126.

177   Y.Y. Hoyano and R.E. Pincock, Can. J. Chem., 58 (1980) 134.

178   F.A. Cotton and G. Wilkinson, Advanced Inorganic Chemistry, Wiley, New York, 4th edn., 1980, Chap. 11.

179   J. Bertin, H.B. Kagan, J.-L. Luche and R. Setton, J. Am. Chem. Soc., 96 (1974) 8113.

180   J.P. Alazard, H.B. Kagan and R. Setton, Bull. Soc. Chim. Fr., (1977) 499.

181   H.B. Kagan, J. Bertin, J.L. Luche and R. Setton, Fr. Demande 2,288,079 (1976); Chem. Abstr., 86 (1977) 88454k.

182   F. Helfferich, Ion Exchange, McGraw-Hill, New York, 1962, Chap. 11.

183   V. Gold and C.J. Liddiard, J. Chem. Soc. Faraday Trans. 1, 73 (1977) 1119.

184   V. Gold and C.J. Liddiard with J.L. Martin, J. Chem. Soc. Faraday Trans. 1, 73 (1977) 1128.

185   R.P. Bell, The Proton in Chemistry, Chapman and Hall, London, 2nd edn., 1973, Chaps. 9, 10.

186   V. Gold, C.J. Liddiard and G.D. Morgan, in E. Caldin and V. Gold (Eds.), Proton-Transfer Reactions, Chapman and Hall, London, 1975, Chap. 13.

187   A.J. Kresge, H.L. Chen, Y. Chiang, E. Murrill, M.A. Payne and D.S. Sagatys, J. Am. Chem. Soc., 93 (1971) 413.

188   R.G. Wilkins, The Study of Kinetics and Mechanism of Reactions of Transition Metal Complexes, Allyn and Bacon, Boston, 1974.

189   J. Bjerrum, Metal Ammine Formation in Aqueous Solution, Haase, Copenhagen, 1957, Part C, Chap. XI.

190   J. Bjerrum and J.P. McReynolds, in W.C. Fernelius (Ed.), Inorganic Syntheses, Vol. II, McGraw-Hill, New York, 1946, Chap. 8.

191   J.C. Bailar Jr. and J.B. Work, J. Am. Chem. Soc., 67 (1945) 176.

192   K.W. Ellis and N.A. Gibson, Anal. Chim. Acta, 9 (1953) 275.

193   D.S. Rustad, J. Inorg. Nucl. Chem., 39 (1977) 2231.

194   A. Tomita and Y. Tamai, J. Colloid Interface Sci., 36 (1971) 153.

195   A. Tomita and Y. Tamai, J. Phys. Chem., 75 (1971) 649.

196   R. Mureinik, J. Catal., 50 (1977) 56.

197   F.P. Dwyer and A.M. Sargeson, Nature (London) 187 (1960) 1022.

198   K.A. Burke and F.G. Donnan, Z. Phys. Chem., 69 (1909) 148.

199   G. Senter, J. Chem. Soc., 97 (1910) 346.

200   J.M. Austin, O.D. E.-S. Ibrahim and M. Spiro, J. Chem. Soc. B. (1969) 669.

201   E.F.G. Barbosa and M. Spiro, J. Chem. Soc. Chem. Commun., (1977) 423.

202   M.S. Santos, E.F.G. Barbosa and M. Spiro, J. Chem. Soc. Faraday Trans. 1, 84 (1989) 4439.

203   T.C. Franklin and M. Iwunze, J. Am. Chem. Soc., 103 (1981) 5937.

204   T.C. Franklin, D. Ball, R. Rodriguez and M. Iwunze, Surf. Technol., 21 (1984) 223.

205   J.H. Fendler and E.J. Fendler, Catalysis in Micellar and Macromolecular Systems, Academic Press, New York, 1975, Chap. 5. J.H. Fendler, Membrane Mimetic Chemistry, Wiley, New York, 1982, Chap. 12, Table 12.1.

206   M. Ohag and P.G.N. Moseley, private communication, 1975.

207   L.G. Hutchins and R.E. Pincock, J. Org. Chem., 45 (1980) 2414.

208   L.G. Hutchins and R.E. Pincock, J. Org. Chem., 47 (1982) 607.

209 L.G. Hutchins and R.E. Pincock, J. Catal., 74 (1982) 275.

210 O. Ito and M. Hatano, Chem. Lett., (1976) 39.

211 M.F.M. Post, J. Langelaar and J.D.W. Van Voorst, Chem. Phys. Lett., 46 (1977) 331.

212 A. Voet, J.B. Donnet, P.A. Marsh and J. Schultz, Carbon, 4 (1966) 155.

213 D.E. Bergbreiter and J.M. Killough, J. Am. Chem. Soc., 100 (1978) 2126.

214 R.E. Pincock, private communication, 1981.

215 W. Pigman and H.S. Isbell, Adv. Carbohydr. Chem., 23 (1968) 11. H.S. Isbell and W. Pigman, Adv. Carbohydr. Chem., 24 (1969) 13.

216 K. Tanabe, A. Nagata and T. Takeshita, J. Res. Inst. Catal. Hokkaido Univ., 15 (1967) 181.

217 T.D.J. Dunstan and R.E. Pincock, J. Phys. Chem., 88 (1984) 5684.

218 T.D.J. Dunstan and R.E. Pincock, J. Org. Chem., 50 (1985) 863.

219 K. Tanabe, in J.R. Anderson and M. Boudart (Eds.), Catalysis: Science and Technology, Vol. 2, Springer Verlag, Berlin, 1981, Chap. 5.

220 H.E. Swift and B.E. Douglas, J. Inorg. Nucl. Chem., 26 (1964) 601.B.E. Douglas and S.-M. Ho, J. Inorg. Nucl. Chem., 26 (1964) 609 and earlier papers.

221 D. Sen and W.C. Fernelius, J. Inorg. Nucl. Chem., 10 (1959) 269.

222 R.J. Mureinik and M. Spiro, J. Chem. Soc. Dalton Trans., (1974) 2480.

223 R.J. Mureinik and M. Spiro, J. Chem. Soc. Dalton Trans., (1974) 2486.

224 P.D. Totterdell and M. Spiro, J. Chem. Soc. Dalton Trans., (1979) 1324.

225 R.E. Test and R.S. Hansen, Atomic Energy Comm. Document IS-341 (Chemistry), Iowa State University, U.S.A., 1961.

226 H. Morawetz and B. Vogel, J. Am. Chem. Soc., 91 (1969) 563. H. Morawetz and G. Gordimer, J. Am. Chem. Soc., 92 (1970) 7532.

227 L. Johansson, Acta Chem. Scand., 25 (1971) 3752.

228 M.L. Tobe, Acc. Chem. Res., 3 (1970) 377.

229 G.H. Searle, F.R. Keene and S.F. Lincoln, Inorg. Chem., 17 (1978) 2362.

230 A. Hammershøi and E. Larsen, Acta Chem. Scand. A, 32 (1978) 485.

231 H. Bartelt and H. Skilandat, J. Electroanal. Chem., 23 (1969) 407.

232 M. Spiro, J. Chem. Soc., (1960) 3678.

233 J.M. Austin, T. Groenewald and M. Spiro, J. Chem. Soc. Dalton Trans., (1980) 854.

234 C. Wagner and W. Traud, Z. Elektrochem., 44 (1938) 391.

235 K.J. Vetter, Electrochemical Kinetics, Academic Press, New York, 1967, Chaps. 2, 3.

236 A.J. Bard and L.R. Faulkner, Electrochemical Methods, Wiley, New York, 1980, Chaps. 1, 3, 8.

237 M. Spiro, J. Chem. Soc. Faraday Trans. 1, 75 (1979) 1507.

238 P.L. Freund and M. Spiro, J. Chem. Soc. Faraday Trans. 1, 79 (1983) 481.

239 M. Spiro and P.L. Freund, J. Electroanal. Chem., 144 (1983) 293.

240 M. Spiro and A.M. Creeth, unpublished work.

241 W. Jaenicke and F. Sutter, in W. Eichler, H. Frieser and O. Helwich (Eds.), Wissenschaftliche Photographie, Helwich, Darmstadt, 1958, p.386.

242 W. Jaenicke and F. Sutter, Z. Elektrochem., 63 (1959) 722.

243 I.R. Jonasson and D.R. Stranks, Electrochim. Acta, 13 (1968) 1147.

244 C. Wagner, Electrochim. Acta, 15 (1970) 987.

245 Z. Takehara, Electrochim. Acta, 15 (1970) 999.

246 M.D. Birkett, A.T. Kuhn and G.C. Bond, Catalysis, Specialist Periodical Report, Vol. 6, Royal Society of Chemistry, London, 1983, p.61.

247 D.V. Sokol'skii, Hydrogenation in Solutions, Israel Scientific Translations, 1964.

248 M. Grätzel, in R.E. White, J. O'M. Bockris and B.E. Conway (Eds.), Modern Aspects of Electrochemistry, No. 15, Plenum Press, New York, 1983, Chap. 2.

249 S. Fronaeus and C.O. Östman, Acta Chem. Scand., 10 (1956) 769.

250 S. Fronaeus, Acta Chem. Scand., 10 (1956) 1345.

251 R. Greef and H. Aulich, J. Electroanal. Chem., 18 (1968) 295.

252 J.R. Jonasson, Ph.D. Thesis, University of Adelaide, 1967, pp. 219–223.

253 G.M. Waind, Discuss. Faraday Soc., 29 (1960) 135.

254   S.M. Hasany, Ph.D. Thesis, University of Adelaide, 1971.

255   S.M. Hasany and D.R. Stranks. Proceedings of the Conference on Industrial Uses of Electrochemistry, Islamabad, Pakistan, 1982, p. 97.

256   M.I. Bellavance and B. Miller, in A.J. Bard (Ed.), Encyclopedia of Electrochemistry of the Elements, Vol. IV, Dekker, New York, 1975, Chap. IV-4.

257   S.M. Hasany and D.R. Stranks, Abstracts of the 4th National Conference of Pure and Applied Physical Chemistry, Bucharest, Roumania, 1974.

258   W.E. Lee and E.R. Brown, in T.H. James (Ed.), The Theory of the Photographic Process, Macmillan, New York, 4th edn., 1977, Chap. 11.

259   J.R. Thirtle, L.K.J. Tong and L.J. Fleckenstein, in T.H. James (Ed.), The Theory of the Photographic Process, Macmillan, New York, 4th edn., 1977, Chap. 12.

260   K.S. Bagdasaryan, Acta Physicochim. USSR, 19 (1944) 421.

261   W.E. Lee, in T.H. James (Ed.), The Theory of the Photographic Process, Macmillan, New York, 4th edn., 1977, Chap. 14 II.

262   G. Just, Z. Phys. Chem., 63 (1908) 513.

263   M. Spiro, R.R.M. Johnston and E.S. Wagner, Electrochim. Acta, 3 (1961) 264.

264   K. Schwabe and W. Schwenke, Electrochim. Acta, 9 (1964) 1003.

265   A.T. Hubbard, R.A. Osteryoung and F.C. Anson, Anal. Chem., 38 (1966) 692.

266   G.M. Waind, Chem. Ind., (1955) 1388.

267   R. Greef, R. Peat, L.M. Peter, D. Pletcher and J. Robinson, Instrumental Methods in Electrochemistry, Ellis Horwood, Chichester, 1985, Chap. 10.

268   M. Spiro, Faraday Discuss. Chem. Soc., 77 (1984) 309.

269   P.L. Freund and M. Spiro, J. Phys. Chem., 89 (1985) 1074.

270   A.E. Reynolds, J.S. Foord and D.J. Tildesley, Surf. Sci., 166 (1986) 19.

271   R.W. Murray, in A.J. Bard (Ed.), Electroanalytical Chemistry, Vol. 13, Dekker, New York, 1984, p. 191.

Chapter 3

Kinetics of Crystallisation of Solids from Aqueous Solution

W.A. HOUSE

1. Introduction

This chapter is concerned with the growth of crystals from aqueous systems. The approach is to present the fundamental concepts which are vital to the understanding of growth processes. This is followed by a discussion of the most important techniques available to the experimentalist and how the growth may be interpreted using current theories of crystal growth. Finally, the effects of impurities on the growth kinetics are examined.

The growth of crystals is of interest in a number of areas of science. These range from industrial manufacturing processes in which crystalline powders are prepared in bulk, to processes in which crystallisation is one part of a complex array of reactions that are occurring. This is true of many biological mineralisation reactions such as shell, bone, and teeth growth. In such systems, the epitaxial growth of one phase onto another may be a crucial step in the mineralisation. In environmental applications, the precipitation of mineral phases such as calcite, aragonite, calcium hydroxyapatite, iron oxides, and various silicas are often found to be important in determining natural water compositions and describing chemical pathways. The accompanying coprecipitation reactions of nutrients such as phosphate, nitrate, organic acids, and trace elements affects their biological availability and the water quality. The factors which control the stability of aqueous solutions in different conditions are also important in many analytical procedures.

2. Fundamental concepts

2.1 SURFACE STRUCTURE

An ideal crystal surface of orientation (hkl) is an imaginary surface formed when all the atoms, ions or molecules on one side of a plane of orientation (hkl) inside the bulk crystal are removed. Such a surface is termed atomically smooth. However, an ideal surface is unstable because the asymmetry of the interatomic forces in the surface region leads to surface relaxation. This usually manifests itself by the movement of the crystal components in a direction normal to the surface plane to enable a reduction

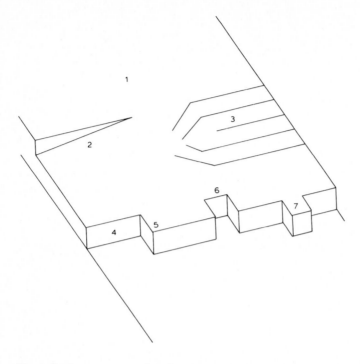

Fig. 1. Surface structure often found on low-index crystal faces. 1, A terrace; perfectly flat crystal face. 2, An emerging screw dislocation. 3, The intersection of an edge dislocation with a terrace. 4, A ledge or monatomic step, 5. A kink; a step in a ledge. 6, A vacancy in a ledge. 7, An adsorbed growth unit on a ledge.

in the total free-energy of the crystal. Of course, the ions, molecules or atoms within the crystal are not a rigid lattice but vibrate about some equilibrium position. It is found that in metals, the atomic vibrations have a larger amplitude at the surface than in the bulk material, particularly in the direction of the surface normal. The vibrations have been measured using the temperature dependence of the intensity of LEED spots.

A real surface will contain many defects; indeed, a surface which differs only slightly in orientation from an atomically smooth surface will be composed of terraces, steps or ledges, and kinks (see Fig. 1). Other defect structures such as emerging screw disclocations or edge dislocations are also possible. The description of surfaces in terms of terraces, steps, and kinks is usually referred to as the Kossel–Stranski model of real surfaces. Not all these types of defects will be necessarily present on a surface in equilibrium conditions. It is possible that, on an annealed single crystal surface, the number of dislocations emerging at the surface is of the order of 10^4–10^6 cm^{-2}. These defects are stable for kinetic reasons. In the case of kinks and steps, the probability of formation can be estimated using the Boltzmann factor.

(i) Kink formation. If the addition of a growth unit at a step leads to a

free-energy change, ΔG_{AS}, and is regarded as the formation of two kinks, then the free-energy of formation of a kink is $\Delta G_{AS}/2$. Because, at equilibrium, $\Delta G_{AS}/2 = d^2\gamma$, where γ is the interfacial free-energy (surface tension) and d the lattice dimension, then the probability of formation of a kink along a step is

$$P_k = \exp(-d^2\gamma/kT) \tag{1}$$

For the growth of calcite at 25°C and with $\gamma = 97\,\mathrm{mJ\,m^{-2}}, P_k \simeq 6 \times 10^{-5}$, i.e. 0.006%. This amounts to approximately 10^5 kinks per metre of step. Nielsen and Christoffersen [1] have estimated that, for low molecular weight, sparingly soluble electrolytes, P_k is in the range of 8–0.7%. Hence, an equilibrium surface at this temperature is expected to have a significant number of kink sites along a step.

For a crystal in contact with a supersaturated solution, the kink site density is expected to be larger than at equilibrium. Since the movement of kinks along a step is energetically neutral, the kinks may be regarded as analogous to a one-dimensional gas. Hence, the density of kinks along a step is

$$\frac{1}{x_o} = \frac{\exp(-\Delta G_{AS}/2kT)}{d} \tag{2}$$

where x_o is the kink distance. At a saturation ratio S (see Sect. 2.3), ΔG_{AS} is related to S and γ through the equation

$$\Delta G_{AS} = 2d^2\gamma - kT\ln S \tag{3}$$

leading to a kink density of

$$\frac{1}{x_o} = \frac{S^{1/2}\exp(-d^2\gamma/kT)}{d} \tag{4}$$

(ii) Step formation. If the crystal surface contains no dislocations, then a step can either start and finish at the surface boundary or form a closed loop. In the case of two-dimensional nucleation on a plane surface, the free-energy of formation of a nucleus of n units and circumference L is

$$\Delta G_s = L\gamma d - nkT\ln S \tag{5}$$

$$\frac{d\Delta G_s}{dn} = \gamma d\frac{dL}{dn} - kT\ln S \tag{6}$$

$$\frac{d\Delta G_s}{dn} = \gamma da_m\frac{L}{2A} - kT\ln S \tag{7}$$

where A is the surface area of the nucleus and $a_m = A/n$.

When $d\Delta G_s/dn = 0$, the critical free-energy of formation is obtained from

$$\Delta G_s^* = \frac{L^*\gamma d}{2} \tag{8}$$

References pp. 230–231

Writing the shape factor $\xi = L^2/4A$

$$\Delta G_s^* = \frac{\xi \gamma^2 d^2 a_m}{kT \ln S} \tag{9}$$

For a square nucleus, $\xi = 4.0$ and

$$\Delta G_s^* = \frac{4\gamma^2 d^2 a_m}{kT \ln S} \tag{10}$$

If $a_m = d^2$ then

$$\Delta G_s^* = \frac{4\gamma^2 d^4}{kT \ln S} \tag{11}$$

Thus, the probability of formation of a critical square nucleus on a particular crystal surface is

$$P_s = \exp[-4\gamma^2 d^4/(kT)^2 \ln S] \tag{12}$$

In the case of calcite, this amounts to a probability of 10^{-1060} at $S = 1.2$, i.e. at 20% supersaturation.

Crystal growth occurs at lower supersaturations than might be expected if surface nucleation is the sole source of steps on a crystal surface. For growth to occur continuously by this mechanism, new steps need to be formed on each layer because once the step advances over the surface, further nucleation is needed for growth to continue. Frank [2] overcame this difficulty by suggesting that screw dislocations (shown in Fig. 1) terminating in the surface provide the necessary steps for growth. The equilibrium distribution of kinks on such a step is adequate to permit growth to progress producing a growth spiral centred on the dislocation. In this case, the step does not disappear but grows continuously, even at very low supersaturations. In reality, very complicated spirals may be formed when several screw dislocations interact on the surface, e.g. a series of cooperating growth spirals resulting from a grain boundary. The growth spiral forms the basis of the Burton, Cabrera and Frank (BCF) theory of crystal growth which is discussed in Sect. 5.1.

Other surface structures are likely to affect growth. Edge dislocations produce regions in the surface in which the potential energy field is perturbed leading to variations in the interaction energy between the growth unit and the surface. This may aid the formation of two-dimensional nuclei and so generate steps. Similarly, the substitution of foreign ions, i.e. non-lattice ions, into the lattice causes defect structures and heterogeneities which may have important effects on the nucleation ability of the surface, particularly at high supersaturations. It is difficult to quantify the precise effects of such structure on the growth kinetics. Some progress has been made in quantifying the surface heterogeneity of crystalline powder surfaces using gases such as nitrogen, argon, and krypton as surface probes [3]. The adsorption

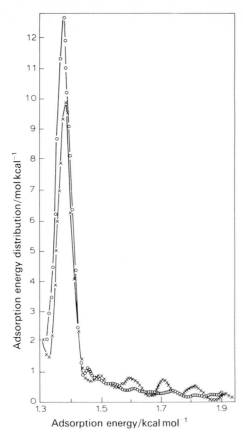

Fig. 2. Surface heterogeneity of a powder sample of sodium chloride computed from the adsorption of argon at 76.1 K using a two-dimensional gas model of adsorption. ×, Sample prepared by electrostatic precipitation of a NaCl aerosol; ○, same sample but after annealing at 310–315°C in a dry nitrogen atmosphere at 600 mm Hg. (Reproduced from ref. 4 by courtesy of Academic Press, Inc.)

of gases depends on the energy of interaction between the gas and the crystal surface. Different crystal faces produce differences in the interaction energy. The measurement of the adsorption isotherm at low temperatures permits the calculation of the distribution of adsorption energies which may reflect heterogeneities on particular crystal faces or the distribution of exposed crystal planes in a particular powder sample. Figure 2 illustrates some results obtained with sodium chloride crystals. These techniques may, in the future, be useful for comparing samples of non-porous crystalline material of the same substance which exhibit differences in their growth kinetics.

2.2 SOLUBILITY PRODUCT

The thermodynamic equilibrium between a solution and crystalline material is usually defined in terms of a solubility product. This is developed for the general case of a solid of composition $A_x B_\beta$ as follows. If one molecule of the solid dissolves according to the equation

$$A_x B_\beta \rightleftharpoons \alpha A^{a+} + \beta B^{b-} \tag{13}$$

to form a total of $\upsilon (= \alpha + \beta)$ ions in solution, then the chemical potential of the ions in solution may be expressed as

$$\mu_{A^{a+}} = \mu^\circ_{A^{a+}} + kT \ln a_{A^{a+}} \tag{14}$$

$$\mu_{B^{b-}} = \mu^\circ_{B^{b-}} + kT \ln a_{B^{b-}} \tag{15}$$

where μ°_i and a_i are the standard chemical potential and activity of the ith ion, respectively, defined on the basis that when $a_i = 1$ then $\mu_i = \mu^\circ_i$. Similarly, for a molecule in the solid phase

$$\mu_{A_x B_x} = \mu^\circ_{A_x B_x} + kT \ln a_{A_x B_x} \tag{16}$$

The chemical potential energy change associated with crystallisation is then

$$\Delta \mu = \alpha \mu_{A^{a+}} + \beta \mu_{B^{b-}} - \mu_{A_x B_\beta} \tag{17}$$

At equilibrium, $\Delta \mu = 0$ and

$$\Delta \mu^\circ = -kT \ln \left[\frac{a^\alpha_{A^{a+}} a^\beta_{B^{b-}}}{a_{A_x B_\beta}} \right] \tag{18}$$

so that

$$\Delta G^\circ = -RT \ln \frac{a^\alpha_{A^{a+}} a^\beta_{B^{b-}}}{a_{A_x B_\beta}} = -RT \ln [K_s] \tag{19}$$

defines the solubility product K_s.

Thus, the thermodynamic solubility product is defined in terms of the activity of the lattice ions in solution and in the solid phase. It is usual to express the activity of the solid phase in terms of the mole fraction of the component in the phase, X_i, and the rational activity coefficient, λ_i, e.g.

$$a_{A_x B_\beta} = \lambda_{A_x B_\beta} X_{A_x B_\beta(c)} \tag{20}$$

so that when $X_{A_x B_\beta(c)} = 1$, the activity coefficient becomes $\lambda_{A_x B_\beta} = 1$. In this case, the definition of K_s simplifies to

$$K_s = a^\alpha_{A^{a+}} a^\beta_{B^{b-}} \tag{21}$$

2.2.1 The effects of other dissolved substances

The solubility product depends on the activity of the lattice ions in solution. If the solution contains other ions or molecules either as impurities, additional chemicals or the result of reactions within the solution phase, then this must be taken into account in the evaluation of K_s. At least four separate effects on the solubility may be identified.

(i) The additional chemicals are not incorporated into the lattice or complex in solution with the lattice ions but only affect the solution phase activities. The effect is usually taken into account by calculating the activity coefficients from the ionic strength of the solution. The most common example of this is the addition of a background electrolyte such as KCl during crystallisation reactions. If the ion activities are determined by direct experimental measurements using specific ion electrodes, then the background electrolyte may not be important.

(ii) The additional material forms complexes in solution with the lattice ions and so alters the free-ion activities needed in the evaluation of the solubility product. When the dissociation constants for such complexes are known, then the free-ion activities may be calculated.

(iii) The solution composition changes because of chemical reactions. If one of the lattice ions is involved in reactions in solution, then any additional material which affects the equilibrium, either directly or through changes in the ionic strength, must be taken into account. The simplest example of this effect is in precipitation reactions in water which involve either H^+, HCO_3^- or CO_3^{2-} ions. In such instances, the reaction of atmospheric CO_2 with the solution perturbs the bicarbonate system equilibria.

(iv) The additional material can incorporate into the lattice of the crystal. In this case, the solubility product may be affected because the solid phase is no longer pure and $a_{A_\alpha B_\beta} \neq 1$. In such situations, it is important to assess whether a new crystalline phase separates or merely that the lattice is only changed slightly to accommodate the additional ions. Solid solutions are only likely to form if the size of the minor component is similar to that of the lattice ion it is replacing. In such cases, the lattice structure is only slightly distorted and the activity of the crystalline material may be estimated using regular solution theory, e.g. for a solid solution $A_\alpha I_\beta$ in $A_\alpha B_\beta$, the activity coefficient is

$$\lambda_{A_\alpha B_\beta} = \exp[X^2_{A_\alpha I_\beta} W/RT] \tag{22}$$

where $X_{A_\alpha I_\beta}$ is the mole fraction of I in the crystal and $2W$ is the interchange energy defined as the total increase in energy caused by exchanging $A_\alpha I_\beta$ and $A_\alpha B_\beta$ in the crystal lattice starting with two pure crystals of $A_\alpha I_\beta$ and $A_\alpha B_\beta$. The interchange energy is assumed to be constant and independent of the mole fraction of $A_\alpha I_\beta$ in the lattice.

2.2.2 The effects of particle size

The solubility product is defined for a semi-infinite plane surface where the interfacial energy between the crystal and the solution makes a negligible contribution to the free-energy of formation of the particle. The definition also necessitates that the solid phase is homogeneous in structure and that a chemical potential may be assigned to the components irrespective of their position within the solid. When the crystals are small this may not be true because the imbalance of interionic forces at the surface produce structural pertubations which may extend through the particle. This means that, for some small particles, both the interfacial free-energy and the solid density depend on the particle size.

A rigorous solution to this problem has not been achieved. The classical approach is to neglect variations in the interfacial energy and particle density with size and write the free-energy of formation of a spherical particle of radius r, in the form

$$\Delta G_{p} = 4\pi r^{2}\gamma - \frac{4\pi r^{3}\Delta\mu}{3V_{m}} \tag{23}$$

where γ is the interfacial energy between the crystal and solution, $\Delta\mu$ is the chemical potential change accompanying the incorporation of a growth unit into the crystal, and V_{m} is the molecular volume. Hence, a critical particle radius $r *$ can be calculated such that $(d\Delta G_{p}/dr)_{T,P} = 0$

$$r^{*} = \frac{2\gamma V_{m}}{\Delta\mu} = \frac{2\gamma M}{RT\rho\ln\Omega} \tag{24}$$

where M is the molecular mass, ρ the solid density, and Ω the ratio defined as the degree of saturation in Sect. 2.3. According to this equation, a $1\,\mu m$ radius particle of gypsum ($CaSO_{4} \cdot 2H_{2}O$) has a solubility $\simeq 2.2\%$ higher (based on Ω) than expected. For smaller particles, the increase is larger, e.g. for a $0.1\,\mu m$ radius particle, a solubility increase of $\simeq 24.8\%$ is predicted. Solids with a lower interfacial energy show less of an effect, e.g. for calcite with $\gamma \simeq 97\,mJ\,m^{-2}$, it is predicted that the solubility increase is $\simeq 2.9\%$ for $0.1\,\mu m$ radius particles.

2.3 DRIVING FORCE FOR CRYSTALLISATION

The driving force is the chemical potential energy change associated with crystallisation [see eqn. (13)].

$$\Delta\mu = RT\ln\left[\frac{a_{Aa+}^{\alpha}\,a_{Bb-}^{\beta}}{K_{s}}\right] = RT\beta \tag{25}$$

where β is defined as the growth affinity

$$\beta = \sum_{i}\frac{\Delta\mu_{i}}{RT} \tag{26}$$

The growth affinity is the difference between the chemical potential of one mole of growth units of the crystallising substance in its supersaturated solution and in its crystalline form divided by RT. The growth unit is defined according to the stoichiometry of the crystal.

Other definitions of the saturation state of a solution are also in use. If we consider the average chemical potential energy change per ion ($\Delta\mu_{ion}$) then

$$\frac{\mu_{ion}}{RT} = \frac{\beta}{v} = \beta_{ion}$$

where $v = \alpha + \beta$. This leads to the definition of the saturation ratio, S

$$\ln S = \beta_{ion} = \ln\left(\frac{a_{Aa}^a \, a_{Bb}^\beta}{K_s}\right)^{1/v} \tag{27}$$

and relative supersaturation, σ

$$\sigma = S - 1 \tag{28}$$

The definition of the saturation ratio originates from the description of the driving force for precipitation from stoichiometric solutions in which $S = a/a_e \simeq c/c_e$ where a and a_e are the mean ion activities in a supersaturated and equilibrium solution, respectively, and c and c_e are the corresponding concentrations, e.g. the saturation ratio of a silver chloride solution is

$$S = \left(\frac{a_{Ag} \, a_{Cl^-}}{K_s}\right)^{1/2} \simeq \frac{c_{AgCl}}{c_{e(AgCl)}}$$

and

$$\beta = \ln\left(\frac{a_{Ag} \, a_{Cl}}{K_s}\right) = 2 \ln S$$

Other methods of expressing the degree of supersaturation are in use. Of these, the degree of supersaturation, Ω, and saturation index, SI, are most encountered.

$$\Omega = \frac{a_{Aa}^z \, a_{Bb^-}^\beta}{K_s} \tag{29}$$

$$SI = \log \Omega \tag{30}$$

2.4 RATE OF CRYSTALLISATION

The most fundamental and unambiguous way of determining the rate of growth of a crystal is to measure the rate of linear growth perpendicular to a crystal face and relative to a fixed point within the crystal. This method is possible in growth experiments with good quality single crystals. However, it does become impractical when measuring the growth of a batch of crystals in suspension. In this situation, it is convenient to measure the changes in the solution composition and relate this to the growth of the crystals, i.e.

$$\frac{dv}{dt} = \frac{VM}{\rho}\frac{dc}{dt} \tag{31}$$

where v is the volume of the crystals, V the volume of the crystal suspension,

M the mass of one mole of the crystal, ρ the density of the crystal, and *c* the concentration of the solute. Equation (31) assumes that the change in concentration of the solution can be solely attributed to the growth of a single characterised solid. The mean linear growth rate of the suspension, $\bar{R}$, may then be defined as

$$\bar{R} = \frac{dv/dt}{A(t)} \tag{32}$$

where $A(t)$ is the total surface area of the growing crystals at time, *t*. If it is assumed that each individual crystal remains geometrically similar to its form at the start of the process (the assumption of persistent similarity), then

$$A(t) = A(0)\left(\frac{v(t)}{v(0)}\right)^{2/3} \tag{33}$$

If the extent of crystallisation is very small, then eqn. (32) becomes:

$$\bar{R} \simeq \frac{dv/dt}{A(0)} \tag{34}$$

A more general treatment of growth in suspensions accounts for the differences in the growth rate on different crystal faces. If each crystal face of type $\{hkl\}$ has a linear growth rate R_j which is only dependent on the growth affinity, then

$$\frac{dv}{dt} = \sum_j R_j A_j(t) \tag{35}$$

where A_j is the total surface area of crystal faces of type *j*. Van Oosterhout and van Rosmalen [5] have made further progress with the assumptions

(i) that $A_j(t)$ is a function of $v(t)/v(0)$ only and does not depend on the concentration path during the reaction, i.e. $A_j(t) = f_j[v(t)/v(0)]$ and

(ii) that all the linear growth rates are linearly related, i.e. have the same β dependence, $R_j = p_j R(\beta)$ where p_j is a constant.

Hence

$$\frac{dv/dt}{v(0)} = R(\beta) \sum_j p_j f_j[v(t)/v(0)]$$

or, defining $g[v(t)/v(0)] = \Sigma_j p_j f_j[v(t)/v(0)]$, we have

$$\frac{dv/dt}{v(0)} = R(\beta)g[v(t)/v(0)] \tag{36}$$

The function $R(\beta)$ is related to the mean linear growth rate by

$$\bar{R} = R(\beta) \sum_j p_j A_j(t)/A(t) \tag{37}$$

In Monte Carlo calculations of crystal growth, the growth rate is often expressed in a dimensionless form as the stick fraction, *s*. This is related to the linear growth rate of a particular face by the equation

$$s_j = \frac{R_j}{k^+d} \tag{38}$$

in which k^+ is the total flux of growth units impinging on the surface (the deposition rate) and d is the lattice spacing in a direction normal to the growth.

3. Nucleation

Homogeneous nucleation is the spontaneous nucleation of crystals in a solution. The nucleation occurs because the driving force for crystallisation is greater than the increase in interfacial energy caused by the formation of a three-dimensional nucleus in solution. Homogeneous nucleation does not require an existing surface to occur but results from the cluster of growth units (sometimes called embryos) growing to sufficient size that a critical nucleus is formed which is stable and able to grow. In contrast, heterogeneous nucleation requires a surface on which nucleation may occur. The term, "heterogeneous", is derived from the fact that the surface is of a different material from the crystallising phase and is usually uncharac-

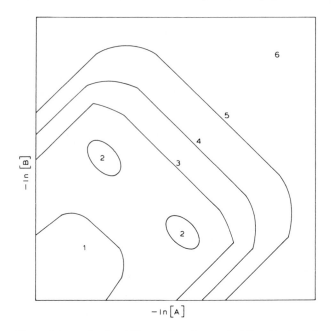

Fig. 3. Schematic of stability regions for a salt AB. 1, Region of precursor formation; 2, colloid stability regions; 3, boundary line for homogeneous nucleation; 4, boundary line for heterogeneous nucleation; 5, boundary line of the metastable region (solubility line); 6, region of undersatuartion.

References pp. 230–231

terised, e.g. dust particles within the solution or the material of the containing-vessel walls.

The division between homogeneous and heterogeneous nucleation is not always clear. However, various solution compositions may be identified according to the magnitude of the ion activity product. For a simple 1:1 electrolyte, the possible stability regions are illustrated in Fig.3. The regions are defined in terms of boundaries expressed in terms of pa_A and pa_B ($pa_i = -\ln a_i$) or pC_A and pC_B. At sufficiently high supersaturations, homogeneous nucleation occurs. In some instances, a precursor phase may form which subsequently recrystallises to a more stable crystalline form of the substance. Also, according to the ionic composition, the nucleated particle may be stabilized by the preferential adsorption of the anion or cation to produce a colliod. At lower supersaturations, heterogeneous nucleation occurs accompanied by the slow loss of lattice ions from solution. In some systems, there is a range of solution compositions which are supersaturated but remain stable in composition over long periods.

3.1 HOMOGENEOUS NUCLEATION

The process of formation of a critical cluster is largely unknown but it is likely that the structure and interfacial properties are different from the material in macroscopic crystals. It is possible that the clusters are formed by a series of reactions of the type

$$U + U \rightleftharpoons U_2$$

$$U_2 + U \rightleftharpoons U_3$$

.
.
.

$$U_{n^*-1} + U \rightleftharpoons U_{n^*}$$

where U is the growth unit and n^* the number of growth units in the critical cluster. The free energy of formation of a critical nucleus, ΔG_n^*, may be calculated using the method described for the formation of a two-dimensional nucleus and discussed in Sect. 2.1. In brief, the free-energy of formation of a cluster of size n with a surface area A is

$$\Delta G_n = \gamma A - nkT \ln S$$

or   (39)

$$\Delta G_n = \gamma A - nkT\beta$$

depending on the choice of the growth unit. Equation (39) leads to a critical free-energy of formation

$$\Delta G_n^* = \frac{\xi \gamma^3 V_m^2}{(kT \ln S)^2}$$ (40)

TABLE 1

Shape factors for different geometry crystals

Shape	Shape factor, ξ
Sphere	16.755
Regular dodecahedron	22.201
Regular octahedron	27.713
Cube	32.000
Parallelepiped (4:2:1)	50.815
Regular tetrahedron	55.426
Rod (10:1:1)	109.760
Plate (10:10:1)	204.800

where the shape factor $\xi = 4A^3/27v^2$, V_m is the molecular volume of the growth unit and v the total volume of the crystal. The critical nucleus size is

$$n^* = \frac{2\xi\gamma^3 V_m^2}{(kT \ln S)^3} \tag{41}$$

For a sphere, the shape factor is $16\pi/3$ and so

$$\Delta G_n^* = \frac{16\pi\gamma^3 V_m^2}{3(kT \ln S)^2} \tag{42}$$

and

$$n^* = \frac{32\pi\gamma^3 V_m^2}{3(kT \ln S)^3} \tag{43}$$

The shape factors for a number of other geometries are tabulated in Table 1.

The rate of homogeneous nucleation has been derived in a number of classical papers [6, 7]. The final result may be expressed in the form

$$J = B \exp(-\Delta G_n^*/kT) \tag{44}$$

where J is the number of nuclei formed per unit volume and time and B is a constant at constant temperature. Nielsen [8] has approximated the pre-exponential factor

$$B = \frac{D}{d^5}\left(\frac{2 \ln S}{3\pi n^*}\right)^{1/2} \tag{45}$$

where D is the diffusion coefficient and d is the mean molecular diameter of the growth unit. This leads to

$$J = B \exp\left[-\frac{16\pi\gamma^3 V_m^2}{3(kT)^3(\ln S)^2}\right] \tag{46}$$

References pp. 230–231

Equation (46) shows that the nucleation rate is an exponential function of the supersaturation. Hence it is expected that J will be negligible until a certain critical supersaturation is achieved after which homogeneous nucleation will be extremely fast.

With $B(\simeq D/V_m^{5/3})$ expected to be of the order of 10^{30}, Nielsen [8] has shown that $\ln J$ versus $(\ln S)^{-2}$ plots produce straight lines for a range of values of γ and V_m. In this region, the approximation

$$J \simeq k_n S^n \tag{47}$$

is valid. In eqn. (47), n is often called the order of nucleation and may be derived from

$$\frac{\mathrm{d}\ln J}{\mathrm{d}\ln S} = \frac{32\pi\gamma^3 V_m^2}{3(kT\ln S)^3} = n^* \tag{48}$$

Hence, in this approximation, the critical nucleus size may be estimated from the slope of plots of the logarithm of the nucleation rate as a function of the logarithm of the supersaturation.

Nucleation data have also been used to estimate γ, the interfacial energy at the nucleus–solution surface. This may be done using eqn. (46) with $J = 1\,\mathrm{cm}^{-3}\mathrm{s}^{-1}$, i.e.

$$\gamma = \left[\frac{3\ln B\,(kT)^3(\ln S^*)^2}{16\pi V_m^2}\right]^{1/3} \tag{49}$$

where S^* is the critical supersaturation corresponding to critical nucleus formation. B may be estimated from $D/V_m^{5/3}$ or, with a knowledge of n^*, from eqn. (45). Alternatively, γ may be calculated from the slope of a plot of $\ln J$ against $(\ln S)^{-2}$ [see eqn. (46)].

The empirical equation

$$J = k_n \Delta c_{max}^n \tag{50}$$

has also been used to relate the nucleation rate to the metastable zone width Δc_{max}^n. This is the concentration difference, measured with respect to the equilibrium concentration which is realized before nucleation occurs. The use of concentration in this respect is unfortunate and parameters based on activities are preferred. However, Δc_{max}^n has been used to characterise nucleation processes. The metastable zone width can be determined from undercooling data obtained by slowly cooling a saturated solution until nucleation occurs or from measurements involving solvent evaporation. Ananikyan et al. [9] have used the latter method to correlate Δc_{max} with various solution properties of potassium iodate, potassium pentaborate, and sodium pentaborate such as osmotic pressure, density, and surface tension. It was discovered that the most transparent and best facetted crystals were grown from solutions having pH values corresponding to the maximum osmotic pressure with solutions with the lowest osmotic pressure producing the worst crystals.

Another parameter often used to characterise nucleation is the induction time or period, τ. This is defined as the time taken for the formation of crystals after creating a supersaturated solution. Hence, the measured induction period does depend upon the sensitivity of the recording technique. It is generally assumed that τ is inversely proportional to the nucleation rate, i.e.

$$J = \frac{B'}{\tau}$$

or

$$\tau = \frac{B'}{B} \exp\left[\frac{16\pi\gamma^3 V_m^2}{3(kT)^3 (\ln S)^2}\right] \tag{51}$$

so that, for homogeneous nucleation, a plot of log τ against $T^{-3}(\ln S)^{-2}$ should be linear. Mullin and Ang [10] have used this method to find that, for nickel ammonium sulphate at $S > 1.8$, eqn. (51) is obeyed. As shown in Fig.4, deviations from the classical behaviour described by eqn. (51) occur at low supersaturations where heterogeneous nucleation becomes the dominant mechanism.

3.2 HETEROGENEOUS NUCLEATION

As shown in Fig.3, a range of solution composition may exist in which the

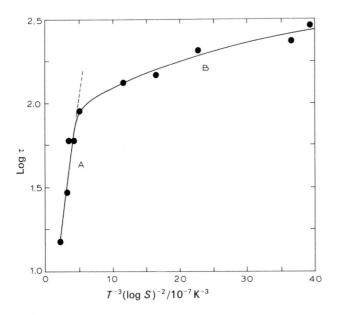

Fig. 4. Results from the nucleation kinetic studies on nickel ammonium sulphate. A, Homogeneous nucleation; B, heterogeneous nucleation; ●, experimental data. (Reproduced from ref. 10 by courtesy of The Royal Society of Chemistry.)

References pp. 230–231

solution supersaturation is insufficient to enable homogeneous nucleation but enough for heterogeneous nucleation on an existing surface.

If we consider two-dimensional nucleation on a substrate x, then eqn. (5) is modified because of the difference in interfacial energy between the substrate and solution, γ_{xs}, and the interfacial energies at the substrate–crystal and crystal–solution interfaces, i.e. γ_{xc} and γ_{cs}, respectively. Hence, eqn. (5) contains an additional term: $A(\gamma_{xs} - \gamma_{xc} - \gamma_{cs})$. The ease of nucleation relative to two-dimensional nucleation on a pure crystal depends on the sign of the additional term. Heterogeneous nucleation produces far fewer particles than homogeneous nucleation and leads, in general, to larger induction periods, e.g. see Fig. 4. The sharp increase in the particle density with increasing supersaturation is characteristic of the onset of homogeneous nucleation.

Because of the difficulties associated with the characterisation of heteronuclei in solution, few studies have attempted to explain experimental results in a quantitative way. If it is assumed that, once nucleation occurs, the particles grow without recrystallisation, then it is possible to get information about the particle density from a consideration of the geometry of the particles and the growth kinetics. One approach is to add heteronuclei to supersaturated solutions and measure the crystallisation kinetics and, from the data obtained, estimate the surface area of the growing crystals. In this way, it is feasible to obtain information about the nucleation capability of different heteronuclei and the effects of pretreatments on the nucleation capability. An example of such an application will be discussed in Sect. 5.4.

4. Experimental methods

A variety of techniques have been used to study crystal growth kinetics in aqueous systems. The most popular are briefly described below.

4.1 DIRECT MIXING

This method usually involves mixing two or more solutions so that the resulting mixture is supersaturated with respect to the crystalline phase and precipitation commences. It is most widely used to prepare crystalline material and to study the kinetics of homogeneous nucleation. In most circumstances, the maximum supersaturation attained will depend on the method of mixing. The possibility of the formation of precursor phases and their subsequent recrystallisation must also be considered. For example, the crystalline products obtained by mixing $Ca(NO_3)_2$ and Na_2CO_3 depends on the temperature, concentration of the reagents and on the digestion time. Major disadvantages of the method for studying growth kinetics after nucleation arise because of the poor characterisation of both the crystal form and morphology.

The method has been used to study the kinetics of homogeneous nuclea-

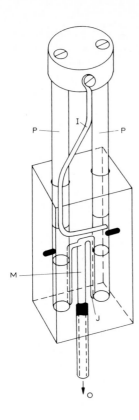

Fig. 5. Direct mixing apparatus after the design of Nielsen [12]. I, Position indicator; J, inflow jets; M, mixing chamber; O, outflow tube; P, pistons.

tion at constant temperature. Typically, a mixing cell of the type illustrated in Fig.5 may be used for fast reactions. The induction period may be determined from turbidity measurements in the mixing chamber. The particles formed can be counted by a number of methods (see, for example, the review by Lieberman [13]) and observed by light microscopy. In this way, the effects of changes in the initial supersaturation on the crystalline product and particle density can be examined in some detail.

4.2 SEEDED GROWTH METHODS

The disadvantages associated with the direct mixing method for measuring growth kinetics can be overcome by ensuring that the initial solution composition is within the metastable region illustrated in Fig. 3 and then inducing growth by the addition of seed material. This should be as well characterised as possible. In the extreme case, the seed could be a single crystal exposing a single face (as discussed in Sect. 4.2.4). However, the majority of research has employed polydisperse powder samples of defined specific surface area, crystal form, and morphology.

References pp. 230–231

When powder is used as the seed, it is inevitable that the analysis of the results depends on the heterogeneity of the seed material and, in particular, the distribution of different crystal faces on the powder sample. Changes in the morphology or in the relative distribution of different crystal faces will also complicate the interpretation of the results (see Sect.2.4). In spite of this, the method is popular because of the simplicity of the experimental technique and the relevance of the experimental conditions to the solution of practical problems, e.g. in batch reactor design. In studies involving inhibiting agents, it is often desirable to investigate the interaction between the crystal and inhibitor without crystal growth, i.e. measurement of adsorption isotherms. This only becomes practical, by conventional methods, if the surface area of the adsorbent is large enough to produce sufficient adsorption to permit measurable changes in the solution composition. Hence, powders are ideally suited to research in which the seed dispersion may be used for both growth and adsorption studies.

Ideally, the following seed characteristics should be known.

(a) Crystal form and morphology by X-ray diffraction, light microscopy, scanning electron microscopy, or high-resolution transmission electron microscopy.

(b) Particle size distribution by various methods [13].

(c) Specific surface area, Σ, measured by gas adsorption (krypton, nitrogen, or argon at low temperature) or radiotracer exchange techniques.

(d) Impurity abundance by wet analysis or surface-sensitive techniques such as electron microprobe analysis and secondary ion mass spectrometry. This characterisation will be of particular importance when comparing growth data obtained using different preparations of the same seed substance. In such circumstances, the adsorption heterogeneity distributions discussed in Sect.2.1 may be of value in understanding differences in growth behaviour. After growth, the crystals may be compared with the original seed material to evaluate changes in the crystal form, morphology, size distribution, or impurity abundance. In this respect, it is important that the solutions are not agitated in such a way that the crystalline material is fractured, e.g. the rotation of a magnetic stirrer bar may be sufficient to grind some brittle crystals. Some care is also necessary to establish that sufficient growth has occurred to enable the identification of the precipitated material. This is only normally necessary when there is a possiblity of different crystal phases nucleating on the seed or when the presence of an inhibitor influences the formation of a new phase.

The various techniques used to study growth kinetics by the seeded growth method are now discussed. These include methods which have been used to study not only powder samples but also single crystals.

4.2.1 Free-drift method

The solution composition is monitored potentiometrically using ion-selective electrodes or conductiometrically using a platinum electrode. No attempt is made to control the solution composition.

The advantages of the method may be summarised by

(i) the growth kinetics may be studied over a range of supersaturations in one experiment and

(ii) the approach of the system to thermodynamic equilibrium may be studied in detail.

The measurement is suited to automation using a laboratory microcomputer connected via an analog-to-digital converter to an amplifier circuit. In complex solutions, a single reference electrode may be combined with various ion-selective electrodes and the computer used to switch the signal from the appropriate electrode pair to the computer interface. In situations where conductance and ion-selective electrodes are used together with a common power circuitry, it is necessary to isolate the electrodes from each other when taking measurements. Also in such situations, it may be necessary to correct all conductivity readings because of the leakage of a reference solution from the liquid junction of the cell. The automation enables the solution composition and growth kinetics to be calculated during the experiment. The response time of the electrode system must be faster than the reaction rate. The major disadvantage of the technique using ion-selective electrodes stems from the problems associated with calibrating the electrodes for applications involving dilute or varying composition solutions. In dilute solutions, the electrode pair may produce substantial fluctuations caused by turbulence and flow around the liquid junction and ion-sensitive surfaces.

In some situations, it may not be possible to use methods based on chemical sensors but necessary to extract a volume of solution for analysis. It is important to ensure the complete separation of the solution from the crystalline phase. Typically, this situation arises when investigating the coprecipitation and inhibition effects of microcomponents, e.g. trace metals and low molecular weight organics and also in monitoring solutions for which suitable electrodes have not been developed.

4.2.2 Constant composition method

After the crystal seed addition, the composition of the reaction solution is maintained constant by the addition of the appropriate lattice ions to the solution. Again, this method is conveniently automated using ion-selective electrodes to control the solution composition. This is done by using the signal from the sensor to control motor-driven burettes so that the solution composition is maintained as near constant as possible. The advantages of this method are:

(i) The growth kinetics can be studied at a fixed supersaturation over long periods. This is particularly useful when studying the effects of inhibitors on the growth kinetics.

(ii) The chemical sensor is only having to respond to small perturbations from a single solution composition in which it is in equilibrium.

In most systems, it is necessary to couple two or more burettes mechanic-

References pp. 230–231

186

ally. This is necessary to ensure that the stoichiometry of the titrant solutions is correct. For instance, in the recent study of the growth of strontium fluoride in potassium nitrate solution [15], the fluoride-specific ion electrode was used to control the addition of two titrant solutions (1) strontium nitrate + potassium nitrate and (2) potassium fluoride. Because the signal from the fluoride electrode was effectively constant throughout the precipitation experiment, the fluoride ion activity in solution was constant. It is also possible to perform reactions in specific conditions, such as constant pH, using potentiostat assemblies. In this type of experiment, only a single ion activity is maintained constant.

Some caution is necessary in evaluating the method for specific applications. It is not necessary to calibrate the electrode because, once a steady signal is obtained before seed addition, an extract from the solution can be analysed. However, it is necessary to check for changes in the electrode performance during the growth experiment. This is possible by sampling the solution at intervals and analysing. In some circumstances, it may be necessary to use two similar electrodes and compare the signals continuously to assess whether a problem occurs.

The use of conductance as a measure of the solution composition is limited unless some other control of the composition is possible. Figure 6 illustrates a schematic of the apparatus used to follow the crystal growth of calcite from calcium bicarbonate solutions under constant composition conditions. In this case, the carbon dioxide concentration is kept constant by the flow of a water-saturated mixture of carbon dioxide and nitrogen gas. The conductance is monitored using an a.c. bridge with the analog signal

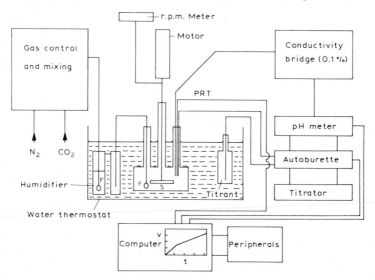

Fig. 6. Schematic of the apparatus for the measurement of growth kinetics in constant composition solution conditions. Both the CO_2 concentration and electrical conductivity are kept constant by the addition of a gas mixture and titrant solution.

connected to a conventional pH-stat assembly. A metastable solution of calcium bicarbonate is used as titrant. In this type of experiment, it is essential to have good control over the temperature and to ensure that the gas transfer is sufficient to maintain a constant concentration of carbon dioxide in solution.

4.2.3 Fluidised bed method

This is included as an example of the techniques used in pilot scale studies of industrial crystallisers. A schematic of a fluidised bed system is shown in Fig.7. The apparatus could be operated in free-drift or constant composition conditions as described above.

The crystals are maintained in suspension by adjusting the flow in the fluidation tube, B, (Fig.7) using the by-pass valve, C. The temperature in the system is controlled via a cooling coil, E, and the flow rate measured using the meter, D. The required supersaturation may be attained by adjusting the temperature of the solution before the seed suspension is added. The major advantage of the design is that the crystal growth may be studied under reproducible hydrodynamic conditions similar to those in real fluidised-bed crystallisers. If the technique is used for the purpose of improving the design of industrial reactors, it is necessary that the effects of supersaturation,

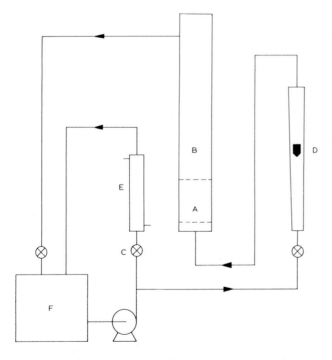

Fig. 7. Diagram of a fluidized bed crystalliser. A, Calming section to reduce large eddies; B, reaction zone; C, by-pass valve; D, flow meter; E, cooling coil; F, solution reservoir.

temperature and solution velocity on the crystal growth be known in detail. These factors are probably best studied in simpler laboratory systems described in Sects. 4.2.1, 4.2.2 and 4.2.4.

4.2.4 Single crystal methods

Ideally, the study of growth kinetics requires both the solution and crystals to be completely characterised. In practice, the most difficult problems are often associated with our lack of understanding of the surface structure of crystals. The kinetics of growth in suspensions represent an average relationship between growth and supersaturation for exposed crystal faces (Sect. 2.4). The use of single crystals greatly reduces the uncertainty because the geometry of the crystal and exposed faces can be controlled. This permits the linear growth rate to be measured very precisely on specific crystal faces. The crystal should normally have a well-developed smooth face of sufficient size to enable study and a minimum number of dislocations.

The major disadvantage with using single crystals is that individual crystals are not easily characterised at the microscopic level. Experiments with individual crystals of the same size and exposed lattice may produce different kinetics depending on the surface structure, e.g. defect density, on the individual crystals. Thus, it may become necessary to pretreat the exposed faces in some way, e.g. annealing in an inert gas atmosphere or polishing, so as to produce a more uniform surface. Alternatively, measurements on a number of individual crystals may permit a better understanding of the role of surface morphology on the kinetics. This approach should provide a convenient link with studies on suspensions. The methods which have been used to study the growth of single crystals may be divided according to whether the reaction is surface-controlled or controlled by the transport of ions from the solution to the crystal surface.

(a) Surface controlled reactions

The methods discussed above in Sects. 4.2.1 and 4.2.2 are equally applicable to studies of the growth of single crystals. The limited surface area available for the reaction leads to correspondingly small changes in the solution composition. Therefore, it becomes more difficult to measure reaction rates from concentration changes and the majority of studies have used other techniques. These range from weighing a suspended crystal [16] to various optical methods designed to measure the linear growth rate of a particular crystal face [17, 18]. The simplest method of doing this is by mounting the crystal in a thermostated flow-cell and controlling the solution composition, i.e. degree of supersaturation, in a separate vessel.

One way of controlling the supersaturation is based on the two-vessel Walker–Kohman crystalliser [19]. A more recent design of this type of reactor is illustrated in Fig.8. The principle is to dissolve a polycrystalline sample in vessel V2 at temperature T_2 and permit growth on a single crystal in another vessel, V1, at temperature, T_1. The temperature difference between the two vessels is achieved using heater H2. Optical distortions are minimised by construction of the vessels using suitable organic glasses.

Various modifications of this design have been suggested, including more complex arrangements involving three vessels (see, for example, ref. 20).

Briefly, the main features necessary for this type of study are

(1) control over the solution composition near the crystal surface and good control of the temperature in the crystalliser;

(2) good mixing of the solution in the vicinity of the crystal surface to ensure temperature homogeneity; and

(3) suitable mounting of the crystal to permit measurement of the growth rate by optical measurements or other means.

(b) Transport controlled reactions

If the growth kinetics are controlled by the transport of ions from the bulk solution to the surface of the crystal, it is essential to perform experiments in well-defined hydrodynamic conditions. This is possible using the rotating

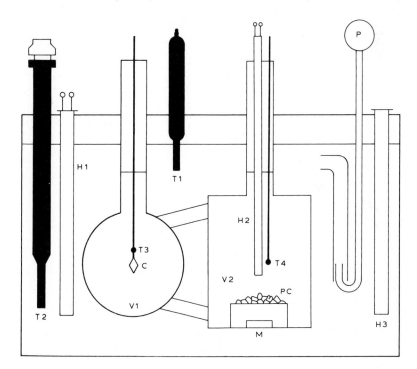

Fig. 8. Diagram of a two-vessel crystalliser. C, Single crystal; H1, H2, and H3, heating elements; M, magnetic stirrer; P, pump; PC, polycrystalline material; T1 and T2, thermometers, typically Hg in glass or platinum resistance thermometers (PRT), T1 is used to control heaters; T3 and T4, miniature thermometers, i.e. thermister or PRT; V1, crystallisation vessel; V2, dissolution vessel.

disk electrode as illustrated in Fig.9. The crystal is usually mounted centrally in an epoxy resin disk, polished using diamond grit, and washed. It is also necessary to mask the crystal so that a disk portion is exposed to the solution.

The rate of transport of ions by diffusion and convection to the surface of the rotating disk has been described by Levich [21]. The flux of ions to the surface is determined by the rotation speed of the electrode. Thus, at constant supersaturation, the linear growth rate of the exposed face depends upon the rotation speed. With this technique, it is not easy to measure the growth rate of the crystal directly. The rate is usually determined using the free-drift or constant composition methods as described above. However, other methods are being developed which allow the crystal to remain stationary and the solution is forced to flow over the surface. As an example, the channel electrode first described by Meyer et al. [22] offers the possibility of studying the growth kinetics in defined hydrodynamic conditions using the methods described above. Thus, a single apparatus with the crystal mounted in a channel would permit an extensive investigation of both transport- and surface-controlled reactions.

5. Models of crystal growth

Research on crystal growth in aqueous systems has been predominantly concerned with improving our understanding of the processes which deter-

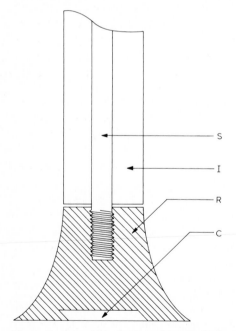

Fig. 9. Rotating disk. S, Main drive shaft connecting to motor and controller; I, inert sleeve, usually constructed from PTFE; R, resin disc; C, single crystal mounted in resin with disk exposed.

mine the growth kinetics. In particular, how the growth rate depends on the crystal morphology (or specific surface area), growth affinity or supersaturation of the solution and temperature. Unfortunately, the complex processes which occur during crystallisation make it impossible to suggest a universal model for growth and so results are often described in terms of empirical or semi-empirical rate equations. In some instances, it has been possible to develop models, based on molecular mechanisms of crystallisation, which explain the rate equations. In fact, as discussed below, the same rate equation may be derived on the basis of different mechanisms. In these situations, it may be impossible to distinguish between the models using growth data obtained at a single temperature.

The basis of our present understanding of growth kinetics stems from the identification of rate-determining steps. The description of the growth kinetics over a range of growth affinities may involve a combination of these steps.

(1) The transport of the growth unit(s) from the bulk solution through the hydrodynamic boundary layer to a region adjacent to the adsorption layer of the crystal. This is often referred to as bulk transport-controlled, volume diffusion-controlled, or simply transport-controlled.

(2) The adsorption of the growth unit(s) into the surface adsorption layer. This step is likely to involve the partial dehydration of the growth unit(s). The adsorption layer is a region immediately adjacent to the surface of thickness < 1000 nm. It is that part of the electrical double layer of ionic crystals, termed the Stern layer, in which specific adsorption occurs.

(3) The diffusion of the growth unit(s) across the crystal surface to a step.

(4) The incorporation of the growth unit(s) into a kink on a step. This process may be accompanied by further dehydration either at the time of incorporation or later. The step may be on a two-dimensional surface nucleus or a dislocation growth spiral.

The slowest step in this series of processes will become rate-limiting. However, if a parallel process, such as the direct incorporation of growth units from solution into kinks, is the fastest, then this step will become rate-limiting.

A schematic of the surface processes is shown in Fig. 10. The states may be defined as follows.

1. Hydrated growth unit in the aqueous phase indicated as $A(H_2O)_k$.

1a. Activated partially dehydrated growth unit for entry into the adsorption layer. Shown in Fig. 10 as $A(H_2O)_l$.

2. Adsorbed growth unit. The potential energy field of the crystal produces barriers to surface migration.

2a. An activated dehydrated growth unit for entering a kink position on a step, $A(H_2O)_m$.

3. The partially hydrated growth unit incorporated at a kink. It seems likely that the growth unit will completely dehydrate at this stage.

This leads to the hydration numbers $k \gtrsim l \gtrsim m$. The relaxation times, τ,

192

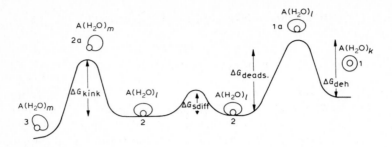

Fig. 10. Schematized states of hydrated ions and potential barriers that must be overcome in order that an ion can enter into a kink. (Reproduced from ref. 23 by courtesy of North-Holland Publishing Co.)

are given by the Eyring theory

$$\tau = \left(\frac{h}{kT}\right) \exp\left(\frac{\Delta G_i}{kT}\right) \tag{52}$$

where the four activation energies, ΔG_i, are shown in Fig. 10.

Perhaps the most rigorous attempts to model crystal growth developed

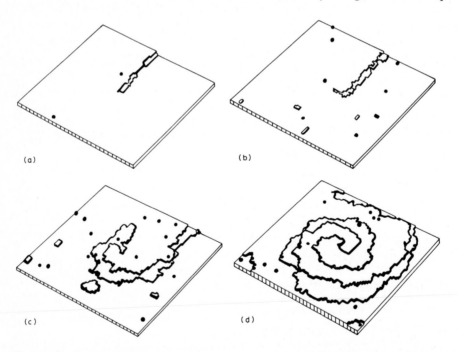

Fig. 11. The formation of a spiral step pattern on a simple cubic {100} face. A single screw dislocation intersects the surface segment at the centre and at this point two steps originate. In equilibrium, the two steps are essentially straight but, when the driving force is applied, they advance and can eventually provide a dense array of edge sites over the crystal face. (Reproduced from ref. 25, © 1980 by the American Association for the Advancement of Science.)

from the Burton et al. (BCF) theory [24]. This will now be considered in some detail.

5.1 GROWTH THEORY OF BURTON, CABRERA AND FRANK

As discussed in Sect. 2.1, Frank [2] suggested that a screw dislocation emerging at a crystal surface provided the necessary steps for growth, even at low supersaturation when two-dimensional nucleation is improbable. When the solution in contact with a crystal becomes supersaturated, steps are generated which wind into spirals about the centre of the dislocation. A number of such spirals have been found [26] and, indeed, the process of formation of a spiral step pattern has been simulated by computer [25]. The results, shown in Fig. 11, illustrate how two associated steps wind up into a double spiral that covers the entire crystal face. Two-dimensional nuclei are also observed, some of which are incorporated into the advancing spiral.

The shape of the growth spirals may be quite complex and only approximated by analytical functions such as the Archimedean spiral employed by Burton et al. [24] and shown in Fig. 12. The relationship between the velocity of a straight step and curved step may be derived using the approach of Nielsen [27]. This enables the steady-state shape of a spiral in contact with a fixed supersaturation solution to be calculated.

Following the method discussed Sect. 2.1, the free-energy of formation of a two-dimensional disk nucleus of radius r on the crystal is

$$\Delta G_{\mathrm{d}} = 2\pi r d\gamma - \frac{\pi r^2 d\Delta\mu}{V_{\mathrm{m}}} \tag{53}$$

where d and V_{m} are the lattice distance and molecular volume, respectively.

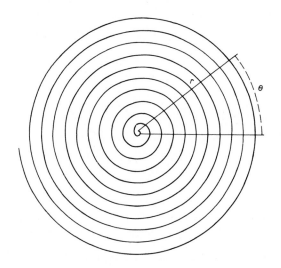

Fig. 12. An approximation to a single growth spiral (an Archimedean spiral).

Thus, the critical radius when $(d\Delta G_d/dr)_{T,P} = 0$ is given by

$$r^* = \frac{\gamma V_m}{\Delta \mu} = \frac{\gamma V_m}{kT\beta} \quad \text{or} \quad \frac{\gamma V_m}{kT\ln S} \tag{54}$$

If the current of growth units into a kink is assumed to be proportional to the saturation, S, then the net current is $(S - S_R)$ where S_R is the equilibrium saturation at a step of radius of R, i.e.

$$S_R = \exp(\gamma V_m/kTR) = [\exp(\gamma V_m/kTr^*]^{r^*/R}$$

$$S_R = S^{r^*/R} \tag{55}$$

If the kink density is proportional to f(S) [see eqn. (4)], then the lateral velocity of a straight step is

$$u_\infty = k'(S - 1)\text{f}(S) \tag{56}$$

and for a curved step

$$u_R = k'(S - S_R)\text{f}(S) \tag{57}$$

where k' is a constant. Eliminating k' using eqns. (56) and (57) and substituting eqn. (55) leads to

$$u_R = \frac{u_\infty(S - S^{r^*/R})}{S - 1} \tag{58}$$

Writing $S^{r^*/R} = [1 - (S - 1)]^{r^*/R}$ and expanding in a binomial series leads to

$$u_R = u_\infty\left(1 - \frac{r^*}{R}\right)\left[1 + \frac{r^*(S - 1)}{2R} + \ldots\right] \tag{59}$$

which, for $S \simeq 1$, reduces to

$$u_R = u_\infty\left(1 - \frac{r^*}{R}\right) \tag{60}$$

the equation used by Burton et al. In this case, when the supersaturation is increased, the spiral will wind itself until the radius of curvature at the centre reaches a critical value r^* [from eqn.(54).], at which point $u_R = 0$ and the whole spiral rotates with a stationary shape and fixed step distance, y_0 (see Fig. 12). Equations such as (59) and (60) ignore the strain energy caused by the presence of a dislocation (see, for example, ref. 28).

The density of steps, $1/y_0$, or step distance, y_0, at the stationary state is an important parameter in the BCF theory. The step distance may be evaluated from $y_0 = 2\pi u_\infty/\omega$ where ω is the angular velocity for a rotating spiral. Using polar coordinates, (r, θ), as shown in Fig. 12, the angular velocity may be determined in terms of u_∞ and r^*, from the solution of the three equations for u_∞ [e.g. eqn. (58) or (60)], the radius of curvature, R, at point (r, θ) and the angular velocity, i.e.

$$R = \frac{(1 + r^2\theta'^2)^{3/2}}{2\theta' + r^2\theta'^3 + r\theta''} \tag{61}$$

and

$$\omega = \frac{u_R}{r}[1 + (r\theta')^2]^{1/2} \tag{62}$$

where $\theta' = d\theta/dr$ and $\theta'' = d^2\theta/dr^2$.

The crudest approximation to the solution is an Archimedean spiral (Fig. 12).

$$r = 2r^*\theta \tag{63}$$

$$\omega = \frac{u_\infty}{2r^*}$$

Burton et al. pointed out that these equations are a good approximation for equidistant steps which are not too close to the centre of the spiral. In this approximation

$$y_o = 2r^*(\theta + 2\pi) - 2r^*\theta = 4\pi r^* \tag{64}$$

A more rigorous solution presented by Burton et al. using eqns. (60)–(62) leads to an angular velocity

$$\omega = \frac{3^{1/2}u_\infty}{2r^*(1 + 3^{1/2})} \tag{65}$$

and step distance

$$y_o = \frac{4\pi r^*(1 + 3^{1/2})}{3^{1/2}} \simeq 19.8r^* \tag{66}$$

Even though these solutions are approximate, the equations derived using u_z from eqn. (60) are expected to produce excellent results for low supersaturations, e.g. $1 < S \leq 1.3$, in which they have been applied. However, the majority of work with aqueous systems involves crystal growth at much

TABLE 2

Ratio of the step distances from eqns. (66) and (68) calculated according to eqn. (69)

Saturation ratio, S	Factor, δ
1.0	1
1.5	0.933
2.0	0.893
2.5	0.865
3.0	0.845
3.5	0.830
4.0	0.817
4.5	0.807
5.0	0.798
10.0	0.750

higher supersaturations. Nielsen [27] recognized this limitation in the estimation of y_o and extended the BCF treatment to higher supersaturations using eqn. (58) and following the original mathematical derivation of Burton et al. The final result may be expressed as

$$\omega = \frac{u_\infty}{r^*[2 + 2(3S)^{-1/2}]} \tag{67}$$

$$y_o = 4\pi r^*[1 + (3S)^{-1/2}] \tag{68}$$

Equations (67) and (68) predict that, at a given supersaturation, the angular velocity will be greater and the step distance, y_o, smaller than values derived from eqns. (65) and (66) for $S > 1$. The prediction of the increased density of steps with supersaturation shown in eqn. (68) may have repercussions in the interpretation of growth data. The ratio of the step distances from eqns. (68) and (66), i.e.

$$\delta = \frac{1 + (3S)^{-1/2}}{1 + 3^{-1/2}} \tag{69}$$

are shown in Table 2 for S between 1 and 10. In more complex configurations of growth spirals, the interaction between them produces changes in y_o. In the case of several spirals of the same sign (i.e. rotation), the step distance is given by Burton et al. as $y_o = 19.8r^*/\varepsilon$ where ε is the number of cooperating spirals, which is probably usually between 5 and 10. In more complex situations, when the spirals emerge from a grain boundary, uncertainties arise in the evaluation of y_o. The density of steps then depends on the length of the grain boundary and number of cooperating spirals [29]. However, the dependence on the supersaturation remains the same.

5.1.1 Surface-diffusion model

The further development of the BCF approach and application to growth from solution follows closely the methods described by Bennema and Gilmer [29]. The expressions for the step density will be called upon later in other modelling treatments.

Having established relationships for the step density, or at least the dependence of y_o on S, it is possible to consider in more detail the mechanisms described in Sect.5. For the moment, only the diffusion of growth units across the surface to steps and their subsequent incorporation will be considered. In a later section, the process of bulk diffusion will be considered in more detail.

In order to have a mathematically treatable problem, it is necessary to make some simplifying assumptions.

(1) The spiral steps are considered to be non-moving sinks into which growth units are incorporated.

(2) The steps are treated as straight steps which are far from the centre of the growth spiral.

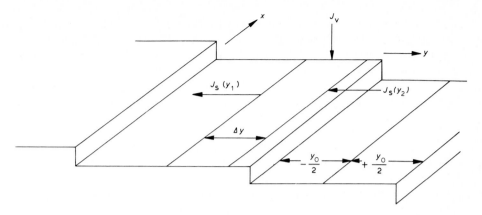

Fig. 13. Section of the surface between growth steps.

The flux of growth units in a strip on the terrace of the surface close to a step, may be divided into two main components (illustrated in Fig. 13)

(1) a flux of units, J_v, from the bulk solution to the surface adsorbed layer and

(2) a flux of growth units, J_s, from the strip towards the step. The concentration of units decreases near the step. In the steady-state

$$\frac{dJ_s(y)}{dy} - J_v = 0 \tag{70}$$

and according to Fick's law

$$J_s(y) = D_s c_e \frac{d[\alpha - \alpha_s]}{dy} = D_s c_e \frac{d\psi}{dy} \tag{71}$$

where D_s is the surface diffusion coefficient, c_e the equilibrium concentration of units cm^{-2}, α and α_s the supersaturations far from the step (as in the bulk) and on the surface at some position y, respectively, i.e. $\alpha_s = c_s/c_e$ where c_s is the surface concentration. ψ is a potential function defined as $\psi = \sigma - \sigma_s(y)$ where σ is the relative supersaturation defined in Sect. 2.3. The net flux from the solution may be written as the difference between the flux entering the surface layer and leaving, i.e.

$$J_v = \frac{c_e[\alpha - \alpha_s(y)]}{\tau_{\text{deads.}}} = \frac{\psi c_e}{\tau_{\text{deads.}}} \tag{72}$$

where the deadsorption relaxation time, $\tau_{\text{deads.}}$, is defined in eqn. (52) (see also Fig. 10). Substituting eqns. (71) and (72) into eqn. (70) gives the BCF differential equation

$$D_s \tau_{\text{deads.}} \frac{d^2\psi}{dy^2} = \psi \tag{73}$$

$$\lambda_s^2 \nabla^2 \psi = \psi$$

where the quantity $\lambda_s = (D_s \tau_{\text{deads.}})^{1/2}$ is the mean surface displacement. The general solution to this equation is

$$\psi = a \exp(y/\lambda_s) + b \exp(-y/\lambda_s) \tag{74}$$

The evaluation of a and b requires boundary conditions. Four cases were considered by Burton et al.

(1) $\lambda_s \gg x_0$ and one single step, where x_0 is the kink distance defined in Sect. 2.1.

(2) $\lambda_s \gg x_0$ and a parallel sequence of equidistant steps.

(3) $\lambda_s < x_0$ without diffusion in a step.

(4) $\lambda_s \ll x_0$ with diffusion in a step.

Case (2), which is most relevant to the growth of a spiral, leads to the boundary conditions

$$\begin{aligned} y &= y_0/2 & \psi &= \xi\sigma \\ y &= -y_0/2 & \psi &= \xi\sigma \end{aligned} \tag{75}$$

where $\xi = (\sigma - \sigma_{\text{step}})/\sigma$ is termed the retardation factor. When the relative supersaturation at the step, σ_{step}, equals 0, then $\xi = 1$. The solution to eqn. (74) is

$$\psi = \frac{\xi\sigma \cosh(y/\lambda_s)}{\cosh(y_0/2\lambda_s)} \tag{76}$$

The retardation factor may be evaluated from a consideration of the incorporation rate of growth units into a kink site of a step. If the average number of growth units per unit length of step that are within a jump distance, a, of a step is ac_{step}, then the number entering the step (from one side) per unit step in unit time is $ac_{\text{step}}/\tau_{\text{kink}}$ where τ_{kink} is the relaxation time for entry into a kink as defined in Sect. 5 (see Fig. 10). The net flux, J_s, must be zero at equilibrium, therefore

$$J_s = \frac{ac_{\text{step}}}{\tau_{\text{kink}}} - \frac{ac_e}{\tau_{\text{kink}}} \tag{77}$$

$$J_s = ac_e(1 - \xi)\sigma/\tau_{\text{kink}} \tag{78}$$

Also, according to eqns. (71) and (76), the flux into a step is

$$J_s = D_s c_e \frac{d\psi}{dy} = \frac{D_s c_e \sigma\xi \sinh(y/\lambda_s)}{\lambda_s \cosh(y_0/2\lambda_s)} \tag{79}$$

and solving eqns. (78) and (79) for ξ at $y = y_0/2$

$$\xi = \left[1 + \frac{D_s \tau_{\text{kink}}}{a\lambda_s} \tanh(y_0/2\lambda_s)\right]^{-1} \tag{80}$$

The linear growth rate is simply

$$R = \frac{u_\infty d}{y_0} \tag{81}$$

where u_∞ is the velocity of advance of a straight step and d the lattice spacing perpendicular to the crystal face. The velocity of the step is

$$u_\infty = 2J_s a_m \tag{82}$$

where a_m is the surface area occupied by a growth unit. The factor of 2 arises because J_s in eqns. (78) and (79) was calculated for growth units entering one side of the step. Thus, eqn. (82) assumes that the rate of entry is the same on each side of the step. Substituting eqn. (79) into eqn. (82)

$$u_\infty = \frac{2D_s c_e \xi \sigma a_m \tanh (y_o/2\lambda_s)}{\lambda_s} \tag{83}$$

and so, with y_o from eqn. (66) and r^* from eqn. (54) written in the form $r^* = \gamma V_m/kT\beta \; (\simeq \gamma V_m/kT\sigma$ at low $\sigma)$

$$R = \frac{\xi D_s c_e a_m d\sigma^2}{\lambda_s^2 \sigma_1} \tanh (\sigma_1/\sigma) \tag{84}$$

where $\sigma_1 = 9.8\gamma V_m/kT\lambda_s$. Thus, the BCF surface diffusion model may be written in the form

$$R = C \frac{\sigma^2}{\sigma_1} \tanh (\sigma_1/\sigma) \tag{85}$$

with $C = \xi D_s c_e a_m d/\lambda_s^2$.

The solution of the Burton et al. differential equation (73) has been done by Burton et al. for all the cases mentioned above. In particular, for a parallel sequence of steps and case (3), i.e. $\lambda_s < x_o$ without diffusion along a step

$$\psi = \frac{\xi c_o \sigma \cosh (y/\lambda_s)}{\cosh (y_o/2\lambda_s)} \tag{86}$$

with $c_o = (\pi\lambda_s/x_o) \ln [2\lambda_s/(1.78a)]$. For $\lambda_s \ll x_o$ with diffusion along a step, a similar expression is obtained but with c_o in a more complex form and dependent on y_o.

Bennema and Gilmer [29] have examined the conditions when assumption (1), i.e. neglect of the advance velocity of step in the diffusion equation, is valid, i.e. when $u_\infty \ll \lambda_s/\tau_{\text{deads.}}$. Substituting for u_∞ from eqn. (83) and with $\lambda_s = (D_s \tau_{\text{deads.}})^{1/2}$

$$\frac{u_\infty}{\lambda_s/\tau_{\text{deads.}}} = 2c_e \xi a_m \sigma \tanh (y_o/2\lambda_s) \gg 1 \tag{87}$$

Hence, because $\xi \leqq 1$, $\tanh (y_o/2\lambda_s) \leqq 1$, and $c_e/a_m \leqq 1$, eqn. (87) is expected to be valid at low supersaturations when $\sigma \ll 1$.

The upper limit of σ is difficult to establish in practice. It is also unfortunate that the saturation state of the system is expressed as the relative supersaturation, σ, rather than $\beta \; (= \Delta\mu/RT)$, the growth of affinity. Apart from the basic assumptions concerning the step velocity and geometry of the step, approximations in the mathematical derivation are made, the effects of which have not been investigated in any detail. For example, as shown by Nielsen [27], at higher supersaturations, y_0 is thought to be more accurately approximated by eqn. (68) rather than eqn. (66).

The BCF surface diffusion model, eqn. (85), may be simplified under the conditions

(1) $\sigma \ll \sigma_1$ then

$$R = \frac{C\sigma^2}{\sigma_1} \quad \text{(parabolic)}$$

(2) $\sigma \gg \sigma_1$

$$R = C\sigma \quad \text{(linear)}$$

where C is a constant.

It is interesting to compare the (R,σ) curves predicted from the model for

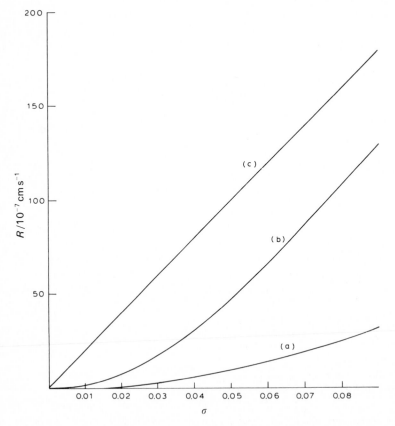

Fig. 14. Predictions of BCF theory eqn. (85) for $C = 2 \times 10^{-4}$ cm s^{-1}. (a) $\sigma_1 = 0.5$; (b) $\sigma_1 = 0.1$; (c) $\sigma_1 = 10^{-3}$.

σ_1 in the range 10^{-3} to 0.5 and with an arbitrary $C = 2 \times 10^{-4}\,\mathrm{cm\,s^{-1}}$. These are shown in Fig. 14. The transformation from linear to parabolic kinetics occurs at $\sigma = \sigma_1$. A number of studies of the application of the surface diffusion model to growth in aqueous systems has been reported for σ in the range 0–0.5, e.g. see the tabulation of Bennema et al. [30]. An example of one such application of the model to the interpretation of (R,σ) data is illustrated in Fig. 15 for sodium chlorate crystallisation. These data were obtained using a large single crystal suspended from an analytical balance and with σ between 0 and 2×10^{-3}. The optimum agreement between eqn. (85) and the data produced a C value of $1.9\,\mathrm{cm\ s^{-1}}$ and $\sigma_1 = 0.6 \times 10^{-3}$.

5.1.2 Volume diffusion model

When a concentration gradient appears in the bulk solution adjacent to the crystal surface, the volume diffusion process will become rate-limiting. Gilmer et al. [31] have examined the coupling of volume and surface dif-

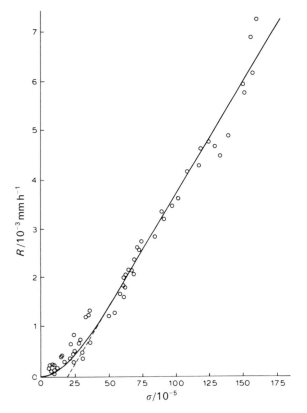

Fig. 15. Application of the BCF diffusion model to the crystal growth of the {100} face of sodium chlorate. O, Experimental $(R\sigma)$ data;——, best-fit BCF curve [eqn. (85)] obtained by a least-squares method. (Reproduced from ref. 16 by courtesy of Akademie Verlag.)

fusion when the concentration in the bulk solution is allowed to depend on position.

Burton et al. have shown that, for a spherical diffusion field around a kink

$$\frac{d\sigma(r)}{dr} = Br^{-2} \tag{88}$$

where B is a constant. With the boundary conditions $\sigma(r) = \sigma(x_o)$ for $r = x_o$ and $\sigma(r) = \sigma(\lambda)$ for $r = \lambda$

$$B = \frac{\sigma(x_o) - \sigma(\lambda)}{(1/\lambda) - (1/x_o)} \tag{89}$$

Upon substituting eqn. (89) into eqn. (88) and introducing the retardation factor, $\xi = [\sigma(x_o) - \sigma(\lambda)]/\sigma(x_o)$ and assuming that $x_o \gg \lambda$

$$\frac{d\sigma(r)}{dr} = \left(\frac{\lambda}{r^2}\right)\xi\sigma(x_o) \tag{90}$$

The flux of growth units entering a kink is then determined by multiplication by $2\pi r^2$ (the surface around a kink) and by Dc_o (where D and c_o are the diffusion coefficient and concentration in the bulk).

$$J_k = 2\pi\lambda\xi Dc_o\sigma(x_o) \tag{91}$$

The total flux into a step per unit length is

$$J_{step} = \frac{2\pi\lambda\xi Dc_o\sigma(x_o)}{x_o} \tag{92}$$

so that $u_x(= J_{step} a_m)$ can be calculated. Hence the linear growth rate $(R = u_x d/y_o)$ is

$$R = C\sigma\sigma(x_o) \tag{93}$$

where $C = 2\pi\lambda\xi Dc_o a_m dkT/(19 V_m x_o \gamma)$

$$\tag{94}$$

and $y_o = 19\gamma V_m/kT\sigma$ [see eqns. (54) and (66)].

The relative supersaturation at a distance $r = x_o$ from a kink is given by [24]

$$\sigma(x_o) = \sigma\left[1 + \frac{2\xi\lambda\pi(\delta - y_o)}{x_o y_o} + \frac{2\xi\lambda}{x_o} \ln(y_o/x_o)\right]^{-1} \tag{95}$$

where δ is the thickness of the unstirred layer given by [32]

$$\delta = \left[\frac{2}{3}\left(\frac{\eta}{\rho D}\right)^{1/3}\left(\frac{\rho q}{\eta X}\right)^{1/2}\right]^{-1} \tag{96}$$

where η is the viscosity of the solution, ρ is the density of the solution, q the rate of flow of the solution relative to a crystal face, and X is the characteristic length of the crystal. If $\xi = 1$, the second term in eqn. (95) vanishes when

$$\frac{2\pi\lambda\delta}{x_o y_o} \ll 1 \tag{97}$$

i.e. substituting for $y_o(\simeq 19\gamma V_m/kT\sigma)$, this becomes

$$\sigma \ll \frac{19\gamma V_m x_o}{2\pi kT\lambda\delta} \tag{98}$$

or

$$\sigma \ll \sigma_c$$

Because the third term in eqn. (95) is not sensitive to σ, the equation breaks down into two regions.

(1) Parabolic region when

$$\sigma \ll \sigma_c, R = C'\sigma^2 \tag{99}$$

where

$$C' = C\left[1 + \frac{2\xi\lambda}{x_o} \ln \frac{(19\gamma V_m)}{kTx_o\sigma}\right]^{-1}$$

(2) Linear region when $\sigma \gg \sigma_c$. This condition leads to

$$R = C\sigma \tag{100}$$

where $C = DC_o a_m d/\delta$.

Thus, both the surface diffusion and volume diffusion models produce similar (R,σ) curves. The application of the volume diffusion model has been somewhat limited (see, for example, the recent review by Bennema [33]).

5.2 CHEMICAL MECHANISTIC MODELS

The BCF theory discusses crystal growth in terms of the physical features of the molecular processes rather than in terms of the chemical changes which occur. If the adsorption interaction between the bulk ions and the crystal surface is slow, this may be the rate-determining step and the surface layer or adsorption layer may be considered to be in equilibrium with the crystal. In this situation, it is the chemical reaction between the solvated ions and the crystal surface which determines the rate of growth.

Ions in aqueous solution are surrounded by a shell of water molecules in tetrahedral or octahedral co-ordination that are relatively immobile because of the intensity of the electric field in the vicinity of the ion. Three regions of solvent around an ion may be labelled. In region 1 (see Fig. 16), all the water molecules are aligned by the field of the central ion forming a solvation shell. Between the distances b_j and R_j is region 2, known as the Gurney co-sphere, in which the water structure is modified by the presence of the central ion. Outside the radius R_j, region 3, the water molecules are considered to be unaffected by the central ion and retain their bulk properties, namely dielectric constant and viscosity, whereas these properties are con-

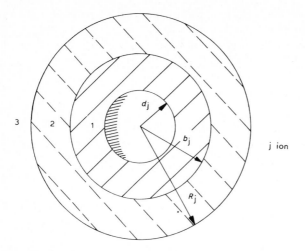

Fig. 16. Organisation of water around an ion.

sidered to be modified in regions 1 and 2. If the approach of such an ion to a crystal surface is considered, it is evident that the charge distribution over the surface of the co-sphere will be influenced by short-range interactions between the distributed charge and the lattice ions in the region of the step or terrace. Hence, even though the growth affinity is high, short-range interactions between the lattice ions and the water of hydration may cause such resistance to ion incorporation, that the process is slow and has a large activation energy.

As discussed by Doremus [34], the growth unit may either be individual ions or an ion-pair or molecule. In the growth of ionic salts, the ions of opposite charge must combine stoichiometrically at some stage in the crystallisation process to establish electrical neutrality in the crystal lattice. The ability of ions to associate in pairs, or sometimes in larger groups, depends on a balance between coulombic, thermal, and solvation forces. When attractive electrostatic forces between ions are high, ion-pairs or even triple-ions tend to be formed and are reasonably stable, remaining as a single entity for long periods compared with the time taken for a single Brownian displacement. The formation of ion-pairs is assumed when the co-spheres of the two ions overlap. However, in some circumstances, solvent-separated ion-pairs may be formed when the ions are separated by two or three solvent molecules. The formation of ion-pairs in aqueous solution may provide a kinetic pathway for crystallisation, of lower activation energy than is feasible for the individual ions. This is likely to be particularly important for small, highly charged ions. Unfortunately, however, no rigorous calculations have been made of the energy barriers associated with partial dehydration at a surface or, indeed, of potential energy barriers for surface migration of partially hydrated ions or ion-pairs.

5.2.1 Adsorption model

For a symmetrical 1:1 electrolyte, the reaction between the ions in a supersaturated solution and the crystal surface may be written as

$$A + B \underset{k_2}{\overset{k_1}{\rightleftharpoons}} AB(s) \underset{fast}{\overset{K_1}{\rightleftharpoons}} AB(c) \tag{101}$$

where (s) and (c) indicate an adsorbed surface state and crystalline state, respectively. The second reaction, involving diffusion and integration, is fast in comparison with the first reaction. The reaction rate per unit mass of seed is

$$\frac{1}{W}\frac{d[A]}{dt} = k_1 a_A a_B - k_2 a_{AB(s)} \tag{102}$$

where W is seed mass.

At equilibrium, $d[A]/dt = 0$ and so

$$k_2 = \frac{k_1 a_A a_B}{a_{AB(s)}} \tag{103}$$

and the solubility product is

$$K_s = \frac{a_A a_B}{a_{AB(c)}}$$

Because the equilibrium constant, K_1, is defined as

$$K_1 = \frac{a_{AB(c)}}{a_{AB(s)}}$$

the rate constant k_2 may be expressed as

$$k_2 = k_1 K_s K_1 \tag{104}$$

Hence substituting eqn, (104) into eqn. (102) gives

$$\frac{1}{W}\frac{d[A]}{dt} = k_1 a_A a_B - k_1 K_s K_1 a_{AB(s)} \tag{105}$$

or

$$\frac{1}{W}\frac{d[A]}{dt} = k_1 K_s a_{AB(c)} \left[\frac{a_A a_B}{a_{AB(c)} K_s} - 1 \right] \tag{106}$$

which, according to the definition of K_s and β [see eqns. (19) and (25)] reduces to

$$\frac{1}{W}\frac{d[A]}{dt} = k_1 K_s a_{AB(c)} [\exp \beta - 1] \tag{107}$$

In pure crystals, $a_{AB(c)} = 1$ by definition and eqn. (107) becomes

References pp. 230–231

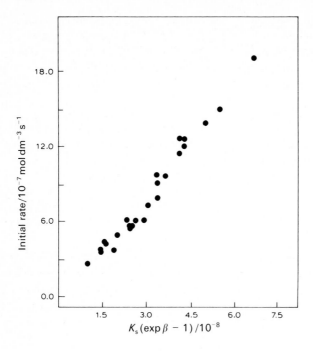

Fig. 17. Test of eqn. (108) for calcite growth. (Reproduced from ref. 36, © 1985 by Pergamon Journals Ltd.)

$$\frac{1}{W}\frac{d[A]}{dt} = k_A[\exp\beta - 1] \tag{108}$$

where k_A is the adsorption model rate constant. This is sometimes referred to as the Nancollas and Reddy equation.

A similar equation has been used to model the growth of calcite from a seeded solution [35]. This was written in the form

$$\frac{d[Ca_T^{2+}]}{dt} = -kA_s\left\{[Ca^{2+}][CO_3^{2-}] - \frac{K_s}{\gamma^2}\right\} \tag{109}$$

where k is the rate constant, A_s the total surface area of the seed crystals, and γ the divalent activity coefficient. This equation produced good agreement with experimental growth data. More recently, Inskeep and Bloom [36] have tested eqn. (108) for calcite precipitation at pH > 8 and partial pressures of CO_2 < 0.01 atm and, as shown in Fig. 17, obtained reasonable agreement. Some deviations at the start of the experiment at high supersaturations were attributed to nucleation on the crystals [37]. The results from these studies on calcite growth established two important points.

(1) The crystallisation reaction is surface-controlled rather than volume diffusion-controlled. The rate constant was found to be independent of the stirring speed and the apparent activation energy for growth was deter-

mined as $\simeq 40\,\mathrm{kJ\,mol^{-1}}$[38], consistent with a surface-controlled mechanism.

(2) The reaction rate was found to be proportional to the mass of seed in the suspension and so the rate constant is independent of the seed concentration (at least over the range of initial supersaturations studied). At low supersaturations, the rate constant may be found to be dependent on the initial supersaturation.

Another approach to studying reactions in which the adsorption step is rate-limiting has been suggested by Nielsen [39]. This treatment expresses the supersaturation in terms of the relative supersaturation, σ, defined in Sect.2.3 and the theory is developed for a fixed stoichiometry electrolyte. For a solution of concentration c, the concentration at the surface, c_s, may be approximated by

$$c_s = Kc_e \tag{110}$$

where c_e is the concentration of solute in thermodynamic equilibrium with the surface. If the assumption is made that the flux in and out of the adsorbed layer occurs in jumps of length a, then the forward flux is $A_s acv_{ad}$, where A_s is the surface area and v_{ad} the adsorption frequency. Similarly, the desorption flux is $A_s ac_s v_{ds}$ where v_{ds} is the desorption frequency. Thus, the linear growth rate, R, is

$$R = V_m a(cv_{ad} - Kc_e v_{ds})$$
$$R \simeq aV_m v_{ad} c_e \sigma = k_A \sigma \tag{111}$$

where V_m is the molecular volume and k_A the rate constant. Equation (111) is a linear rate law. It may be used to calculate the adsorption frequency $[v_{ad} = k_A/(aV_m c_e)]$from the (R,σ) plots with the parameter a being taken as the molecular diameter of the growth unit. This has been done for $Ba(NO_3)_2$, $MgSO_4 \cdot 7H_2O$, $Ni(NH_4)_2(SO_4)_2 \cdot 6H_2O$, $Na_2S_2O_3 \cdot 5H_2O$, and KNO_3, yielding v_{ad} between 9×10^3 and $4 \times 10^4\,\mathrm{s^{-1}}$[39]. A value of $v_{ad} = 10^4\,\mathrm{s^{-1}}$ leads to an activation energy, $\Delta G_{ads} \simeq 50\,\mathrm{kJ\,mol^{-1}}$, which is about three times the activation energy of the normal diffusion step of a small ion in an aqueous solution.

5.2.2 MODELS INVOLVING ASSOCIATED IONS

The incorporation of an ion-pair into the lattice should be easier in comparison with a free-ion and thus lead to a kinetically favourable crystallisation pathway. This type of mechanism has been suggested by Doremus [34] and Gunn [11]. In the following, two examples will be discussed in more detail.

Firstly, the crystal growth of calcium sulphate may be described by the mechanism [11]

$$Ca^{2+} + SO_4^{2-} \rightleftharpoons CaSO_4^0 \tag{112}$$

$$CaSO_4^0 + Ca^{2+} \rightleftharpoons [CaSO_4 \cdot Ca]^{2+} \tag{113}$$

$$CaSO_4^0 + SO_4^{2-} \rightleftharpoons [CaSO_4 \cdot SO_4]^{2-} \tag{114}$$

The associated ions then react with either negatively or positively charged kinks, e.g.

$$[CaSO_4 \cdot Ca]^{2+} + [-] \rightarrow CaSO_4(c) + Ca^{2+} \tag{115}$$

$$[CaSO_4 \cdot SO_4]^{2-} + [+] \rightarrow CaSO_4(c) + SO_4^{2-} \tag{116}$$

where the symbols $[-]$ and $[+]$ represent kink sites. Because reactions (112)–(114) are considered to be instantaneous, the surface reactions (115) and (116) are expected to be rate-limiting reactions. Hence, the ion incorporation rate is given by

$$\frac{1}{A_s} \frac{dn_{Ca_T}}{dt} = k_1 [CaSO_4 \cdot Ca]^{2+} [-] \tag{117}$$

$$\frac{1}{A_s} \frac{dn_{SO_4}}{dt} = k_2 [CaSO_4 \cdot SO_4]^{2-} [+] \tag{118}$$

For a neutral crystal, $dn_{Ca_T}/dt = dn_{SO_4}/dt$ and substituting the equilibrium conditions [eqns. (112)–(114)] leads to

$$\frac{[Ca^{2+}]}{[SO_4^{2-}]} = K\frac{[+]}{[-]} \tag{119}$$

At equilibrium

$$CaSO_4(c) \rightleftharpoons Ca^{2+} + SO_4^{2-} + [+] + [-] \tag{120}$$

so that

$$\frac{[Ca^{2+}][SO_4^{2-}][+][-]}{CaSO_4(c)} = \text{constant}$$

Thus, $[+][-] = \text{constant}$. If this condition holds for growth, then

$$[+] = c_1 \left\{ \frac{[Ca^{2+}]}{[SO_4^{2-}]} \right\}^{1/2}$$

$$[-] = c_2 \left\{ \frac{[SO_4^{2-}]}{[Ca^{2+}]} \right\}^{1/2} \tag{121}$$

and therefore the precipitation rate is given by

$$\frac{dn_{Ca_T}}{dt} = kA_s \{[Ca^{2+}][SO_4^{2-}]\}^{3/2} \tag{122}$$

Other mechanisms may be formulated. As an example, if the ion-pair $CaSO_4^0$ reacts directly with a $[+]$ or $[-]$ kink, the reaction is

$$CaSO_4^0 + [\pm] \rightarrow CaSO_4(c) \tag{123}$$

and the rate is given by

$$\frac{dn_{Ca_T}}{dt} = kA_s[Ca^{2+}][SO_4^{2-}] \tag{124}$$

with $[\pm]$ constant. It is important to remember that only the forward precipitation reactions are considered in the above treatment. The reverse dissolution reactions will become significant near equilibrium.

The results obtained by this method have been used to predict the order of reactions for $CaSO_4$, $Mg(OH)_2$, and $MgNH_4PO_4$ [11].

The second example is for the precipitation of calcium carbonate from $Ca(HCO_3)_2$ solutions. The mechanistic model was derived by Plummer et al. [40] from a kinetic study of the dissolution of Iceland spar crystals. The approach has more recently been extended to the interpretation of crystal growth [41]. The model is based upon the reaction between a partially dehydrated $[Ca-HCO_3]^+$ complex or ion-pair and possible anionic reaction sites within the adsorbed layer.

$$[Ca-HCO_3]^+ + CO_3^{2-}(s) \xrightarrow{k_4'} CaCO_3(c) + HCO_3^-(s) \tag{125}$$

$$[Ca-HCO_3]^+ + HCO_3^-(s) \xrightarrow{k_4''} CaCO_3(c) + H_2CO_3(s) \tag{126}$$

$$[Ca-HCO_3]^+ + OH^-(s) \xrightarrow{k_4'''} CaCO_3(c) + H_2O(s) \tag{127}$$

The corresponding dissolution reactions are

$$CaCO_3(c) + HCO_3^- \xrightarrow{k_1'} Ca^{2+} + CO_3^{2-} + HCO_3^- \tag{128}$$

which, at low pH, is equivalent to

$$CaCO_3(c) + H^+ \xrightarrow{k_1} Ca^{2+} + HCO_3^- \tag{129}$$

$$CaCO_3(c) + H_2CO_3 \xrightarrow{k_2} Ca^{2+} + 2HCO_3^- \tag{130}$$

$$CaCO_3(c) + H_2O \xrightarrow{k_3} Ca^{2+} + HCO_3^- + OH^- \tag{131}$$

where the rate constants are labelled according to Plummers system. Considering each of the reactions separately, the following rate equations may be derived.

(i) *Reaction between $[Ca-HCO_3]^+$ and $OH^-(s)$*

Combining reactions (127) and (131) leads to the precipitation rate G $[= (dn_{Ca_T}/dt)/W$ where W is the mass of seed and is directly proportional to $A_s]$.

$$G = k_3 a_{H_2O} - k_4''' a_{Ca^{2+}} a_{HCO_3^-} a_{OH^-(s)} \tag{132}$$

At equilibrium, $R = 0$ and

$$k_4''' = \frac{k_3 K_2}{(K_s K_w)} \tag{133}$$

References pp. 230–231

if $a_{\text{OH}^-(\text{s})} = a_{\text{OH}^-}$ where K_2 is the second dissociation constant of carbonic acid, K_s is the solubility product of calcite and K_w is the ionic product of water. Substituting eqn. (133) into eqn. (132) and using the relationship

$$\frac{K_2 a_{\text{Ca}^{2+}} a_{\text{HCO}_3^-}}{K_s} = a_{\text{H}^+} \exp(\beta) \tag{134}$$

leads to

$$G = k_3 a_{\text{H}_2\text{O}} \left[1 - \left(\frac{a_{\text{H}^+}}{a_{\text{H}^+(\text{s})}} \right) \exp \beta \right] \tag{135}$$

where it has been assumed that $a_{\text{H}_2\text{O}} = a_{\text{H}_2\text{O}}(\text{s})$.

(ii) Reaction between $[\text{Ca–HCO}_3]^+$ and HCO_3^- (s) The rate may be expressed as [see eqns. (126) and (130)]

$$G = k_2 a_{\text{H}_2\text{CO}_3} - k_4'' a_{\text{Ca}^{2+}} a_{\text{HCO}_3^-} a_{\text{HCO}_3^-(\text{s})} \tag{136}$$

Thus, $k_4'' = k_2 K_2/(K_s K_1)$ if $a_{\text{HCO}_3^-(\text{s})} = a_{\text{HCO}_3^-}$ at equilibrium. K_1 is the first dissociation constant of carbonic acid. Equation (136) becomes

$$G = k_2 a_{\text{H}_2\text{CO}_3} - k_2 a_{\text{H}_2\text{CO}_3(\text{s})} \left(\frac{a_{\text{H}^+}}{a_{\text{H}^+(\text{s})}} \right) \exp \beta \tag{137}$$

and with the assumption $a_{\text{H}_2\text{CO}_3} = a_{\text{H}_2\text{CO}_3(\text{s})}$, this becomes

$$G = k_2 a_{\text{H}_2\text{CO}_3} \left[1 - \left(\frac{a_{\text{H}^+}}{a_{\text{H}^+(\text{s})}} \right) \exp \beta \right] \tag{138}$$

(iii) Reaction between $[\text{Ca–HCO}_3]^+$ and CO_3^{2-} (s)

Combining eqns. (125) and (129) leads to the expression

$$G = k_1 a_{\text{HCO}_3^-} - k_4' a_{\text{Ca}^{2+}} a_{\text{HCO}_3^-} a_{\text{CO}_3^{2-}(\text{s})} \tag{139}$$

and $k_4' = k_1/K_s$ if $a_{\text{CO}_3^{2-}(\text{s})} = a_{\text{CO}_3^{2-}}$ at equilibrium. Substituting for k_4' and $a_{\text{CO}_3^{2-}} (= K_2 K_1 a_{\text{H}_2\text{CO}_3}/a_{\text{H}^+}^2)$ in eqn. (139) gives

$$G = k_1 a_{\text{HCO}_3^-} \left[1 - \left(\frac{a_{\text{H}^+}^2 a_{\text{H}_2\text{CO}_3(\text{s})}}{a_{\text{H}^+(\text{s})}^2 a_{\text{H}_2\text{CO}_3}} \right) \exp \beta \right] \tag{140}$$

and with $a_{\text{H}_2\text{CO}_3(\text{s})} = a_{\text{H}_2\text{CO}_3}$, this becomes

$$G = k_1 a_{\text{HCO}_3^-} \left[1 - \left(\frac{a_{\text{H}^+}^2}{a_{\text{H}^+(\text{s})}^2} \right) \exp \beta \right] \tag{141}$$

Equation (135) has been applied to the crystal growth of Iceland spar with the rate constant k_3 determined from dissolution experiments [40, 43]. Plummer et al. [40] found that the dissolution rate could be described by an equation of the form

$$G = k_1 a_{\text{H}^+} + k_2 a_{\text{H}_2\text{CO}_3^*} + k_3 a_{\text{H}_2\text{O}} - k_4 a_{\text{Ca}^{2+}} a_{\text{HCO}_3^-} \tag{142}$$

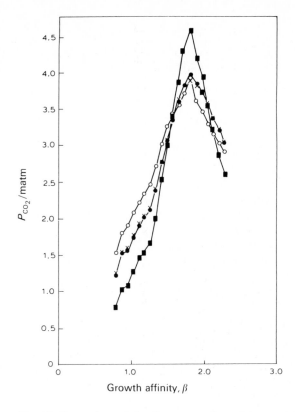

Fig. 18. Comparison of the P_{CO_2} curves for calcite precipitation at 25°C. ●, P_{CO_2} in the bulk solution; ×, $P_{CO_2(s)}$ obtained using the CO_3^{2-} (s) mechanism [eqn. (141)]; ○, $P_{CO_2(s)}$ obtained using the OH^- (s) mechanism [eqn. (135)]; ■, $P_{CO_2(s)}$ obtained using the HCO_3^- (s) mechanism [eqn. (138)]. (Reproduced from ref. 41 by courtesy of The Royal Society of Chemistry.)

where k_1, k_2, and k_3 are defined in eqns. (129)–(131), respectively and the rate constant for the back reaction (crystal growth) is k_4, i.e.

$$k_4 = k_4' a_{CO_3^{2-} (s)} + k_4'' a_{HCO_3^- (s)} + k_4''' a_{OH^- (s)} \qquad (143)$$

Hence, given the value of k_3 and with $a_{H_2O} \simeq 1$, the growth data may be used to evaluate $a_{H^+}/a_{H^+ (s)}$ or $P_{CO_2(s)}$ from eqn. (135). The results indicated that the $P_{CO_2(s)}:P_{CO_2}$ ratio increased to a maximum value and then decreased with decreasing β to converge at a value of 1.

Alternatively, the three mechanisms [eqns.(125)–(127)] may be examined individually to determine values of k_1, k_2, and k_3 by optimising agreement between $P_{CO_2(s)}$ and P_{CO_2}. The results obtained by this method, for the seeded growth of calcite, are illustrated in Fig.18. As shown, the assumption that $p_{CO_s(s)} = p_{CO_2}$ is best for the CO_3^{2-} (s) model with the OH^- (s) model producing reasonable agreement. A comparison of the rate constant, k_3, obtained from crystal growth and dissolution experiments is shown in Table 3. Good agreement between all but one of the values is shown. The value of k_3 obtained in

TABLE 3

Experimental values obtained for k_3 using various techniques and for different samples [43] Powder (1) and powder (2) are calcite samples of BET specific surface areas 0.22 and 5.65 $m^2 g^{-1}$, respectively.

Method	Sample	k_3/mol cm^{-2} s^{-1}
Rotating disk (dissolution)	Polished Iceland Spar crystals	1.41×10^{-10}
Free-drift (dissolution)	Crushed Iceland Spar	1.20×10^{-10}
Free-drift + pH stat (dissolution)	Powder (1)	1.34×10^{-10}
Free-drift (precipitation)	Powder (1)	1.35×10^{-10}
Free-drift (precipitation)	Powder (2)	3.61×10^{-11}

the rotating disk experiments was found to be dependent on the polishing grit size with the rate constant increasing with increasing grit size. The value shown in Table 3 is the rate constant obtained by extrapolation to zero grit size. The difference between powder (1) and (2) is thought to be caused by morphological differences between the samples. The powder (1) consisted of rhombohedral particles $< 5\mu$ in size with no distinct flaws, whereas the powder (2) material was a much higher specific surface area sample composed of elliptically shaped particles.

5.3 THE PARABOLIC RATE LAW

For a number of substances, the growth rate may be expressed by a parabolic expression

$$R \propto (c - c_e)^2 \tag{144}$$

where c is the concentration of the crystallising substance in solution and c_e the equilibrium concentration. Typical plots are illustrated in Fig. 19. This empirical relationship has been known for some time, e.g. for the growth of potassium sulphate [44] and silver chloride [45].

Davies and Jones [45] made an attempt to explain the parabolic dependence based on a double layer model that allows for a difference in concentration of the adsorbed ions and ions in the bulk of solution. For a (2:2) electrolyte, the concentration of lattice ions in a supersaturated solution is

$$[A](s) = [A] \exp(-2\psi/kT)$$
$$[B](s) = [B] \exp(2\psi/kT) \tag{145}$$

where ψ is the electric potential at the surface and (s) denotes the surface adsorption layer. In a supersaturated solution, the concentration of A which is available for growth is

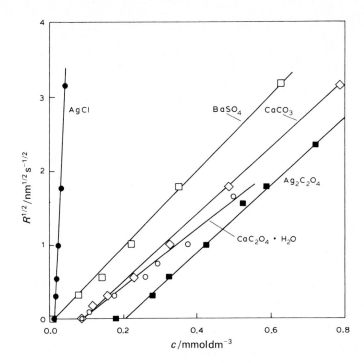

Fig. 19. Examples of electrolytes which follow the parabolic rate law. ■, $Ag_2C_2O_4$; ○, $CaC_2O_4 \cdot H_2O$;◇, $CaCO_3$;□, $BaSO_4$; ●, AgCl. (Reproduced from ref. 47 by courtesy of North-Holland Publishing Co.)

$$[A] \exp(-2\psi/kT) - [A]_e(s) \tag{146}$$

If the concentration of the ions on the surface is equal, then

$$[A]_e(s) = \frac{K_s^{1/2}}{\gamma_2} \tag{147}$$

where γ is the divalent ion activity coefficient. Hence, eqn. (146) becomes

$$[A] \exp(-2\psi/kT) - \frac{K_s^{1/2}}{\gamma_2} \tag{148}$$

If $[A](s) = [B](s)$, then, from eqn. (145)

$$\exp(2\psi/kT) = \left\{\frac{[A]}{[B]}\right\}^{1/2} \tag{149}$$

The concentration of A ions available for growth may be calculated using eqns. (149) and (146). Similarly, the concentration of B ions may be calculated and the growth rate determined from the product $[A][B]$, i.e.

$$G = k_p \gamma_2^2 \left\{A^{1/2}B^{1/2} - \left(\frac{K_s}{\gamma_2^2}\right)^{1/2}\right\}^2 \tag{150}$$

References pp. 230–231

where k_p is the rate constant. With $[A] = [B] = c$, this reduces to the form given in eqn. (144). Similarly, for such systems $(c - c_e)^2 = c_e^{1/2}\sigma^2$, which is in the form predicted by the BCF surface diffusion theory [eqn. (85)] when $\sigma \ll \sigma_1$ and volume diffusion [eqn. (99)].

The Davies and Jones derivation makes some fundamental assumptions concerning the surface concentrations of the lattice ions and the BCF theory is only applicable to very small supersaturations. Thus, both theories have limitations which affect the interpretation of the results of growth experiments. Nielsen [27] has attempted to examine in detail how the parabolic dependence can be explained in terms of the density of kinks on a growth spiral and the adsorption and integration of lattice ions. One of the factors, $\sigma = S - 1$, comes from the density of kinks on the spiral [eqns. (4) and (68)] and the other factor is proportional to the net flux per kink of ions from the solution into the lattice. Nielsen found it necessary to assume that the adsorption of equivalent amounts of constituent ions occurred and that the surface adsorption layer is in equilibrium with the solution. Rather than eqn. (145), Nielsen expresses the concentration in the adsorption layer in the form of a simple adsorption isotherm equation

$$[A](s) = K_{AS}[A]$$
$$[B](s) = K_{BS}[B]$$
(151)

Defining n_A and n_B as the number of growth sites in which A and B can fit, respectively, with $n = n_A + n_B$, then the net flux of A and B into the kinks is

$$J_A = k_A[A]n_A - k'_A n_B$$
$$J_B = k_B[B]n_B - k'_B n_B$$
(152)

i.e. eqns. (151) are valid at all times during growth. For a crystal AB, the two currents J_A and J_B must be equal and so by substituting for n, eqn. (152) becomes

$$J = J_A + J_B = \frac{2n(k_A k_B[A][B] - k'_A k'_B)}{k_A[A] + k_B[B] + k'_A + k'_B}$$
(153)

This may be further simplified by using the equilibrium condition $J_A = J_B = 0$ to obtain a relationship between $k_A k_B$ and $k'_A k'_B$, i.e.

$$[A]_e[B]_e = \frac{k'_A k'_B}{k_A k_B} = K_s$$
(154)

and so the total flux [eqn. (153)] becomes

$$J = \frac{2n k_A k_B([A][B] - K_s)}{k_A[A] + k_B[B] + k'_A + k'_B}$$
(155)

Because J/n is the flux per kink, eqn. (155) is not in the form to predict the parabolic dependence. Nielsen has suggested [27] that the electrical charges at the interface must exert an important effect on the composition of the interface that has not been taken into consideration in the derivation of eqn. (155). This amounts to an additional constraint on the composition of the surface layer as defined according to eqn. (151). Nielsen suggested that, because there is no great net electric charge, the ions must be adsorbed in equivalent amounts (as described by the Davies and Jones theory above)

$$[A](s) = [B](s) \tag{156}$$

Hence, applying eqn. (155) to the adsorbed layer gives

$$J = \frac{2nk_A k_B \{[A]^2(s) - [A]_e^2(s)\}}{(k_a + k_B)[A](s) + k'_A + k'_B} \tag{157}$$

$$J = \frac{2nk_A k_B}{(k_A + k_B)} \frac{\{[A](s) + [A]_e(s)\}\{[A](s) - [A]_e(s)\}}{[A](s) + [(k'_A + k'_B)/(k_A + k_B)]} \tag{158}$$

which is in the correct form, i.e. $J \propto [A]^{1/2}[B]^{1/2} - K_s^{1/2}$ if

$$\frac{(k'_A + k'_B)}{(k_A + k_B)} = [A]_e(s) \tag{159}$$

and

$$\frac{[A](s)}{[A]_e(s)} = \frac{[A]^{1/2}[B]^{1/2}}{K_s^{1/2}} = S \tag{160}$$

From eqn. (152) and the equilibrium condition, eqn. (159) is equivalent to $n_{Ae} = n_{Be}$ where the attached subscript e denotes the equilibrium values. Also, with $[A](s) = [B](s)$, eqn. (160) may be transformed to

$$[A](s)[B](s) = K_a[A][B] \tag{161}$$

where

$$K_a = \frac{[A]_e^2(s)}{K_s} = \frac{[A]_e(s)[B]_e(s)}{[A]_e[B]_e} \tag{162}$$

is an adsorption constant. Equation (161) expresses the equilibrium condition between the surface adsorption layer and the solution. Under these conditions, i.e. eqns. (159) and (160) are valid, and the total flux equation (158) becomes

$$J = \frac{2nk_A k_B \{[A](s) - [A]_e(s)\}}{k_A + k_B} \tag{163}$$

so that the flux per kink is

$$J_k = k_i V_w [A]_e(s)(S - 1) = k_i V_w K_{ads} c_e(S - 1) \tag{164}$$

and the linear growth rate

TABLE 4

Comparison of k_A and k_h for aqueous systems at 25°C calculated from the data given by Nielsen [39] for the parabolic rate law

Electrolyte	S	k_A/s^{-1}	k_h/s^{-1}	$\Delta G_i/\Delta G_h$
AgCl	1–3.5	2×10^9	10^{10}	1.25
$Ag_2M_0O_4$	1–2	2.9×10^8	10^{10}	1.55
KH_2PO_4	1–1.2	3.1×10^6	10^9	1.66
KNO_3	1–1.05	5.8×10^6	10^9	1.59
$KZnF_3$	1–2.5	1.05×10^5	10^9	2.05
K_2SO_4	1–1.2	5.4×10^6	10^9	1.60
CaF_2	1–3	8.9×10^4	1.6×10^8	1.71
$CaSO_4 \cdot 2H_2O$	1–1.16	2.6×10^4	1.6×10^8	1.82
$CaCO_3$	1–3.5	2×10^5	1.6×10^8	1.63
$Fe(CHOHCOO)_2$	1–7	1500	3.2×10^6	1.53

$$R = \frac{V_m}{x_o y_o} k_i V_w K_{ads} c_e (S - 1) \tag{165}$$

where k_i is the integration rate constant, V_w is the molar volume of water, and K_{ads} is the adsorption coefficient, i.e.

$$k_i = \frac{2k_A k_B}{V_w(k_A + k_B)} \tag{166}$$

$$K_{ads} = K_a^{1/2}$$

In summary, the parabolic rate law may be derived using eqn. (152) to determine the fluxes, with the adsorption layer composition determined by the equilibrium relationship (161) with [A](s) = [B](s) for a salt AB. Because the integration step is considered to be the rate-limiting process, it is of considerable interest to evaluate k_i. If the rate constant for the removal of water from the anion is much larger than for the cation, the dehydration of the cation will be rate-limiting. In this instance

$$k_i = \frac{2k_A}{V_w} \tag{167}$$

It has been suggested that k_A should be of a similar magnitude to k_h, the rate constant for the removal of a single water molecule of hydration from the inner sphere. Table 4 summarises some results calculated from the data given by Nielsen [39]. The free-energy changes were calculated according to the Eyring theory [see eqn. (52) with $\tau_A^{-1} = k_A$]. The results are somewhat limited but do suggest that more than one water molecule of hydration is lost during the incorporation step. With the assumption that $k_A \ll k_B$, eqn. (164) may be used to write the linear growth as

$$R = \frac{2k_A V_m K_{ads} c_e (S - 1)^2}{19d^2(\gamma/kT) \exp(\gamma/kT)} \tag{168}$$

where $1/x_o y_o \simeq (S - 1)/19d^2(\gamma/kT)\exp(\gamma/kT)$ [see eqns. (66) and (2)]. Thus, the rate constant depends on k_A, K_{ads}, and γ. The theory requires further detailed experimental investigation with particular attention to the estimation of the quantities which contribute to the rate constant.

5.4 POLYNUCLEAR LAYER MECHANISM

When the rate of crystallisation of a generalised salt $A_\alpha B_\beta$ is expressed in the functional form

$$G = k\{[(A^{a+})^\alpha(B^{b-})^\beta]^{1/v} - K_s^{1/v}\}^n \tag{169}$$

a log plot will enable n to be evaluated. For the linear and parabolic rate laws, $n = 1$ and 2, respectively. When $n > 2$, the crystal growth is generally associated with a polynuclear mechanism. This occurs when the surface nucleation is so fast that each layer of the crystal is the result of the intergrowth of numerous individual nucleated two-dimensional "islands". The rate-determining step is the incorporation of growth units at kink sites but, unlike the theories described above, the surface steps are generated through surface nucleation.

The theory of polynuclear growth has recently been reformulated from the original Hillig [46] treatment for crystallisation from aqueous solutions [39]. The first step in the development of the model is the calculation of the lateral velocity, v_∞, of a straight step assuming that the incorporation of the growth unit at a kink site is the rate-determining process. The number of growth units entering a kink in unit time may be expressed as

$$J_k^+ = 2k_A[A](s)V_m \tag{170}$$

i.e. there are $2V_m[A](s)$ growth units within one jump of the kink. The constant, k_A, is the integration rate constant for the growth unit designated A. If the adsorption layer is in equilibrium with the bulk solution, then

$$[A](s) = K_{ads}[A] = K_{ads}[A]_e S \tag{171}$$

Therefore, eqn. (170) becomes

$$J_k^+ = 2k_A K_{ads}[A]_e V_m S \tag{172}$$

The flux of growth units out of the kink site is equal to J_k^+ when $S = 1$, i.e.

$$J_k^- = 2k_A K_{ads}[A]_e V_m \tag{173}$$

and so the net flux entering a kink is

$$J_k = 2k_A K_{ads}[A]_e V_m (S - 1) \tag{174}$$

The lateral velocity of the step is then given by $v_\infty = d^2 J_k/x_o$ and so, from eqns. (4) and (174), this becomes

$$v_\infty = 2dk_A K_{ads}[A]_e V_m (S - 1)S^{1/2}\exp(-d^2\gamma/kT) \tag{175}$$

The second stage of development is to calculate the nucleation rate of surface "islands", J_i, and so determine the number of "islands" nucleated on an area A_s in a period dt, i.e. $J_i A_s dt$. Nielsen [39] has evaluated J_i to give

$$J_i = 2k_A (K_{ads}[A]_e V_m/d)^2 S^{5/2}(\ln S)^{1/2} \exp(- d^2\gamma/kT)$$
$$\times \exp(- \Delta G^*/kT) \tag{176}$$

where

$$\Delta G^* = \frac{\pi d^4 \gamma^2}{kT \ln S}$$

is the critical free-energy for formation of a circular nucleus [see eqn. (9) with $\xi = \pi$]. If the surface "islands" do not reach the edge of the crystal face but stop growing as they meet other islands in the same layer, then the are a covered at time $t = \tau$ (no nucleation at $t = 0$) is

$$A_\tau = \int_0^\tau \pi[v_\infty(\tau - t)]^2 J_i A_s dt = \frac{\pi v_\infty^2 J_i A_s \tau^3}{3} \tag{177}$$

As an approximation, the time taken to complete one layer is obtained by letting $A_\tau = A_s$ and solving for τ, i.e.

$$\tau = \left(\frac{3}{\pi J_i v_\infty^2}\right)^{1/3} \tag{178}$$

and the linear growth rate is

$$R = d\left(\frac{\pi J_i v_\infty^2}{3}\right)^{1/3} \tag{179}$$

Hence, substituting into eqn. (179) for J_i and v_∞ gives

$$R = k_e S^{7/6}(S - 1)^{2/3}(\ln S)^{1/6} \exp(- K_e/\ln S) \tag{180}$$

where $k_e = 2dk_A (K_{ads}[A]_e V_m)^{4/3} \exp(- d^2\gamma/kT)$

$$K_e = \frac{\pi d^4 \gamma^2}{3k^2 T^2}$$

For a salt AB, then $[A]_e = c_e$.

This growth model predicts a sharper increase in growth rate with increasing S than the parabolic model. This is illustrated in Fig. 20 for the two models. Nielsen and Toft [47] have observed this type of growth for both CaF_2 and SrF_2. For CaF_2, the growth was found to be parabolic at low supersaturations, i.e. $S < 3$ but agreed with the polynuclear layer mechanism for S between 4 and 40. In the case of SrF_2, eqn. (180) described the data between $S = 1$ and 33 with $k_e = 3.4 \times 10^{-8}$ and $K_e = 12.9$. Hamza et al. [48] also found that the crystallisation of CaF_2 obeyed eqn. (169) with the value of $n = 3.7 \pm 0.1$. The addition of Mg^{2+} to the reaction solution did not affect the form of the growth equation. Similarly, Heughebaert and Nancollas [49] found that the seeded growth of octacalcium phosphate followed eqn. (169) with $n = 4$. This contrasts with the crystallisation of dicalcium phosphate

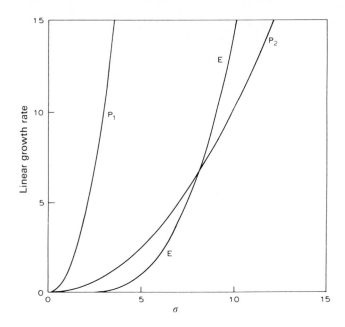

Fig. 20. Comparison of the parabolic and polynuclear layer predictions. P_1, $R = 1.1 (S - 1)^2$; P_2, $R = 0.1 (S - 1)^2$; E, polynuclear layer model with $k_c = 10$, $K_c = 10$.

dihydrate ($CaHPO_4 \cdot 2H_2O$) for which the parabolic rate law was obeyed [50]. The polynuclear layer mechanism has also been applied to the study of the crystallisation of calcite following heterogeneous nucleation onto ground glass seeds [51]. The model suggested the absence of any induction period and the growth data were interpreted in terms of a polynuclear layer mechanism at high supersaturations followed by a parabolic dependence at lower supersaturation. It is interesting to note that the treatment of the glass with alkali led to a large reduction in the nucleation rate.

At high supersaturations, it is expected that the growth rate will be determined by more than one principal mechanism. Also, in view of the discussion in Sect. 2.1, the transition from polynuclear layer growth to other mechanisms will be important when interpreting data obtained over a wide range of supersaturation. Few data have been tested in detail using eqn. (180) and the majority of studies have relied on the value of n from the generalised equation (169) to postulate possible mechanisms. Other theories, such as that of Madsen and Boistelle [52], which describe growth by surface nucleation according to the polynuclear mechanism, differ in form to that of eqn. (180). Further work is still needed to evaluate the applicability of these equations.

5.5 VOLUME DIFFUSION EQUATIONS

The BCF volume diffusion model has already been discussed in Sect. 5.1.2. The classical approach is to consider the diffusion of lattice ions from the bulk solution through a stationary liquid layer adjacent to the surface. In the case of diffusion-controlled growth, the flux of ions to the surface is given by

$$J = D\frac{dc}{dx} \tag{181}$$

where x is the diffusion path. The reaction rate is determined by the rate of diffusion to a spherical surface at a distance r from the centre, i.e.

$$\frac{dn}{dt} = 4\pi r^2 D\frac{dc}{dr} \tag{182}$$

At steady state, eqn. (182) may be integrated between $c = c_e$ at $r = r_1$, i.e. at the surface of the sphere, and c, the bulk concentration at $r = r_2$ where $r_2 > r_1$

$$\frac{dn}{dt} = \frac{4\pi D(c - c_e)r_1 r_2}{(r_2 - r_1)} \tag{183}$$

With the condition that $r_2 \gg r_1$ and evaluating the linear growth rate

$$R = \frac{DV_m(c - c_e)}{r_1} \tag{184}$$

Hence

$$R = k_D(S - 1) \tag{185}$$

i.e. linear rate law is obtained.

The situation is slightly more complex in a non-stoichiometric solution. Examining the general case of $A_\alpha B_\beta$ when $\alpha = \beta$, the fluxes J_A and J_B for ions A and B are

$$J_A = \frac{D_A([A] - [A]_e)}{r}$$
$$J_B = \frac{D_B([B] - [B]_e)}{r} \tag{186}$$

for a particle of radius r. The equilibrium concentration, $[A]_e$, may be evaluated using the relationship $[B]_e[A]_e = K_s$ and the condition that $J_A = J_B$ to preserve electrical neutrality. With the assumption that $D_A = D_B = D$, this leads to

$$[A]_e = \frac{[A] - [B]}{2} + \left\{\left(\frac{[A] - [B]}{2}\right)^2 + K_s\right\}^{1/2} \tag{187}$$

so that the linear growth rate is

$$R = \frac{DV_m}{r}\left[\frac{[A] + [B]}{2} - \left\{\left(\frac{[A] - [B]}{2}\right)^2 + K_s\right\}^{1/2}\right] \tag{188}$$

Equation (187) is only strictly applicable to crystals in a stationary liquid. Particles less than $5\,\mu m$ in size (the value depending on the difference in the densities of the solid and solution) will tend to be carried with the solution during stirring and so grow by a purely diffusion-controlled growth rate. However, for larger particles, the transport of ions to the surface will depend upon the solution movement around the particles and so both convection and diffusion have to be taken into account. In this case, the assumption that $r_2 \gg r_1$ that led to eqn. (184) is no longer valid. Instead, if $(r_2 - r_1) = \delta$ and $r_1 \simeq r_2$, then eqn. (184) becomes

$$R = \frac{DV_m(c - c_e)}{\delta} \tag{189}$$

where δ is the thickness of the effective film for mass transfer or simply the stagnant film thickness. Although δ is not the same thing as the diffusion boundary layer thickness, it is expected to be of the same order of magnitude. The diffusion boundary layer is that region close to the surface in which transport occurs by purely molecular diffusion, whereas δ is a hypothetical layer which allows for the effects of both convection and pure diffusion transport. The value of δ may be estimated from the relationship derived by Mullin [53] given by eqn. (96).

For a sphere sedimenting by gravity, the value of δ may be estimated from the theoretical equation given by Nielsen [54].

$$\delta = \frac{r}{(1 + P_e)^{0.285}} \tag{190}$$

where P_e is the Péclet number given by

$$P_e = \frac{2gr^3\Delta\rho}{9D\eta} \tag{191}$$

where g is the acceleration due to gravity and $\Delta\rho$ the density difference between the solid and solution.

It follows from eqn. (189) that, when convection is important, the rate constant for the linear rate law is given by

$$k_c = \frac{DV_m c_e}{\delta} \tag{192}$$

Hence, δ may be estimated from experimental growth data and compared with theoretical predictions derived by eqn. (190) or eqn (96). Reasonable agreement has been obtained between the theoretical and experimental values of δ for NH_4NO_3 growth [39] but other systems which exhibited linear growth kinetics produced poor agreement, e.g. for $(NH_4)_2SO_4$ the theoretical estimate of δ was $18\,\mu m$, which contrasted with the value derived from the growth data of $66\,\mu m$. Unfortunately, there are no reports of the application of the rotating disk method [see Sect. 4.2.4] to similar systems. Indeed, future research on bulk transport-controlled growth should consider experimental

References pp. 230–231

techniques which permit characterisation of the hydrodynamics in the vicinity of the crystal surface.

6. The effects of inhibitors

Other components in solution may have a significant effect on the crystal growth kinetics. They may exist in solution as impurities at very low concentrations. However, the low density of kink sites coupled with the affinity of the impurity to the crystal surface means that the growth may be very sensitive to such minor components. There are many reports of such impurities, all of which are effective at low concentration, i.e. $10^{-9} < [I] < 10^{-4}$ where [I] is the concentration of the inhibitor. For example, Meyer [55] has recently examined the effects of various additives ranging from organics to a variety of cationic and anionic additives on the growth of calcite. A number of substances produced an 80% reduction in the growth rate at a concentration of the order of $10^{-7} \, \text{mol} \, \text{dm}^{-3}$. The problem, of course, is that, even in pure systems, it is often difficult to ensure against contamination at this level during crystallisation.

Following the discussion in Sect. 2.2.1, any additional substance which affects the solubility product will also influence the reaction rate. In brief, the effects of inhibitors may be classified as follows.

(i) The inhibitor may influence the rate-determining step during growth. It is possible that the inhibitor will so affect a particular mechanism that a parallel process may become rate-determining.

(ii) If the inhibitor is co-precipitated, then the composition of the solid will be altered, e.g. isomorphous substitution in the lattice will lead to the formation of solid solutions.

(iii) The morphology of the crystals may be changed. This may be the result of poisoning selective crystal planes.

(iv) The complexation of the inhibitor with lattice ions in solution will lead to a decrease in the supersaturation of the solution.

The effect of (i) will be discussed in more detail in Sect. 6.1.

The role of the inhibitor in determining the morphology of the solid is related to its ability to block growth sites. In this way, the flow and generation of steps is changed. If the effect of the inhibitor is different on different crystal faces, then the growth of one or more faces will be favoured and the shape of the crystal will be different from that produced from a pure solution (see, for example, the work of Troost [56] on the influence of surface-active agents on the growth of sodium triphosphate hexahydrate). The effect of inhibitors on the morphology has been discussed in detail by Cabrera and Vermilyea [57]. Considerations of morphology changes are important when calculating linear growth rates from data obtained using seed suspensions.

The complexation of the inhibitor is generally considered to be unimportant at low concentrations. However, in complex mixtures such as freshwater or seawater, the concentration of the less effective inhibitors may be quite high and must be taken into consideration when calculating the growth affinity. This is also true of those components which have no direct effect on the growth process but merely change the ionic strength of the solution and thus the activity of the lattice ions in solution.

The incorporation of the inhibitor into the crystal affects the solid phase activity, a_{AB} [eqn.(19)]. The activities in the crystalline phase may be expressed in terms of rational activity coefficients, i.e. for solid solution formation

$$a_{AB(ss)} = \lambda_{AB} X_{AB(ss)}$$

and

$$a_{CB(ss)} = \lambda_{CB} X_{CB(ss)}$$

for a solid AB in the presence of a minor component C. The symbol (ss) refers to a solid solution and X_i is the mole fraction of the ith component in solution. The driving force or growth affinity [eqn. (26)] requires an expression for $a_{AB(ss)}$. The value of $X_{AB(ss)}$ may be obtained from a knowledge of the distribution coefficient. The evaluation of λ_{AB} is more difficult. It may be obtained directly from solubility measurements or indirectly from theoretical models [e.g. regular solution theory, eqn. (22)] predicting the changes in the distribution coefficient with variations in $X_{AB(ss)}$.

6.1 SURFACE ADSORPTION

The increased concentration of the inhibitor in the surface layer relative to the bulk solution is caused by adsorption. The adsorption may be at a kink, step, or terrace site and be effective in influencing the step velocity. Unfortunately, the experimental determination of adsorption isotherms of inhibitors is rare and the approach in the past has been to try and obtain information indirectly from the analysis of crystal growth data.

6.1.1 Adsorption at kinks

The simplest way of considering the effects of inhibitors is to assume a site-blocking mechanism. This mechanism postulates that the adsorption of an inhibitor at a growth site may affect the reaction rate because

(i) its entry into a kink site means that it must be desorbed before a lattice ion can be incorporated. This process effectively increases the activation energy for kink site entry (see, for example, Fig. 10, ΔG_{kink}) and

(ii) the adsorption at a kink site leads to the formation of a new but different type of kink site or pseudo-kink. The incorporation of a lattice ion at this kink position will involve a perturbation in ΔG_{kink} and also co-precipitation of the inhibitor.

If the integration rate constant for a site without and with an adsorbed inhibitor molecule or ion is k_0 and k_1, respectively, then the overall rate constant may be expressed as

$$k = k_0(1 - \theta) + k_1\theta = k_0 + \theta(k_1 - k_0) \tag{193}$$

where θ is the fraction of growth sites occupied by the inhibitor. The value of θ may not correspond with the fractional coverage derived from the adsorption isotherm measurement. This is because the adsorption on terraces and ledges may not influence growth to any extent.

If the rate constant $k_1 = \alpha k_0$, then eqn. (193) becomes

$$\frac{k - k_0}{k_0} = \theta(\alpha - 1) \tag{194}$$

The function θ may be expressed as the Langmuir equation

$$\theta = \frac{[I]K_L}{1 + [I]K_L} \tag{195}$$

where K_L is a constant related to the standard free-energy of adsorption i.e. $K_L = \exp(-\Delta G^0/RT)$. Substituting eqn. (195) into eqn. (194) leads to

$$\frac{k_0}{k_0 - k} = \left\{1 + \frac{1}{[I]K_L}\right\}(1 - \alpha) \tag{196}$$

If $\alpha = 1$, no inhibition occurs and when $\alpha = 0$, complete inhibition occurs. A plot of $k_0/(k_0 - k)$ against $[I]^{-1}$ permits the calculation of both α and K_L.

Some typical results are shown in Fig. 21 for the inhibition of calcium hydroxyapatite growth. In this case, the rates were measured in constant composition conditions so that k_0 and k could be replaced by the appropriate crystallisation rates. The intercept indicates that $\alpha \simeq 0$ and the slope gave $K_L = 8.4 \times 10^4$, i.e. $\Delta G^0 = -28.1\,\text{kJ mol}^{-1}$. The results may be explained by the geometric fit of the P–O–P groups of phytic acid on the {010} face of the hydroxyapatite crystals. Recently, Hamza et al. [48] have tested the applicability of the site-blocking model to the growth of SrF_2 in the presence of Mg^{2+}. The kinetic data were consistent with a polynuclear layer mechanism of growth and produced straight lines when plotted according to the form of eqn. (196) (see Fig. 22). However, it was discovered that only experiments at the same supersaturation produced agreement with eqn. (196). Increasing the supersaturation had the effect of reducing the effectiveness of the inhibitor. Figure 22 implies that the value of K_L is dependent on the supersaturation. This seems very unlikely and the results probably reflect the increased nucleation at higher supersaturations.

6.1.2 Adsorption at steps

So far, the main influence of the inhibitor has centred on its interaction with the kink site and how this subsequently affects the integration of growth units. However, the adsorption on steps may also influence the kinetics in two possible ways.

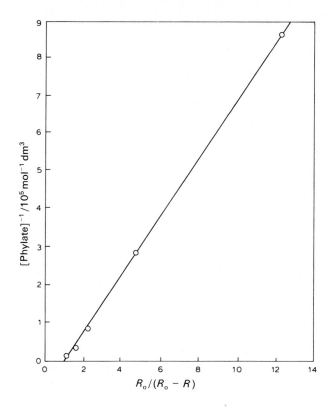

Fig. 21. Example of results analysed using the site-blocking mechanism, eqn. (196), for the growth of calcium hyroxyapatite (HAP) on HAP seeds in the presence of phytic acid. (Reproduced from ref. 58 by courtesy of Academic Press, Inc.)

(i) Berner and Morse [59] have explained the influence of the inhibitor in terms of the energy requirement for growth between two inhibitor ions a distance $2r$ apart in a step (see Fig. 23). The free-energy change associated with the formation of such an "island" is

$$\Delta G_i = \pi r d\gamma - 2r d\gamma - \frac{0.5\pi r^2 d\Delta\mu}{V_m} \tag{197}$$

so that the critical radius when $d(\Delta G_i)/dr = 0$ is

$$r^* = \frac{V_m}{\pi\Delta\mu}(\pi\gamma - 2\gamma) \simeq \frac{1.14 V_m\gamma}{\pi\beta kT} \tag{198}$$

Berner and Morse [59] envisage the two-dimensional nucleation of a disk between the adsorbed inhibitor, i.e.

$$r^* = \frac{\gamma V_m}{\beta kT} \tag{199}$$

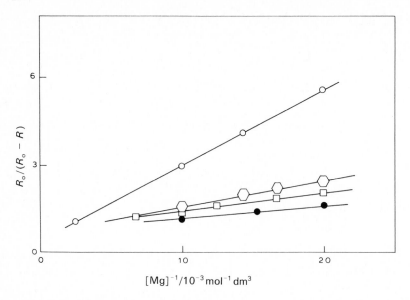

Fig. 22. Test of the site-blocking mechanism for Mg^{2+} inhibition of SrF_2 growth. The results are for different supersaturations. O, $\sigma = 1.86$; $\bigcirc$, 1.56; $\square$, 1.26; $\bullet$, 0.96. (Reproduced from ref. 48 by courtesy of North-Holland Publishing Co.)

If the inhibitor is distributed uniformly over the crystal surface, then the surface density of inhibitor Γ, is attained from

$$\Gamma^{-1/2} = 2r^* \tag{200}$$

Substituting eqn. (199) gives

$$\Gamma = \left[\frac{kT\beta}{2\gamma V_m}\right]^2 \tag{201}$$

or alternatively, using eqn. (198)

$$\Gamma \simeq 7.6\left[\frac{kT\beta}{2\gamma V_m}\right]^2$$

Both equations predict that, given a particular surface density of inhibitor, there is a corresponding critical growth affinity below which growth cannot occur. If this mechanism is operative, the effectiveness of the inhibitor will depend on the initial supersaturation and the extent of the co-precipitation of the inhibitor. If the inhibitor concentration is maintained constant during growth, it is expected that the growth will stop at some critical growth affinity determined by the inhibitor concentration. The main problem with this approach is the uncertainty associated with the density of the inhibitor on the surface. The density on steps and terraces is expected to be quite different. Even along steps, some heterogeneity in density is expected to occur because of the effects of step "bunching", multiple steps, and im-

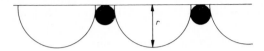

Fig. 23. Critical nucleus of radius r between two inhibitors.

purities. This heterogeneity is expected to lead to a reduction in the measured rate constant, at constant inhibitor concentration, as the initial growth affinity is reduced.

Davey [60] has also considered the same mechanism. In advancing between the inhibitor, each separate length of step must "bow out", thus increasing its curvature and reducing the step velocity according to eqn. (60), which is applicable at low supersaturations, i.e.

$$u_r = u_\infty \frac{r - r^*}{r}$$

where r is the radius of the curved step and u_∞ the velocity of a straight step, i.e. without inhibitor, and far from the centre of the growth spiral. Thus, the linear growth rate is $R = u_r d / y_0$.

(ii) The adsorption of the inhibitor at steps may influence growth in other ways. The presence of the inhibitor reduces the effective length of the step along which kinks can form and may produce steric barriers to the entry of growth units. The overall result of this will be to increase the kink distance, x_0, and reduce the growth rate. If the concentration of the inhibitor is expressed as the number per molecular spacing, $[I]_m$, then the probability of finding a site along a step without an inhibitor will be $1 - [I]_m$. Albon and Dunning [61] have suggested that this mechanism leads to the equation

$$R = R_0[2r^* - 2r^*(1 - [I]_m) + (1 - [I]_m)](1 - [I]_m)^{2r^*} \tag{203}$$

where R_0 is the linear growth rate in the absence of the inhibitor. This model has been found to give good agreement with experimental data from the growth of sucrose in the presence of raffinose. Albon and Dunning [61] related $[I]_m$ to $[I]$ using an isotherm equation of the form of the Freundlich equation. The analysis enables the evaluation of the critical radius, r^*, and hence an estimate of the interfacial energy, γ [from eqn. (200)]. Davey [60] has found good agreement between the value of γ estimated in this way and that from other sources for the {100} face of sucrose and ammonium dihydrogen phosphate and the {100} face of sodium triphosphate hexahydrate.

6.1.3 Adsorption on terraces

Competition between the inhibitor and the growth units will affect the concentration of growth units on the terraces. If the adsorption step is rate-limiting, then this competition can be expected to affect the growth rate in various ways. The area available for adsorption is effectively decreased according to the form of the adsorption isotherm. The isotherm may not only depend upon the temperature but also the solution speciation, e.g. changes in pH during growth may affect the competition between the growth units and inhibitor for surface sites. In this case, the measured rate constant, k_A in eqn. (108), may be expressed as $k_A = k'_A(1 - \theta)$ where k'_A is the rate constant which is independent of the inhibitor concentration, i.e. $k_A = k'_A$ when $\theta = 0$. The adsorption of the inhibitor on steps and kinks may lead to a change in the growth kinetics because an increase in ΔG_{kink} will lead to a reduction in the integration rate constant, which may cause the interaction step to become rate-limiting. Similarly, a change in the energetics of surface diffusion may make this step more important.

In systems in which surface diffusion is rate-limiting, the adsorption of inhibitor molecules onto terraces must reduce the catchment area of the step so that $k = k_0(1 - \theta)$ [62]. If θ is given by the Langmuir equation, then

$$\frac{k_0}{k_0 - k} = 1 + \frac{1}{[I]K_L} \tag{204}$$

which is similar to eqn. (196). Equation (204) predicts no growth as $[I] \to \infty$, i.e. $\alpha = 0$ in eqn. (196). A number of systems have been found to produce agreement with eqn. (204), e.g. {100} ammonium dihydrogen phosphate in the presence of $FeCl_3$ and $AlCl_3$, {100} KBr in the presence of phenol and the carboxylic acids HCOOH, CH_3COOH, C_2H_5COOH, and C_3H_7COOH [60].

The form of eqn. (204) is expected from both adsorption- and surface diffusion-controlled reactions and also from the site-blocking mechanism. Therefore, the agreement with this form of equation is of limited help in understanding the underlying mechanism of growth. This requires a detailed examination of the kinetics without inhibitors.

6.1.4 Two-dimensional nucleation enhancement

If the impurity has a strong affinity for the surface of the crystal, it is expected that steps on the surface will be completely blocked. Therefore if $\alpha = 0$ [eqn. (197)], no growth occurs by the incorporation of growth units at kinks. Thus, providing that the rate is not limited by transport of ions to the surface, the rate-determining step will be the rate of two-dimensional nucleation on the surface of the crystal. Gilmer [63] has proposed that strongly adsorbed impurities will create favourable sites for the adsorption of growth units on the surface. By forming strong bonds to both the solid and the adsorbed molecule, the probability that the molecule will be desorbed is reduced. Thus, the effect of a strongly adsorbed molecule will be to promote two-dimensional nucleation. It follows that, if the inhibitor density on the surface is increased from zero, an optimum concentration will be reached that corresponds to a maximum adsorption enhancement. Above this value, the inhibitor molecules become too close to enable growth to occur between them. At the other limiting case, i.e. $[I] \to 0$, a change in reaction kinetics from two-dimensional nucleation to a parabolic rate law might be expected. It is expected that the rate of two-dimensional nucleation will depend on the growth affinity and the impurity surface density.

Glossary of main symbols

A, A_s	surface area
a_m	molecular area of a growth unit
a_i	activity of ith component
a_e	activity at equilibrium with the crystal
c	concentration of solute
c_e	concentration at equilibrium with the crystal
c_s	concentration at the surface of the crystal
Δc_{max}	metastable zone width
D	diffusion coefficient
D_s	surface diffusion coefficient
d	lattice dimension
G	precipitation rate $[\ = (\mathrm{d}n/\mathrm{d}t)/W]$
ΔG	Gibbs free-energy change
J	flux of growth units or nucleation rate
K_L	Langmuir isotherm constant
K_s	solubility product
K_w	ionic product of water
k_A	rate constant for adsorption-controlled reaction
k_D	rate constant for diffusion-controlled reaction
k_P	rate constant for parabolic rate law
L	circumference of nuclei
M	molecular mass
n	number of growth units
q	flow rate of solution
$\bar{R}$	mean linear growth rate
R	linear growth rate
S	saturation ratio
SI	saturation index
U	symbol for growth unit
u	step velocity
V_m	molecular volume
v	volume of crystal
V	volume of solution or seed suspension
W	mass of seed material or interchange energy
X	characteristic length of crystal
X_i	mole fraction of component i
x_o	kink distance
y_o	step distance

References pp. 230–231

α	supersaturation or inhibition parameter [eqn. (194)]
β	growth affinity
β_{ion}	average growth affinity per ion
γ	divalent activity coefficient or interfacial free energy
δ	stagnant film thickness
ε	number of cooperating spirals
η	solution viscosity
θ	fractional surface coverage of inhibitor
λ_{s}	mean surface displacement
λ_{i}	rational activity coefficient
μ	chemical potential
υ	$\alpha + \beta$ in $A_{\alpha}B_{\beta}$
ξ	shape factor or retardation factor in BCF theory
ρ	density
Σ	specific surface area
σ	relative supersaturation
τ	induction time
ψ	potential function in BCF theory or electric potential at a surface
ω	angular velocity

References

1   A.E. Nielson and J. Christoffersen, in G.H. Nancollas (Ed.), Biological Mineralization and Demineralization, Springer-Verlag, New York, 1982, p. 37.
2   F.C. Frank. Discuss. Faraday Soc., 5 (1949) 48.
3   W.A. House, in D.H. Everett (Ed.), Colloid Science, Specialist Periodical Reports, Vol. 4, The Royal Society of Chemistry, London, 1983, p.1.
4   W.A. House and M.J. Jaycock, J. Colloid Interface Sci., 59 (1977) 252.
5   G.W. Van Oosterhout and G.M. Van Rosmalen, J. Cryst. Growth, 48 (1980) 464.
6   R. Becker and W. Doring, Ann. Phys., 24 (1935) 719.
7   D. Turnbull and J.C. Fisher, J. Chem. Phys., 17 (1949) 71.
8   A.E. Nielsen, Kinetics of Precipitation, Pergamon Press, Oxford, 1964, Chap.2.
9   T.A. Ananikyan, A.G. Nalbandyan and H.G. Nalbandyan, J. Cryst. Growth, 73 (1985) 505.
10   J.W. Mullin and H.M. Ang, Faraday Discuss. Chem. Soc., 61 (1976) 141.
11   D.J. Gunn, Faraday Discuss. Chem. Soc., 61 (1976) 133.
12   A.E. Nielsen, Krist. Tech., 4 (1967) 17.
13   A. Lieberman, AIChE Symp. Ser., 78 (1982) 76.
14   C.G. Inks and R.B. Hahn, Anal. Chem., 39 (1967) 625.
15   S.M. Hamza, A. Abdul-Rahman and G.H. Nancollas, J. Cryst. Growth, 73 (1985) 245.
16   P. Bennema, Phys. Status Solidi, 17 (1966) 555.
17   N. Albon and W.J. Dunning, Acta Crystallogr., 15 (1962) 474.
18   L.N. Rashkovich and A.A. Mkrtchan, J. Cryst. Growth, 74 (1986) 101.
19   C. Walker and G.T. Kohman, Trans. AIEE, 67 (1948) 580.
20   J. Novotny, Krist. Tech., 9 (1971) 343.
21   V.G. Levich, Acta Physicochim. URSS, 17 (1942) 257.
22   R.E. Meyer, M.C. Banta, P.M. Lantz and F.A. Posey, J. Electroanal. Chem., 30 (1971) 345.
23   P. Bennema, J. Cryst. Growth, 1 (1967) 278.
24   W.K. Burton, N. Cabrera and F.C. Frank, Philos. Trans. R. Soc. London, 243 (1951) 299.

25   G.H. Gilmer, Science, 208 (1980) 355.
26   A.R. Verma, Crystal Growth and Dislocations, Butterworths, London, 1953.
27   A.E. Nielsen, Pure Appl. Chem., 53 (1981) 2025.
28   N. Cabrera and M.M. Levine, Philos. Mag., 1 (1956) 450.
29   P. Bennema and G.H. Gilmer, in P. Hartman (Ed.), Crystal Growth: An Introduction, North-Holland, Amsterdam, 1973, Chap. 10.
30   P. Bennema, J. Boon, C. Van Leeuwen and G.H. Gilmer, Krist. Tech., 8 (1973) 659.
31   G.H. Gilmer, R. Ghez and N. Cabrera, J. Cryst. Growth, 8 (1971) 79.
32   P. Bennema, J. Phys. Chem. Solids Suppl., (1967) 413.
33   P. Bennema, J. Cryst. Growth, 69 (1984) 182.
34   R.H. Doremus, J. Phys. Chem., 62 (1958) 1068.
35   G.H. Nancollas and M.M. Reddy, J. Cryst. Growth, 37 (1971) 824
36   W.P. Inskeep and P.R. Bloom, Geochim. Cosmochim. Acta, 49 (1985) 2165.
37   H.N.S. Wiechers, P. Sturrock and G.V.R. Marais, Water Res., 9 (1975) 835.
38   G.E. Cassford, W.A. House and A.D. Pethybridge, J. Chem. Soc. Faraday Trans. 1, 79 (1983) 1617.
39   A.E. Nielsen, J. Cryst. Growth, 67 (1984) 289.
40   L.N. Plummer, T.M.L. Wigley and D.L. Parkhurst, Am. J. Sci., 278 (1978) 179.
41   G.E. Cassford, W.A. House and A.D. Pethybridge, J. Chem. Soc. Faraday Trans. 1, 79 (1983) 1617.
42   W.A. House, J. Chem. Soc. Faraday Trans. 1, 77 (1981) 341.
43   R.G. Compton, P.J. Daly and W.A. House, J. Colloid Interface Sci., 113 (1986) 12.
44   R. Marc, Z. Phys. Chem., 61 (1908) 385.
45   C.W. Davies and A.L. Jones, Discuss. Faraday Soc., 5 (1949) 103.
46   W.B. Hillig, Acta Metall., 14 (1966) 1868.
47   A.E. Nielsen and J.M. Toft, J. Cryst. Growth, 67 (1984) 278.
48   S.M. Hamza, A. Abdul-Rahman and G.H. Nancollas, J. Cryst. Growth, 73 (1985) 245.
49   J.C. Heughebaert and G.H. Nancollas, J. Phys. Chem., 88 (1984) 2478.
50   P.G. Koutsoukos, Z. Amjad, M.B. Tomson and G.H. Nancollas, J. Am. Chem. Soc., 102 (1980) 1553.
51   W.A. House and J.A. Tutton, J. Cryst. Growth, 56 (1982) 699.
52   H.E.L. Madsen and R. Boistelle, J. Cryst. Growth, 46 (1979) 681.
53   J.W. Mullin, Crystallisation, Butterworths, London, 1961.
54   A.E. Nielsen, Croat. Chem. Acta, 53 (1980) 255.
55   H.J. Meyer, J. Cryst. Growth, 66 (1984) 639.
56   S. Troost, J. Cryst. Growth, 3/4 (1968) 340.
57   N. Cabrera and D.A. Vermilyea, in R.H. Doremus, B.W. Roberts and D. Turnbull (Eds.). Proc. Int. Conf. Cryst. Growth, Cooperstown, NY, Wiley, New York, 1958.
58   P.G. Koutsoukos, Z. Amjad and G.H. Nancollas, J. Colloid Interface Sci., 83 (1981) 599.
59   R.A. Berner and J.W. Morse, Am. J. Sci., 274 (1974) 108.
60   R.J. Davey, J. Cryst. Growth, 34 (1976) 109.
61   N. Albon and W.J. Dunning, Acta Crystallogr., 15 (1962) 474.
62   R.J. Davey and J.W. Mullin, J. Cryst. Growth, 26 (1974) 45.
63   G.H. Gilmer, J. Cryst. Growth, 42 (1977) 3.

Chapter 4

An Introduction to Corrosion and its Prevention

P. HAMMONDS

1. What is corrosion?

1.1 DEFINITION

Corrosion may be broadly defined as the degradation of a material due to interaction with its environment. From this broad definition, it is evident that the material need not be man-made; for instance the "weathering" of granite tors on moorlands may be considered as corrosion of their crystal constituents by the atmosphere (rain, sun, wind, etc). In this chapter, how-ever, it is only metallic corrosion that will be considered.

Metals are good conductors of electricity and if the environment with which they are in contact is also conductive, then corrosion will occur via an electrochemical process. Notice that, in every case, the corrosion of the metal occurs due to interaction with its environment. Therefore, for metallic corrosion to occur, the metal or alloy must be unstable in that particular environment. Thus, changing the environment of a metal can greatly affect its rate of corrosion in either direction. For example, wrought iron gates will last indefinitely in a clean urban environment; however, those same gates placed near the coastline would rapidly corrode due to the effects of salt spray from the ocean.

1.2 WHY STUDY CORROSION?

Lack of corrosion control may affect our lives in ways which vary from the inconvenience of a seized nut on our cars to a catastrophe caused by cracking of a combustion chamber in a jet aircraft engine. Thus, the study of corrosion processes helps in choosing designs and materials necessary for maintaining high safety standards. Safety is the primary consideration in preventing corrosion, but economics are also of major importance. As will be seen later, in order to produce metals from their ores an input of energy is required. Energy is an expensive commodity and therefore prevention of corrosion will increase the longevity of a metal article and thus prevent premature replacement costs. It is not only the replacement of an article which costs money, but also lost time and production capability may in-crease the cost of corrosion. Many estimates of the cost of corrosion have been made [1] yet even these may be underestimated as peripheral costs are very difficult to quantify. For example, a relatively minor component such as

References pp. 277–279

a chain link in a can-sterilising unit may corrode and break. This causes the machine to jam and production is halted. Not only is the cost of replacement and labour incurred but also loss of production and several thousand cans of produce have to be discarded. One only has to look around at everyday items to see how ubiquitous corrosion is, from iron railings, car bodies, and exhausts, to spalling concrete from road bridges due to the corrosion of reinforcing bars.

The best way to combat corrosion is at the design stage. It is therefore essential that the designer has a sound knowledge of corrosion principles and a specialised knowledge of environment and materials particular to their field of design. Proper design negates the need for expensive retrofits. Subsequent commissioning and running of plants should also be undertaken keeping the need for corrosion protection in mind. For example, the treatment of cooling waters with appropriate inhibitors or the optimisation of current distribution in impressed current cathodic protection.

2. Electrochemical nature of corrosion

Most corrosion processes are electrochemical in nature or are a combination of physical/chemical and electrochemical processes. Electrochemistry is the study of phenomena associated with the transfer of electrical charge from one phase to another and in itself is an extremely broad subject ranging from batteries as power supplies to propagation of nerve impulses in living organisms. In considering corrosion phenomena, we are concerned first with how stable the metal is in its environment, and secondly how fast the corrosion process will proceed. The stability of a metal is related not only to the environment in which it is placed, but also to the properties of the metal itself and how much energy is required to produce the metal from its ore, i.e. the thermodynamic properties of the system. How fast the corrosion reaction proceeds depends on many factors and need not be related to the thermodynamic stability of the metal. For example, aluminium requires nearly $400\,\text{kcal}\,\text{mol}^{-1}$ to be formed from its oxide, Al_2O_3, bauxite, whereas iron requires only about $200\,\text{kcal}\,\text{mol}^{-1}$ and we would therefore expect aluminium to corrode rapidly. However, many articles are constructed from aluminium and its alloys (saucepans to aeroplanes) and remain quite corrosion free. The reason for this is that aluminium oxidises to form a thin protective oxide film so that the remaining metal is protected from its environment. If this oxide layer is removed by abrasion then, on exposure to air, an oxide film is rapidly reformed; if, however, mercuric chloride solution is rubbed on the metal surface, an amalgam is formed which prevents adhesion of the forming oxide and thus rapid corrosion will continue.

2.1 THERMODYNAMIC CONSIDERATIONS

Many metals are won from their ores by an input of energy. Those that occur "native", i.e. uncombined, are "noble" or untarnished by long exposure to the atmosphere. These metals are easily won from and may be more stable than their compounds (e.g. gold and silver). Many corrosion products are similar to the ores from which the metal was initially won. For example, copper is found as basic carbonate, sulphide, and oxide and these compounds may also be corrosion products of copper. The free energy of formation of such ores/corrosion products is the free energy for the reaction of the metal with the other species both present in their elemental state, for example [2]

$$Zn + \tfrac{1}{2}O_2 = ZnO \qquad \Delta G^\circ = -318.2 \, kJ \, mol^{-1}$$

In an aqueous environment, the "free energy" of a metal is associated with a potential difference between the metal and the solution interface. It is not possible to measure this potential directly because, in order to complete the measuring instrument/metal/solution circuit, it is necessary to dip a conductor into the solution. This immediately introduces a further potential and the reading on the instrument will be the sum of the metal/solution and conductor/solution potentials. The difficulty is overcome by choosing the hydrogen electrode as an arbitrary standard of value O V, all other electrodes may be related to this. Using standard conditions, this gives rise to the familiar electrochemical series. Using the IUPAC sign convention, noble metals have positive potentials relative to base metals, e.g.

$$Zn^{2+} + 2e \rightarrow Zn \qquad E^\circ \quad -0.76 \, V$$

$$Ag^+ + e \rightarrow Ag \qquad E^\circ \quad 0.22 \, V$$

These potentials are standardised, that is, they only apply to systems at unit activity, 25°C, and 1 atm pressure. Deviation from these conditions will result in quite different values being obtained. For example, if two pieces of zinc are dipped into zinc sulphate solutions of different concentration (the solutions being connected via a salt bridge), a difference in potential will be measurable between the two electrodes. The electrode in the more concentrated solution will be the cathode and that in the dilute solution the anode. This is a differential concentration cell, many examples of which are found in corroding systems. Thus, any electrochemical series may be produced for a given environment, that is metals and alloys may be arranged in order of corrosion resistance to that environment. Table 1 illustrates this point for alloys immersed in flowing seawater [3].

Anodes and cathodes need not be separate electrodes but can be areas on the same piece of metal. O'Halloran et al. [4] have developed a technique in which isopotential contours on the corroding electrode may be mapped (see Fig. 1). As the technique involves gathering a large number of data points, a microprocessor is used. A small reference electrode is passed across a corroding specimen close to its surface and the potential differences relative to another fixed reference electrode are recorded. The potential profile reflects the ion current density in the vicinity of the corroding surface and

TABLE 1

Electrochemical series for metals in flowing seawater

Metal	EMF range/V (SCE)
Graphite	0.3/0.20
Platinum	0.24/0.19
Ni–Cr–Mo alloy C	0.09/ − 0.03
Titanium	0.06/ − 0.04
Ni–Cr–Mo–Cu–Si alloy B	0.04/0.02
Nickel–iron–chromium alloy 825	0.04/ − 0.02
Alloy "20" stainless steels, cast and wrought	0.05/ − 0.03
Stainless steel Types 316, 317	0.0/ − 0.10
Nickel–copper alloys 400, K-500	− 0.04/ − 0.13
Stainless steel Types 302, 304, 321, 347	− 0.05/ − 0.10
Silver	− 0.10/ − 0.15
Nickel 200	− 0.10. − 0.20
Silver braze alloys	− 0.10/ − 0.20
Nickel–chromium alloy 600	− 0.14/ − 0.18
Nickel–aluminium bronze	− 0.15/ − 0.22
70-30 Copper–nickel	− 0.17/ − 0.23
Lead	− 0.19/ − 0.25
Stainless steel Type 430	− 0.20/ − 0.27
80–20 copper–nickel	− 0.21/ − 0.27
90–10 copper–nickel	− 0.22/ − 0.28
Nickel silver	− 0.25/ − 0.28
Stainless steel Types 410, 416	− 0.26/ − 0.35
Tin bronzes (G & M)	− 0.25/ − 0.31
Silicon bronze	− 0.26/ − 0.29
Manganese bronze	− 0.27/ − 0.24
Admiralty brass, aluminium brass	− 0.27/ − 0.36
Pb–Sn solder (50/50)	− 0.28/ − 0.36
Copper	− 0.30/ − 0.37
Tin	− 0.31/ − 0.33
Naval brass, yellow brass, red brass	− 0.30/ − 0.40
Aluminium bronze	− 0.31/ − 0.42
Austenitic nickel cast iron	− 0.43/ − 0.54
Low alloy steel	− 0.57/ − 0.62
Mild steel, cast iron	− 0.60/ − 0.71
Cadmium	− 0.70/ − 0.74
Aluminium alloys	− 0.76/ − 1.00
Beryllium	− 0.96/ − 0.98
Zinc	− 0.99/ − 1.02
Magnesium	− 1.60/ − 1.63

thus gives information about the location and magnitude of the corrosion sites. The study also included investigations into the effect of reference electrode design, distance from specimen surface, and type of reference electrode. The contour maps produced by this technique showed good correlation with photomicrographs of the scanned area.

The deviation of equilibrium electrode potential from the standard value

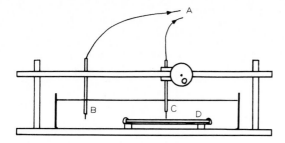

Fig. 1. Apparatus for mapping of the potential across a corroding steel surface. A, monitoring instrumentation; B, stationary reference electrode; C, travelling reference electrode; D, steel electrode in electrolyte.

when non-standard conditions are used is expressed by the Nernst equation

$$E = E^\circ + \frac{RT}{nF} \ln \frac{a_m^{n+} \cdot}{a_m}$$

where E is the potential under non-standard conditions, E° is the potential under standard conditions, R is the gas constant, T is the absolute temperature, F is the Faraday constant, a_m is the activity of metal (unity), and a_m^{n+} is the activity of metal ion in solution. In dilute solutions, the activity may be replaced by concentration. Care must be exercised when using this equation since natural electrode processes rarely consist of a single simple equilibrium. The equation is useful in looking at effects of pH on simple electrode reactions and in the production of Pourbaix diagrams [5]. These diagrams have been produced for many metals and alloys in various environments and provide information on the thermodynamic stability of the metal and any reaction products such as ions, oxides, hydroxides, etc. Certain kinetic data may also be incorporated on these diagrams, e.g. whether the metal will freely corrode or be coated with a passivating oxide. Figure 2 depicts such a diagram for the iron–water system [6].

2.2 KINETIC CONSIDERATIONS

So far, consideration has been given to whether a metal will or will not corrode in a given environment and not how fast this corrosion would proceed. A metal may be used in an environment in which it is unstable provided that its rate of corrosion is very low and allows the structure to perform its task for the required period of time, e.g. steel brake drums on road vehicles.

Loss of metal is due to an anodic reaction at the metal surface; this may occur locally or be spread equally over the metal surface. The manner in which this anodic process occurs dictates the type of corrosion obtained. For instance, if a brass and steel pipe are coupled together, as in many condensate lines, then the brass behaves as the cathode and causes the steel pipe to corrode close to the contact region of the two pipes. This may proceed to

238

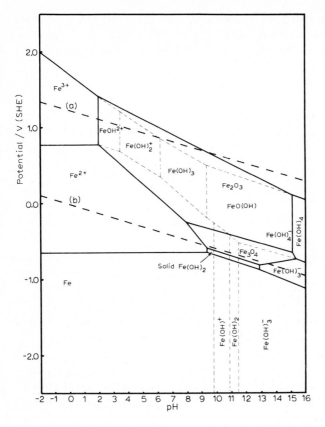

Fig. 2. E.M.F./pH diagram for iron (activities 10^{-6} M). (a) Oxygen evolution/reduction; (b) hydrogen evolution/reduction.

such an extent that the whole of the steel may be oxidised, resulting in leakage of condensate. (These types of corrosion will, however, be discussed more fully in a later section). Thus, the anodic reaction rate is how fast the metal corrodes.

$$Fe \;\rightarrow\; Fe^{2+} + 2\,e$$

The relationship between metal dissolution (or plating) and passage of electric charge is governed by Faraday's laws.

(a) The amount of chemical decomposition is proportional to the quantity of electric current (or charge) passed.

(b) The masses of the different species dissolved or deposited at the electrodes by the same quantity of electric current are in direct proportion to their equivalent weights.

Thus, by measuring the quantity of charge passed at an electrode over a given time period, we would be able to calculate the corrosion rate in terms of weight loss. As mentioned above, however, corrosion occurs on one piece of metal which has both anodic and cathodic reactions occurring on its surface together. That is, all charge passing from anode to cathode occurs within the metal and is not directly measurable externally. The relationship between potential and current of a single electrode reaction was found by Tafel [7] to be

$$E = a + b\log I \tag{1}$$

where E is the electrode potential, I the current density, and a and b constants. This originally empirical equation may be derived theoretically and is a special case of the Butler–Volmer equation [8, 9].

The Butler–Volmer equation relates the effect of anodic or cathodic overpotential to net anodic or cathodic current density for an electrode reaction under activation control; that is, free from mass transport and concentration effects.

$$i = i_o[\exp\{(1 - \beta)Fn/RT\} - \exp\{-\beta F\eta/RT\}] \tag{2}$$

where i is the net current density, i.e. an external current which may be measured by a meter, i_o is the exchange current density (it is the internal current when the anodic and cathodic reactions are occurring at an equal rate), η is the overpotential which is the amount the potential has to be moved from its rest potential in order to produce the net current i, and β is the transfer coefficient or symmetry factor, the value of which depends upon the shape of the reaction-coordinate/energy curves for the electrode process.

Consider the case when the overpotential is negative (i.e. cathodic polarisation) and much less than (RT/F), i.e. a large negative number, then the first term in the Butler–Volmer equation, eqn (2), may be neglected and

$$i = -i_o \exp(-\beta F\eta/RT) \tag{3}$$

If the electrode is polarised anodically such that η is much greater than RT/F, then the second term in eqn (2) may be neglected and

$$i = i_o \exp[(1 - \beta)F\eta/RT] \tag{4}$$

Equations (3) and (4) may be re-written expressing overpotential in terms of net current and exchange current densities.

$$-\eta = -\frac{RT}{\beta F}\ln i_o + \frac{RT}{\beta F}\ln(-i) \tag{5}$$

$$\eta = -\frac{RT}{(1 - \beta)F}\ln i_o + \frac{RT}{(1 - \beta)F}\ln(-i) \tag{6}$$

Compare the form of these equations with the Tafel equation, eqn (1); the slope, b, of the Tafel line is thus $(RT/(1 - \beta)F)$ (anodic) and $(RT/(\beta F)$ (cathodic). Note also that, when $\eta = 0$, $i = i_O$ and the exchange current density may be found from the intersection of the anodic and cathodic Tafel slopes (at $\eta = 0$). This is one method of determining corrosion rates since, at i_o, $i_a = i_c = i_o$. Thus, the anodic exchange current is known and from Faraday's laws, a rate of corrosion in terms of weight loss may be calculated. The drawback to this method is that such large overpotentials may cause physical changes at the metal surface, causing an erroneous corrosion rate to be calculated. However, if η is kept small, say $< 10\,\text{mV}$, then there is less danger of disrupting the interphase structure. For this condition, eqn . (2) may be simplified by expanding the exponential and neglecting higher than first-order terms.

$$i = i_o \left[\left\{ 1 + \frac{(1-\beta)F\eta}{RT} \right\} - \left\{ 1 - \frac{\beta F\eta}{RT} \right\} \right] \qquad \text{for } |\eta| < 10\,\text{mV}$$

$$i = i_o \, F\eta/RT$$

or

$$\frac{\eta}{I} = \frac{RT}{Fi_o} \qquad \text{for } |\eta| < 10\,\text{mV} \tag{7}$$

where η/I is the polarisation resistance for small potentials. Equation (7) is for a single electrode process, e.g. $M \rightarrow M^{z+} + z\text{e}$. The cathodic process in corrosion is not the reverse of the anodic process but may be oxygen reduction or hydrogen evolution.

There are thus at least two electrochemical reactions occurring at a corroding electrode. The potential of such an electrode will lie between the reversible electrode potentials of the two electrochemical reactions at a point where the magnitude of the cathodic current of one is equal to the anodic current of the other.

Let $\vec{i}_z$ be the cathodic current of the species to be reduced, e.g. oxygen, $\overleftarrow{i}_z$ the anodic current for oxidation of, for example, oxygen evolution, $\vec{i}_m$ the cathodic current for metal ions plating out, $\overleftarrow{i}_m$ the anodic current for metal oxidation, i_x a measurable external current, $-\beta_z$ the Tafel slope for $\vec{i}_z$, and β_n the Tafel slope for $\overleftarrow{i}_m$.

Then, for cathodic polarisation

$$i_x = (\vec{i}_z + \vec{i}_m) - (\overleftarrow{i}_z - \overleftarrow{i}_m)$$

If $\vec{i}_m \ll \vec{i}_z$ and $\overleftarrow{i}_z \ll \overleftarrow{i}_m$, then

$$i_x = (\vec{i}_z + \overleftarrow{i}_m)$$

At $\vec{i}_z = \overleftarrow{i}_m = i_{corr}$, $i_x = 0$.

Measuring potential relative to E_{corr}, i.e. $E_{corr} = 0$, then

$$i_x = i_{corr}(e^{-\eta/\beta_z} + e^{-\eta/\beta_m})$$

At low overpotentials

$$i_x = i_{corr}\eta \left(\frac{1}{\beta_z} + \frac{1}{\beta_m} \right)$$

$$i_x = i_{corr} \eta \left(\frac{\beta_z + \beta_n}{\beta_z \beta_m} \right)$$

As $\eta \to 0$, then

$$\frac{di}{d\eta} = i_{corr} \left(\frac{\beta_z + \beta_m}{\beta_z \beta_n} \right) \tag{8}$$

Equation (8) was derived by Stern and Geary [10] and is the theoretical principle upon which many commercial corrosion rate meters operate.

3. Environmental effects on corrosion

Corrosion is the result of the interaction between the metal and the environment in which it is placed. It follows, therefore, that a change in environment may result in a change in the rate of corrosion. It should also be noted that a change in environment may cause a previously protected metal to corrode. For example, a piece of steel may be cathodically protected at $-800\,mV$ S.C.E. in oxygenated seawater, but if it is buried in sand containing sulphate-reducing bacteria, then this potential may be insufficient to prevent corrosion in the acid sulphide environment. Similarly, many chemical corrosion inhibitors have a range of pH over which they are effective, if the environmental pH strays beyond this range, then corrosion may result.

In discussing environment, we can look at its effect on a macro scale, e.g. in the atmosphere, in the ocean, etc. and also examine effects on a micro scale, i.e. what is happening on the metal surface or over short distances. Due to the great variety of environments in which metals are put to use, the range of corrosion problems are equally numerous. Often, similar types of corrosion occur in many environments and may stem from similar mechanisms; these have been given specific names which indicate how the corrosion has occurred. For example, under-deposit corrosion and crevice corrosion are related, both being due to oxygen concentration cells.

3.1 MACRO ENVIRONMENTS

Corrosion of a particular metal may change in extent and mechanism by changing the environment. Although we may quote a given environment in discussing rates of corrosion, in reality environments may be a continuum. For example, the environment experienced by a car component may vary from salt spray to urban atmosphere to polluted industrial atmosphere or, at any one time, may be a "mix" of environments. For an environment may not fit into a single compartment and may also vary in time and space.

These latter reasons make it essential in corrosion failure analysis to know the "history" of the failed article.

References pp. 277–279

3.1.1 Exposure to high temperature

High-temperature conditions such as those occurring in gas turbines or rocket propulsion units present severe conditions of operation to any metallic component. In such conditions, the hot gases attack the metal surface causing oxide formation (in the case of oxygen attack). In order for the metal to be serviceable under these conditions, oxidation must cease immediately a thin adherent film has formed on the metal surface, i.e. the film is protective. What factors govern the rate of scale formation in metal/hot gas reactions?

Initially, a clean metallic surface may be presented to the hot gases and the rate of oxidation is proportional to the exponent of the activation enthalpy, which will be related to the heat of formation of the scaling components. However, once this initial layer of scale has been formed, the metal will be separated, to some degree, from the hot gaseous environment and other factors must now come into play in controlling oxidation rate.

In order for further reaction to occur, the metal must migrate through the scale layer and/or the oxidising molecules must migrate in the opposite direction, towards the metal surface [11]. Within the scale phase, the gaseous molecules will be initially chemisorbed, e.g. as O^{2-}, and thence migrate under chemical/electrostatic influence towards the metal surface. Defects in the scale phase allow both cation and anion migration, the process being analogous to electron migration in semi-conductor materials.

Other important factors are the thermodynamic stability of the resultant scale and its structure relative to that of the underlying metal. For example, an oxide which has a far greater volume than the metal crystal from which it was produced, and little bonding with the metal, is liable to produce a thick, loosely adherent scale which easily spalls from the metal surface, thereby leaving the metal surface open to further attack.

Combinations of the above factors may thus lead to very different rates of attack on metals. The most common method of studying high-temperature oxidation of metals is to analyse the pattern of film (scale) growth and then assess which physical/chemical mechanisms would fit those rate laws. In this way, the effects of adding alloying elements to the metal can clearly be seen.

High-temperature oxidation studies are of increasing importance as metal components are required to perform under increasingly severe conditions as design of engines and turbines progress to greater efficiencies.

3.1.2 Atmospheric corrosion

Atmospheric corrosion is probably that which is most evident to the layman. The pleasant green patina formed on the roofs of many buildings is due to the even corrosion of the copper sheeting underlying the adherent corrosion products, which consist of basic copper carbonate and sulphates. Atmospheric corrosion occurs electrochemically and is due to the joint action of thin films of water together with dissolved gases and salts and solid particulate matter [12,13]. Thus, the factors which affect the rate of atmospheric corrosion are the relative humidity, critical humidity, concentration of dissolved gases and pollutants, and the time of wetness. The presence of the aqueous layer is controlled by the shape of the surface and its chemical nature. A convex liquid surface, i.e. a droplet which has no surface wetting action, is stable only at high relative humidity and the vapour pressure above the surface is therefore high. A surface film of water, i.e. one which wets the metal surface and has a flat interface with the atmosphere, is more stable and the equilibrium water vapour pressure is lower than that of the convex liquid surface. The most stable, and therefore most persistent, water layer is that with a concave surface toward the atmosphere; this is stable at the lowest relative humidity and has the lowest equilibrium water vapour pressure. A concave liquid surface will exist where there are crevices in the wetted metal surface or under solid deposits such as soot, sand grains, or loose corrosion products.

If corrosion rate is plotted against humidity, then a curve such as that in Fig.3 would be obtained. Here, corrosion rate is low until, over a narrow range of humidity, the rate suddenly begins to increase. This point is termed the critical humidity and its value will depend upon the metal and nature of any dissolved species in the water film. For example, iron in a sulphur dioxide polluted atmosphere will have a critical relative humidity of above 75%, whereas a copper surface polluted with iodide will reach a critical relative humidity of about 35%.

The "time of wetness" of a surface will depend upon its geometry and presence or absence of surface deposits. It will also depend on how long the relative humidity remains above the critical level. In warmer climates, this is most likely to occur at late evening and through to early morning.

The products of atmospheric corrosion may be protective or may enhance corrosion. For example, zinc in an urban atmosphere will form a protective basic carbonate layer. However, if sulphur dioxide is present, then this layer is disrupted and corrosion proceeds. The degree of film breakdown will depend upon the concentration of sulphur dioxide in the atmosphere [12].

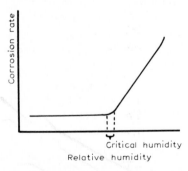

Fig. 3. Response of the corrosion rate to relative humidity.

Seashore environments are particularly aggressive. Humidity is often high and the wind carries droplets of brine, salt crystals, and abrasive sand grains, all of which reduce or destroy protective films on exposed metal surfaces.

The geometry and the orientation of metal structures will also affect their corrosion resistance. For instance, portions of structures where there is limited air flow such as the raised base of a steel tank, are likely to corrode faster than the top of a tank, this being due to longer periods of high humidity and lack of sun or wind to dry the base surface.

Away from the seashore, the aggressive character of the atmosphere to a metal surface is associated with the concentration of pollutants in the air, especially acid gases from car exhausts and chimney stacks. It is increased public awareness of atmospheric pollution, due to acid rain and its effect on life in lakes, on forests, and on buildings, that has brought about pressure for changes in legislation concerning the discharge of these materials. This, in turn, leads to a need for further research into the corrosive effects of pollutant gases (especially oxides of nitrogen and sulphur dioxide) and methods of prevention, i.e. gas scrubbers, electrostatic precipitators, etc. The old method of building taller chimneys to "throw" the waste into another area is no longer acceptable [14,15].

3.1.3 Reducing environment

Reducing conditions are found in anoxic soils, muds, and in oil and gas wells. In fact, for the formation of coal, oil, and gas, anaerobic conditions are necessary otherwise the carbon from the plant and animal remains would have been oxidised by oxygen from the air and lost as carbon dioxide. The cathodic reaction in reducing environments is not, therefore, oxygen reduction but hydrogen evolution enhanced by bacterial action, particularly sulphate-reducing bacteria. For example, sheet steel piling exposed to sulphate reducing bacteria (SRB) in marine and estuarine muds may corrode at a higher rate than that exposed directly to the oxygenated water. The detailed mechanism of corrosion by SRB on steel is still not certain [16] but a likely mechanism is the utilisation of sulphate, carbon source, and hydrogen by the bacteria for its energy requirement and the production of sulphide, acetate, carbon dioxide, and water as waste products. The evolved hydrogen is subsequently utilised by the bacteria, thus depolarising the cathodic regions of the steel. At anodic sites, iron is dissolved in the ferrous state, which combines with the sulphide produced by the bacteria and causes ferrous sulphide to precipitate on the steel surface. These precipitates become local cathodes, thus aiding further depolarisation of the cathodic reaction and subsequent increase in corrosion (anodic reaction) rate.

3.1.4 Industrial waters

In industry, many environments are encountered, many of them being "man-made". In the chemical industry, metals are required to withstand extremes in pH, ranging from concentrated acids to concentrated caustic substances. The metal may be in the form of a pipe, a storage container, or a reaction vessel. High temperatures are also encountered, such as in the pumping of liquid sodium through an atomic reactor core to act as a coolant. High pressures are also common both at low temperatures (gas liquefaction) and high temperatures (polymerisation processes). Each of these environments require detailed investigation as to its effect on metals and their corrosion resistance.

One of the main requirements of industry is water. This may be used as a solvent, a cleaner, potential power (hydroelectricity), lubricant, or coolant. In most of these cases, the water will come into contact with metals and the corrosivity toward a given metal will depend on the nature and amount of dissolved and suspended material in that water.

As an example, water used for cooling purposes will differ in quality depending on the location of the industry and the amount of waste heat it is required to remove from a process. Cooling waters may range from seawater, brackish estuarine waters, hard bore hole water, and lime-softened water to demineralised water. These waters will be recycled and removal of waste heat from the cooling water is often achieved by evaporation of a portion of the water. This warm water is passed down a tower packed with baffles and a natural or forced draft of air is passed counter-current up the tower. The latent heat of evaporation is removed from the water, thus causing a cooling effect. Due to the loss of water from the system, the concentration of salts in the water increases and also further water (also containing dissolved salts) must be added to maintain the volume. If nothing were done, the salts would become so concentrated that precipitation would occur and eventually the cooling system would become blocked. However, before this state is achieved, the less soluble salts such as calcium carbonate would have already precipitated, causing loss of heat transfer in the cooling system and hence loss of efficiency. Thus, in order to maintain a workable concentration of salts in the system, a portion of the water is continually "blown down", i.e. sent to drain. As cooling water costs money, then higher cycles of concentration (CF) will be more economic. The main ways of gaining higher cycles of concentration are either to acid-dose the water to maintain a pH less than 8 (thus preventing calcium carbonate precipitation) and/or the use of chemicals to prevent deposition of salts within the cooling system. Acid dosing lowers the system pH and also removes or prevents the formation of the calcium carbonate film which provides protection against corrosion of the underlying metal, particularly steel. The method of preventing corrosion is then to use a chemical inhibitor added to the cooling water. Cooling water systems may contain many metals, particularly brass, steel, copper, and aluminium. These metals are often in direct contact and, consequently, galvanic corrosion may result. Contaminants from the process side may also gain entry into the cooling water. These may be aggressive, such as acids, or indirectly harmful, such as nutrients for bacterial slime which restricts surface access to corrosion inhibitors.

Water is also used with oil and other additives in the form of an emulsion as a coolant and lubricant in cutting processes. These fluids must be non-flammable, as a spark from the cutting tool or electrical equipment would cause a fire, and they should not cause irritation to the skin or produce harmful vapours. Furthermore, for purposes of protection from corrosion, they should not cause corrosion of the cutting equipment or the material being cut and should give short-term protection from atmospheric corrosion.

3.1.5 Marine environment

The ocean covers about 70% of the Earth's surface and the chemistry of seawater varies from region to region depending upon local geography [17]. Near the mouth of a large river such as the Mississippi, the amount of dissolved solids will be low, whereas in a land-locked sea such as the Dead Sea, the dissolved solids concentration will be very high due to surface evaporation. The large oceans, however, have a fairly constant chemical composition (see Table 2). The salinity is about 35 parts per thousand (p.p.t.) and the major constituents are sodium and chloride ions. Thus, the high conductivity and chloride ion content make seawater an extremely corrosive medium and, because few metals are resistant to seawater, further methods of protection are generally employed, such as cathodic protection and painting of ships. Seawater is slightly alkaline, about pH 8.2, due to carbonate/bicarbonate equilibria, and thus the cathodic reaction is that of

TABLE 2

Average composition of seawater (major ionic species) (35 ppt) [64,65]

Species	Concentration	
	$g\,kg^{-1}$	$mol\,kg^{-1}$
Na^+	10.77	468.5
K^+	0.399	10.21
Mg^{2+}	1.290	53.08
Ca^{2+}	0.4121	10.28
Sr^{2+}	0.0079	0.090
Cl^-	19.354	545.9
Br^-	0.0673	0.842
F^-	0.0013	0.068
HCO_3^-	0.140	2.30
SO_4^{2-}	2.712	28.23
$B(OH)_3$	0.0257	0.416

oxygen reduction. Dissolved oxygen concentration, therefore, has a major effect on the corrosion rate, particularly of steel. Long-term exposure to seawater, such as in the case of wrecked ships, does not, however, lead to total destruction of the metal. The thick layer of rust and scale reduces the corrosion rate and this is the reason why articles such as cannons from ancient ships have been recovered and restored. This slowing down of corrosion rate is probably due to a combination of the thick rust layer, deposition of adherent calcium carbonate [18] and partial or total burying in silts. The calcium carbonate is deposited due to secretion by marine organisms, but is also due to the higher alkalinity at cathodic sites on the metal surface.

3.2 MICRO ENVIRONMENTS AND CORROSION TYPES

Changes of environment over small distances on metal surfaces can lead to specific types of corrosion some of which, once propagated, may be self-sustaining. The nature of the metal surface itself can lead to the production of different environments along its surface; for example, an intercrystalline crack or inhomogeneity in the surface will lead to preferential cathodic or anodic reaction which, in turn, will cause changes in the chemistry of the local solute/solvent. These "micro environments" and types of corrosion are closely related and some common occurrences are described below.

3.2.1 Corrosion due to differential concentration cells

Corrosion may be caused by local differences in oxygen concentration at the metal surface. This may be demonstrated by a simple experiment (Fig.4). Two identical mild steel electrodes are placed in two beakers of sodium chloride solution (0.01 M) and electrical contact of the solutions is made via a salt bridge. One solution is sparged with air or oxygen and the other with nitrogen. If an ammeter is connected across the steel electrodes, then a current is observed to flow. The oxygenated electrode is the cathode and the

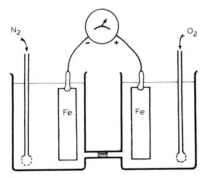

Fig. 4. Apparatus to show the production of electrical current due to the differential aeration of steel electrodes in NaCl solution.

References pp. 277–279

nitrogen sparged electrode the anode. Corrosion products are seen to have formed in the latter beaker.

Thus, any surface geometry or covering which leads to poorer access of oxygen to one part of a metal surface over another will lead to the formation of an oxygen concentration cell and resultant localised corrosion. In such cases, the anodic dissolution occurs as close to the cathodic (aerated) area as possible, especially in poorly conducting media. This is due to the current taking the shortest path, i.e. least resistance in both metal and electrolyte.

3.2.2 Corrosion under deposits

Under-deposit corrosion is a particular type of corrosion caused by differential aeration. If sparingly soluble salts, loosely adherent corrosion products, algal, or other fouling, is deposited on a metal surface, then these areas become depleted in oxygen. Unfouled or less fouled areas have a greater supply of oxygen and hence become cathodic to the fouled areas. Thus, the anodic under-deposit areas will corrode preferentially.

This type of corrosion may be observed on heat exchanger tubes which have been fouled by loosely adherent scale. Beneath the scale, the steel tube begins to corrode, releasing ferrous ions. On seeping from beneath the deposit, these become oxidised to ferric ions which precipitate as oxides and hydroxides (rust). This enlarges the fouled area and fresh corrosion occurs beneath the new rust deposits. If flow is present, then it is generally found that the deposits will develop as long streaks in the direction of flow (Fig.5). The cathodic area may be seen as a light halo in front of the advancing deposit. This region will have higher oxygen transport due to local turbulence set up by the deposit and also a higher hydroxyl concentration due to the reduction of oxygen, hence insoluble hydroxides will be formed from ferric and ferrous ions on entering this region and aid in advancing the deposit.

3.2.3 Crevice corrosion

Crevice corrosion is yet another example of corrosion caused by a difference in oxygen concentration between two areas on the metal surface. In this case, the region of low oxygen concentration lies inside a crevice caused by the overlapping of a piece of metal or other material, e.g. the crevice which exists under a washer pressed onto a metal surface (Fig.6). Even if the washer is insulating (e.g. nylon) as is used for mounting test coupons, corrosion will still occur.

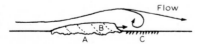

Fig. 5. Advance of under-deposit corrosion on a heat exchanger tube. A, Anodic area; B, corrosion products; C, cathodic area.

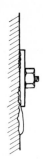

Fig. 6. Crevice corrosion beneath a washer. Note the under-deposit corrosion due to seepage of corrosion products from the crevice.

In fact, when testing corrosion inhibitors, the major amount of corrosion may occur under the mounting washers as fresh inhibitor cannot reach such a confined space and, in addition, the low oxygen concentration leads to corrosion. As in the section above, ferric ions may seep from within the crevice and deposit around the fixture, thus giving rise to under-deposit corrosion.

Crevice corrosion can be particularly harmful to stainless steel. Within the crevice, the passivating oxide layer is broken down, particularly with high chloride concentration. However for renewal, oxygen is required and this is depleted inside the crevice. Thus, the stainless steel within the crevice actively corrodes while that outside, in an oxygenated environment, acts as the cathode.

3.2.4 Pitting corrosion

Many metals are resistant to corrosion due to a compact adherent oxide film which acts as a barrier between the metal and its environment. Aluminium and stainless steels are examples of such metals. Often, the properties of such a barrier film will depend on the environment in which the metal is situated. For example iron in seawater does not produce corrosion products which are protective but in dilute carbonate solutions with no or very low chloride concentrations, a passive complex $Fe(OH)_2/FeCO_3$ layer is formed [19, 20].

A common form of corrosion met with this type of metal is pitting corrosion. Pitting arises when local cells are formed. This may be due to the presence of impurities in the metal such as sulphides, silicates, or noble metals such as copper. These areas will form local cathodes, adjacent weak spots in the oxide film will break down, and the underlying metal corrodes locally. Species within the aqueous medium may also induce pitting; chloride ions are particularly aggressive in this respect [21].

The process of pitting may be divided into initiation and growth. Initiation of pitting may be caused by setting up of local cathodes due to impurities as described above. This causes the local cell potential to rise above the pitting potential. Thus, local weak spots in the oxide film break down and a local anodic area develops. Oxide films may be weakened by the incorporation of chloride ions by diffusion into the lattice. At this stage, the local cell may be stopped if the metal dissolving at the anodic site is able to form protective corrosion products. Growth will continue if repassivation is not possible. Repassivation is prevented by several factors. Chloride ions will migrate into the pit, producing high concentrations which will prevent any adherent oxide film being formed. Chlorides have a further effect in that metal chloride complexes are formed which hydrolyse with water, producing an acid environment within the pit. Corrosion products may also form a semi-permeable cap on the pits, thus reducing the influence of dissolved oxygen. This, together with the acid environment, causes a local rise in passivation potential within the pit, which therefore remains active. The pit does not spread laterally because of the high electrical conductivity of the solution and more anodic potential within the pit; the high current density within the pit prevents the surrounding area from falling below the passivating potential (Fig.7). In terms of weight loss, pitting may not appear to be a serious form of corrosion. However, it is the localised loss of metal which can cause drastic reduction in strength or local perforation of the metal. Perforations will cause leakage of fluids from the metal vessel or pipe causing further damage or at least loss of fluid. Pitting damage to structurally important components such as derrick support legs may lead to collapse and loss of the whole structure. It is therefore essential when choosing a metal for service that it is resistant to pitting attack in the environment to which it will be exposed.

3.2.5 Intergranular corrosion

Another form of localised corrosion is that which occurs along grain boundaries of a metal from the surface into the body of the metal. This is intergranular corrosion. Austenitic stainless steels are prone to this type of corrosion unless suitably alloyed. If such a stainless steel is heated, say during a welding operation, then inhomogeneity from grain to grain boundary is introduced. The grain boundary tends to have a higher carbon content than the grain body; upon heating, chromium from near the grain boundary migrates, combining with the carbon at the boundary to form

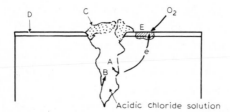

Fig. 7. Cross-sectional view of a pit. A, metal dissolution reaction, e.g. $Al \rightarrow Al^{3+} + 3e$, acid chlorides form, which produce hydrochloric acid and aluminium hydroxide on hydrolysis; B, $2H^+ + 2e \rightarrow H_2\uparrow$; C, porous corrosion product restricting oxygen access; D, passive layer on the metal surface; and E, inclusion acting as a local cathode.

chromium carbide. This leaves the adjacent areas low in chromium and sensitised to attack and thus any breakdown in surface oxide film at these points will not be repaired and corrosion will proceed along the sensitised grain margins. Thus, in determining the cause of failure due to corrosion, it may also be necessary to know the thermal history of the component and not just the bulk chemistry of the metal and its environment.

3.2.6 Mechanical effects on corrosion

Corrosion in conjunction with mechanical factors may also lead to premature failure of metallic components. Certain alloys are particularly vulnerable when placed in an aggressive environment and loaded under tensile stress or cyclical stress loading. Under stressed conditions, the metal begins to crack. If the crack follows grain boundaries, then it is termed intercrystalline stress corrosion cracking or, if it cuts across grains, it is termed transcrystalline stress corrosion cracking (SCC). This is a localised form of corrosion with a mechanism similar to that of pitting.

Stress corrosion cracking differs from corrosion fatigue cracking in that it is specific to certain metal/environment combinations. In corrosion fatigue, any corrosive environment will reduce the limit of endurance of the component. Failure by fatigue occurs when the metal is subjected to cyclical stresses such as those imposed on rotating blades or taut cables vibrating in windy conditions. The endurance limit will depend upon the stress and the number of cycles to which the metal is subjected. The endurance limit is further reduced if the metal is fatigued in a corrosive environment. Corrosion fatigue is accelerated if the metal is anodically polarised and is reduced if the metal is made more cathodic. Fatigue cracks may begin at surface imperfections such as pits, grinding or cutting marks. Stress is a maximum at such sites, i.e. the defects of the surface act as localised stress centres. The imperfections and subsequent cracks tend to be anodic to the rest of the metal surface; this is due to the high stress and also the destruction of protective oxide film at these sites. Cracks induced by corrosion fatigue are usually transgranular and the fracture surface shows a characteristic structure similar to the growth rings on a shell.

Stress corrosion cracking occurs when a susceptible metal is maintained under stress in an environment which is specifically aggressive to this type of corrosion on a metal. Brittle failure of the metal occurs, i.e. there is little reduction in cross-sectional area of the metal. Both transcrystalline and intercrystalline cracking may be noted in a single specimen; changes in the environment may also cause a change in the predominant type of cracking. As in corrosion fatigue, the crack is anodic and its propagation may be aided by concentration cell formation (e.g. oxygen concentration cell). Examples of specific metal/environmental species which may lead to SCC are 18/8 austenitic stainless steel/Cl⁻, carbon steel/caustic, and brass/ammonia vapour. SCC rarely occurs in pure metals and appears to require inhomogeneity in the metal; this increases the "weak spots" in the protective film leading to local cell action. Mechanical factors such as stresses set up during cooling of castings or elongation and flattening of crystals during rolling, may also increase susceptibility to SCC.

Hydrogen embrittlement, which is due to atomic hydrogen collecting within the metal at dislocations and preventing or reducing plastic flow, is another form of cracking. If voids occur within the metal body, then the atomic hydrogen may combine to form dihydrogen (H_2) and the resultant build-up of pressure causes the metal to blister. As the hydrogen content of the metal increases, the strength of the metal decreases. Unlike stress corrosion cracking, hydrogen embrittlement is exacerbated by cathodic protection [22], especially at low potentials where the major cathodic reaction is hydrogen evolution. The presence of such elements as As and S increases the susceptibility of steel to hydrogen embrittlement as they inhibit the combination of hydrogen atoms to form dihydrogen and thus increase the concentration of atomic hydrogen.

Erosion corrosion is caused by the conjoint action of corrosion and mechanical abrasion by a moving fluid or suspended material in the fluid. Turbulent flow or jets of liquid on a metal surface may lead to erosion corrosion. The mechanical action of the fluid removes the protective corrosion deposit, thus exposing fresh metal to the corrosive. As corrosion products build up, they are removed and so the process continues. The surface of a piece of metal exposed to this type of corrosion has a characteristic structure (Fig. 8).

Cavitation corrosion is similar to erosion corrosion in that it is due to the conjoint action of corrosion and the mechanical effect of the fluid. Here, however, the mechanical damage is due to the collapse or implosion of bubbles at the metal surface. The collapse of the vapour bubbles occurs when the internal pressure of the bubbles is lower than the liquid vapour pressure; the collapse leads to strong compression and rarefaction waves at the metal surface. Cavitation corrosion leads to a pitted and roughened metal surface and may be brought about by excessive turbulence at, for example, rotor blades in a centrifugal pump. This type of wear is made use of in ultrasonic cleaning baths where the passage of ultrasound causes similar cavitations at sites of rarefaction which, upon collapse at immersed solid bodies, generates many atmospheres of pressure at their surface.

Fig. 8. Erosion corrosion. Turbulent flow breaks through the passive layer and allows corrosion to proceed.

3.2.7 Dealloying

Corrosion by dealloying is common in brasses; here the zinc component of the alloy is preferentially removed. Brasses with high proportions of the β phase are especially prone to this type of attack. The mechanism appears to be corrosion of both copper and zinc from the metal; the zinc passes into solution but the copper is re-deposited with a porous structure of low strength. Aluminium bronzes also suffer dealloying of the aluminium component if incorrectly heat treated. Other metals which may be preferentially dissolved from their alloys are manganese from copper–manganese, nickel from copper–nickel, copper from either copper–silver or copper–gold, and tin from tin–lead (solders). It is evident from this list that it is the component which is anodic to the alloy which is removed.

A further example of dealloying is graphitization of cast irons. Here, the iron is selectively removed leaving the graphite with corrosion products in the original form of the corroded article. This type of corrosion is common in grey cast iron in which the graphite is in the form of flakes, thus presenting a large surface area for the cathodic reaction.

4. Properties of metals

Modern man is dependent on the use of metals. In the course of a day, it is impossible not to use an article which was either made using a metallic machine or is itself composed in some part of metal. Copper and cast iron carry water to our houses, iron nails support timbers, steel lintels support upper floors, we are transported to work in vehicles composed of numerous metals, our meals are cooked on metal stoves, and we eat using metal cutlery. All metal articles have a particular function to fulfil and must do so with the minimum of wear, corrosion, or loss of strength. It is these factors that govern whether metal is suitable for a particular purpose.

Of all the metals, the alloys of iron are used in the greatest tonnage. Other greatly used metals are the alloys of aluminium, copper, nickel, tin, lead, and zinc. The study of the properties of these metals is the realm of the metallurgist. The properties of metals depend on numerous factors such as elemental composition, heat treatment, mechanical treatment, etc.

4.1 STRUCTURE OF METALS

The surface of a piece of metal, after polishing and etching, viewed through a microscope exhibits a crystalline structure. The size of crystals will vary according to the type of metal and its thermal history. Large crystals of zinc may be easily seen on such articles as galvanised buckets or corrugated iron. The crystals are orderly three-dimensional arrays of metal atoms. The way in which the atoms pack together dictates the crystal structure. The three most common types of packing are face centred cubic (f.c.c.), body centred cubic (b.c.c.), and hexagonal close packed (h.c.p.) (Fig.

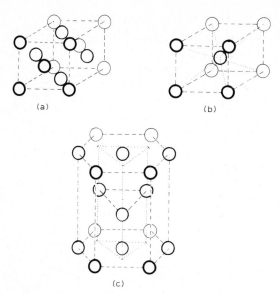

Fig. 9. Unit cell structures of metals. (a) Face centred cubic (f.c.c.); (b) body centred cubic (b.c.c.); and (c) hexagonal close packed (h.c.p.).

9). Thus, metal crystals, or grains, are three-dimensional with axes and planes of symmetry, as are any other crystals. This directional nature causes differences in etching rates and hence produces the relief in micrographs of prepared metal samples. The crystal size and shape depends greatly on the rate at which the metal solidifies from the molten state, the history of working and subsequent heat treatment, and the chemical composition. Metals which are cold-worked show crystals which are distorted and faulted along lines of weakness. Cold working increases the strength and hardness of a metal but reduces its ability to undergo further plastic deformation.

Annealing a cold-worked metal has several effects depending upon the time period over which annealing is performed. The cold-worked material contains a great deal of stress. On annealing, the first effect is to relieve this stress and so the metal recoups some of its ductility, but retains its higher strength. This treatment improves fatigue limit and corrosion resistance. Further annealing leads to nucleation of new crystals at old grain boundaries. These new crystals are free from strain and grow by diffusion of material from old grains. If annealing is continued for long enough, the metal would be completely recrystallised and would again have the ductility and lower strength properties of the pre-cold-worked material (Fig. 10).

In order to obtain the correct physical properties, such as tensile strength, machinability, etc., elemental metals are alloyed together. If the elements are soluble in one another, then an alloy composed of a single phase will be produced but commonly there is a limited solubility and thus multiphased heterogeneous alloys result. Rapid cooling of a solid solution may produce a single-phased metal whereas slow cooling of the same solution may

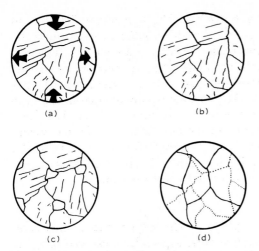

(a)   (b)

(c)   (d)

Fig. 10. Effect of the annealing time on metal grain structure. (a) unannealed, metal contains residual stresses from cold working; (b) annealed for a short time, the residual stresses are removed; (c) increased time of annealing, fresh grains begin to grow at the old grain boundaries; (d) long period of annealing, the old structure is completely obliterated, the metal regains the ductility and lower strength of the pre-cold worked state.

produce more than one phase. This is because the solubility of one metal in another varies with temperature.

The phases within an alloy may have differences in resistance to corrosion, thus leading to relative removal of one phase, i.e. one phase may be cathodic to another. Inclusions such as sulphides in steel may act as local cathodes, thus initiating pitting in the surrounding metal.

4.2 COMPOSITION OF METALS

Of the 106 or so natural and man-made elements, less than 20% are non-metals. At present, few of the metals are used to any large extent by man, either because of their rarity or their instability. The major "tonnage" metals are iron, copper, aluminium, zinc, nickel, and lead. These metals may be alloyed with one another, such as copper and zinc to form brass, and/or may be alloyed with smaller quantities of other metals. Steels are commonly alloyed with chromium, vanadium, molybdenum, or tungsten. Aluminium, for increased lightness and strength, may be alloyed with magnesium.

The range of combinations of metal elements which have been used to form alloys is enormous, only a general outline of the more common alloys can be given here.

4.2.1 Iron and its alloys

Iron is the most important metal in terms of tonnage and usage. Although fairly dense and of poor corrosion resistance, it has tremendous strength and is the fourth most abundant element in the Earth's crust. Iron is produced by the reduction of various ores, e.g. haematite, Fe_2O_3, siderite, $FeCO_3$, and magnetite, Fe_3O_4. Metallic iron is unstable under conditions existing at the Earth's crust and thus is not found native except in meteorites. The corrosion products of iron are variable in composition, consisting of oxides, hydrated oxides, and hydroxides of iron(II) and iron(III). Other constituents may be incorporated into the "rust" from the environment surrounding the metal such as sodium chloride, carbonate, and calcium ions. The composition of the corrosion products will depend on the nature of the environment; highly oxidising conditions will favour iron(III)-type deposits whilst reducing acid conditions will favour iron(II) oxides and hydroxides. The rate of corrosion of iron increases at pH < 5 due to the reaction

$$Fe + 2H^+ = Fe^{2+} + H_2$$

but is independent of pH up to about pH 11. In this region, rate of corrosion is related to the presence of oxidising agents, especially dissolved oxygen.

$$2Fe + O_2 + 2H_2O = 2Fe^{2+} + 4OH^-$$

Other factors are temperature, ionic strength, rate of flow of corrosive fluid, etc. At pH > 11 corrosion rate decreases due to the formation of a protective film of ferric hydroxide/oxide. Although general corrosion rate decreases above this pH, the metal becomes susceptible to intercrystalline attack (at defects in oxide film) and thus fails due to caustic embrittlement.

Iron always contains a small amount of carbon and other impurities unless it is specially prepared, say for spectroscopic purposes. The impurities are dissolved during production, phosphorus and sulphur from the ore, silicon and carbon from the limestone and coke. Silicon and carbon do not greatly reduce the corrosion resistance of iron when present in small quantities. Sulphur, however, may increase corrosion rates if allowed to precipitate as FeS. Crystals of FeS have a fairly high conductivity and can act as local cathodes. Additions of small amounts of manganese to the melt will cause the formation of MnS rather than FeS, the former being a poor conductor.

Carbon has a limited solubility in iron of about 2% by weight. In general, iron containing more than about 2% carbon is termed cast iron and with less than that value, it is steel. The crystal structure of a steel will depend upon the amount of carbon in the steel and also its thermal history. A simplified phase diagram of the Fe/C system is shown in Fig. 11. Slow cooling of an austenitic steel allows a change in lattice structure to occur. As the steel cools below about 700°C, carbon can no longer be held in solid solution and begins to form lamellae of Fe_3C and Fe known as "Pearlite" because of its iridescence when viewed through the microscope. If the carbon content is lower than the eutectoid (0.8%), then a mixture of pearlite and ferrite (Fe) grains will form. If the carbon content is higher, then a mixture of pearlite and cementite (Fe_3C) grains will be formed. Ferrite confers ductility on the alloy, cementite brittleness, and pearlite hardness and strength.

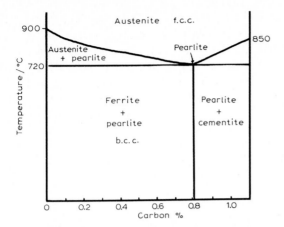

Fig. 11. Solid solution phase diagram for iron/carbon.

The corrosion resistance of steel can be greatly increased by alloying with chromium to form the stainless steels. Figure 12 shows the effect of increasing chromium content on the corrosion rate of steel. At 12–14% Cr there is a dramatic decrease in corrosion rate. The corrosion resistance is due to the formation of a thin adherent layer of chromium oxide on the steel surface [23]. The steel will remain stainless provided the oxide layer remains intact or can be rapidly repaired, i.e. the steel is exposed to oxidising conditions. The precipitation of chromium carbide at grain boundaries will cause disruption of this oxide film (See Sect. 3.2.5) and hence localised corrosion. Precipitation of chromium carbide can be reduced by alloying with elements which form carbides more readily than chromium, e.g. titanium, niobium, and tantalum.

Iron–nickel alloys tend to be of lower corrosion resistance than iron–chromium alloys except towards attack by hot concentrated alkalis. Iron–chromium–nickel alloys are superior to either of the above and are resistant to alkaline and neutral aqueous solutions, atmospheric and seawater attack. Hot non-oxidising acids will cause corrosion, the rate depending on concentration and temperature.

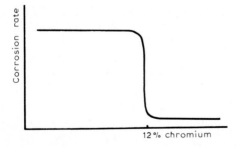

Fig. 12. Effect of alloying with chromium on the corrosion rate of steel.

References pp. 277–279

4.2.2 Copper and its alloys

Copper is a more "noble" metal than iron and is found both native, i.e. in metallic form, and as mixed hydroxide/carbonate (malachite, azurite) or combined with iron and sulphur as copper pyrites.

The metal may corrode by formation of both monovalent and divalent cations, the latter case occurring most readily in corrosion [24]. The passive film on copper is poor and may be broken down by atmospheric pollutants.

Copper is commonly used in sheets as a roofing material when, upon exposure to the atmosphere, the surface becomes covered with a basic film of copper carbonate and/or sulphate. This gives rise to the familiar green patina of such roofs. Heavily polluted atmospheres, however, cause the film to be thicker and less adherent, in which case corrosion rates are high and film spalling will occur, exposing fresh metal surface to the corrosive atmosphere.

Copper with a small amount of phosphorus (about 150 ppm) is used for pipes and tanks in domestic water supplies; it is easy to fabricate, bend and joint.

The most important alloys of copper are the brasses and bronzes, both of which have been used since ancient times. Brass is an alloy of copper and zinc whilst bronze consists of copper and tin.

Below about 36%, zinc brass is a single-phase alloy (α brass), at 36–47% two phases are present (α and β brass), and above this, the β form exists. The single-phase brass is slightly more resistant to corrosion than the two-phase brass. Brasses are resistant to the atmosphere, salt solutions, acid, and alkali. A severe form of corrosion known as dezincification may occur in weakly acidic and neutral solutions [25]. This may be prevented by adding arsenic to the alloy (Sect. 3.2.7). Intercrystalline cracking of brasses may occur, particularly in aggressive environments (ammonia vapour) and where the brass has been cold worked or cast, leaving the metal under stress. The resistance to cracking can be increased by addition of cadmium to the alloy. Many brasses have been produced with higher corrosion resistance than the simple copper–zinc alloy. The increased resistance is achieved by adding such components as nickel (10–30%), aluminium (2%), tin (0.75–1%), and arsenic (< 0.1%).

Bronzes are alloys of copper and tin and are generally more resistant to corrosion than brass, especially toward non-oxidising dilute acids. The bronzes which are used in pump heads and propellers for ships may contain manganese, nickel, and/or aluminium.

Both brasses and bronzes are used in the form of tubes and plates in the fabrication of heat exchangers for use in desalination units and heat exchangers in general.

4.2.3 Aluminium and its alloys

Aluminium is a comparative newcomer due to the difficulty of production from its ore, bauxite. It is the third most abundant element in the Earth's crust but most of it is combined in complex rock-forming silicates. The weathering of such rocks leads to enrichment of aluminium as the oxide. Production of the metal is by electrolytic reduction of the molten oxide [the melting point of which is lowered by addition of cryolite (Na_3AlF_6)].

Aluminium is a light metal with a density about one third of that of steel. It conducts electricity well and may be alloyed with other metals to improve its strength characteristics. The standard electrode potential of aluminium is -1.67 V and would be a highly reactive metal were it not for the protective oxide formed on its surface. If aluminium is immersed in a solution containing species which prevent the oxide film from reforming, then the metal corrodes rapidly with the evolution of hydrogen gas from the cathodic regions.

Aluminium is an amphoteric metal and thus exhibits high corrosion rates in both strong acid and alkaline solutions but is relatively inert at intermediate values of pH (Fig. 13). Aluminium is resistant to waters with low dissolved solid contents such as potable and river water; high chloride levels such as those found in seawater will lead to pitting attack. Hence, the metal is not very resistant to halogen acids. Oxidising salts and acids help in repassivating any damage in the film. The metal is therefore resistant to such species as dichromate, chromate, nitric acid, etc.

Pure aluminium is a fairly corrosion-resistant material. However, the presence of impurities in the metal markedly affects its resistance. This strong effect is due to the highly reactive metal underlying the protective oxide film. If impurities cause breaks in the film, then the metal is open to attack. Moreover, these impurities will act as local cathodes, thus maintaining corrosion and preventing repassivation. Some alloying elements used to increase the strength of aluminium thus reduce its corrosion resistance by the formation of intermetallic compounds, e.g. $CuAl_2$, $FeAl_3$, which are cathodic to pure aluminium. Addition of manganese and magnesium does not reduce the corrosion resistance as the electrode potentials of the intermetallic compounds are about the same and lower than that of aluminium.

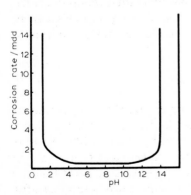

Fig. 13. Effect of pH on the corrosion rate of aluminium.

A common method of maintaining high corrosion resistance of aluminium alloys is to clad the alloy with pure aluminium. Subsequent to cladding, the alloy cannot be heat treated as diffusion of alloying metals into the pure aluminium cladding will again reduce corrosion resistance.

The high electrical conductivity and low density of aluminium make it the most suitable method for transporting electricity in the national grid. The low strength of the metal necessitates the aluminium wires to be wrapped around a high tensile steel core, so that long spans can be achieved. Without aluminium and its alloys, the aircraft industry could not have developed to the present extent. Other industries have also benefited from its low density; the superstructure of ships when constructed of aluminium reduces the weight of the ships, allowing either more cargo or a shallower draft. Other uses of the metal include kitchen ware, wrapping foil, parabolic reflectors, vehicle bodywork, canning, window frames, etc.

A great advantage of aluminium is the ability of its oxide film to absorb dyes, this leads to its use in decorative articles such as pen cases, cigarette lighters, etc. The oxide layer is thickened by "anodising", thus producing a porous structure in the oxide film. The dye is absorbed into the pores which are then sealed by hydrolysing the surface oxide.

One more recent use of aluminium is in cathodic protection. In this application, the alloy sacrificially corrodes, i.e. is anodic and thus prevents corrosion of the structure to which it is electrically connected. In order for these alloys to be efficient, they must undergo little self-corrosion, but corrode uniformly when stimulated by galvanic contact.

5. Testing methods

The corrosion resistance of metals requires testing in order to establish the metals' suitability, or otherwise, for use in specific environments. There may also be a need for testing during use if, for example, unusual conditions arise or unforeseen factors come into play. Test methods should be designed to simulate actual use conditions as closely as possible unless simplifications are required, say for mechanistic investigations. Testing methods can be divided into those conducted on site and those conducted in the laboratory; some methods are applicable to both areas.

5.1 FIELD TESTS

The type and method of a field test will depend upon the environment to which the material is to be exposed. It may not be the metal itself which is under test but the degree of protection afforded by a paint film or other protective. Reliable information is required in the shortest time and thus tests may be made more severe. A standard method of testing paint films for exterior use is to paint metal panels, some of which are scratched after painting. These panels are then exposed on racks to the atmosphere, either marine (close to the shoreline) or industrial (on a roof in an industrial area). The condition of the panels and amount of disbondment of the paint film, especially around the scratch, is noted at intervals. Such effects as fading colour, powdering, crazing, blistering, etc. are also noted. Newly painted pipelines are tested for film continuity. "Holidays" in the paint film may be located electrically (Fig. 14) using a battery, ammeter and sponge soaked in brine, suitably mounted in the form of a "probe".

The corrosion rate in aqueous systems such as cooling water may be assessed by using a side stream line containing pre-weighed metal coupons. The coupons are removed at intervals, cleaned, and re-weighed. Their weight loss, area, and time of immersion are used to calculate the average corrosion rate.

Corrosion rates may be quoted in different ways, as thousandths of an inch penetration per year (mpy), milligrams of weight loss per square decimetre per day (mdd), or millimetres penetration per year (mmpy) are some of the more usual ways. To convert from weight loss to penetration rate, the density of the metal must be known. For example, for mild steel, (density $= 7.8\,\mathrm{g\,cm^{-3}}$)

$$1\,\mathrm{mpy} = \frac{2.54}{1000}\,\mathrm{cm\,py}$$

$$= \frac{2.54}{1000} \times 7800\,\mathrm{mg\,cm^{-2}\,py}$$

$$= \frac{2.54}{1000} \times 7800 \times 100\,\mathrm{mg\,dm^{-2}\,py}$$

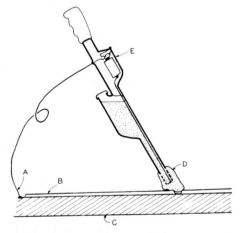

Fig. 14. Apparatus for detecting "holidays" in paint films. A, Electrical connection to pipe; B, paint film; C, metal substrate; D, brine-soaked sponge; E, power and meter section.

$$= \frac{2.54}{1000} \times 7800 \times \frac{100}{365.25} \text{ mg dm}^{-2}\text{day}^{-1}$$

$$= 5.42 \text{ mdd}$$

For aluminium, the factor is 1.79 and for brass 7.69. The reciprocal of these factors will convert mdd to mpy. Some electrochemical techniques may express corrosion rates in terms of an electrical current. In such cases, the anodic reaction must be known so that Faraday's laws may be used in converting to a mass rate loss. Thus, an exchange current density of $8 \mu A \text{ cm}^{-2}$ on mild steel will result in a corrosion rate of about 20 mdd, i.e.

$$\frac{20 \times 2 \times 96\,500 \times 10^6}{60 \times 60 \times 24 \times 56 \times 100} \approx 8 \mu A \text{ cm}^{-2}$$

Many field methods of corrosion rate determination use some form of electrical measurement and these instruments are commercially available. Figure 15 shows some of the probe types used in these measurements.

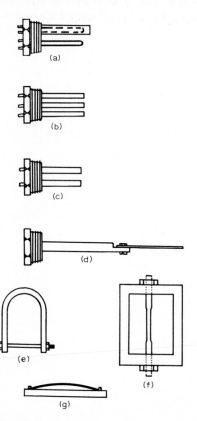

Fig. 15. Corrosion test probes. (a) Electrical resistance; (b) linear polarization; (c) zero resistance; (d) weight loss coupon; (e) U-bend specimen; (f) tensile specimen; and (g) large radius bend specimen.

5.1.1 Corrosion rates from electrical resistance

The electrical resistance of a metal is proportional to its length and inversely proportional to its cross-sectional area.

$$R = \rho \frac{l}{A}$$

where R is the resistance in ohms, l the length in metres, A the area in square metres, and ρ the resistivity in ohm metres. Thus, a metal of interest is formed into a wire and immersed in the corrosive environment. Measurement of the increase in resistance due to decrease in section yields a method of corrosion rate determinations from two successive readings over a period of time.

In order for this instrument to be used successfully, the effect of system temperature fluctuations on resistance of the wire must be nullified. This may be accomplished by having an identical loop of wire in the probe but protected from the corrosive fluid. The difference in resistance of the two loops gives one of the values for evaluation of corrosion rate. This instrument is only useful if corrosion occurs evenly over the wire loop. Pitting will cause excessively high readings and the error becomes increasingly great towards the end of the probe's life.

5.1.2 Linear polarisation technique

In Sect. 2.2, it was shown that the slope of the E versus i curve at small overpotentials was inversely proportional to the corrosion current [27, 28], i.e.

$$R_p = \frac{K}{i_{corr}}$$

where R_p is the polarisation resistance measured at $E_{corr} \pm 20\,mV$, i_{corr} is the corrosion current, and K is a constant which depends on the Tafel slopes of the anodic and cathodic reactions and on the electrode geometry.

Commercial instruments are generally calibrated directly in corrosion rate units and conversion factors are utilised for probes of metals other than that for which the meter is calibrated [29]. Some instruments also have data capture facilities for "unmanned" monitoring. Probes may consist of from two to four elements of which at least one is of the material under test. The higher the solution resistance, the larger the number of elements in the probe, the extra elements are used to assess and nullify the effects of solution resistance.

5.1.3 Zero resistance ammeter

This meter utilises the current which flows when dissimilar metals are short-circuited in an electrolyte. The test material is connected through a specially designed ammeter of zero internal resistance to a more noble metal if the anodic reaction is important or to a more base metal if the cathodic reaction is of interest. The probe is only short-circuited during measurement of the galvanic current. The measured current is proportional to the corrosion rate.

5.1.4 Chemical methods

In some cases, an estimate of corrosion rate may be gained from chemical analysis of the corrosive fluid. For example, the level of copper ions in the discharge from a desalination plant will indicate the rate of metal loss from the system.

Radiotracers [30] may also be used. Metal is exposed to radiation causing a thin layer of radioactive species to be formed in the metal. The specimen is then inserted in the plant and the rate of loss of radioactivity is monitored. After allowance for natural decay, an estimate of corrosion rate can be obtained.

The rate of hydrogen evolution in oxygen-free systems may also be monitored by using thin foils through which the hydrogen can migrate and thus cause an increase in pressure on one side of the foil.

5.2 LABORATORY TECHNIQUES

Laboratory corrosion tests are conducted in order to obtain information on the interaction of a metal with a particular environment. The tests are generally designed to simulate some field situation. Studies in the laboratory are either aimed at obtaining data in a more convenient way and in a shorter time than on-site determinations or are to provide information, either mechanistic or simulated use, before field application. Final testing should therefore correlate closely to the results obtained from field studies [31]. In devising tests, detailed consideration of their applicability is necessary. Short-term laboratory tests are always a compromise and this should be borne in mind when interpreting results. The number and diversity of test results are such that it is only possible to give a brief description of the more common tests.

5.2.1 Combined corrosion and mechanical testing

For testing the susceptibility of a material to stress corrosion cracking, specimens are held in instruments which provide either a constant tensile load or constant rate of elongation. In more sophisticated instruments, the specimens may be subjected to an oscillating force, ranging from compression/tension to tension/increased tension. These latter instruments are for fatigue studies and, generally, are able to operate at variable frequencies. The specimen may be surrounded by air or immersed in a corrosive fluid. The effect of environment on strength, fatigue life, type of failure, etc. can then be determined. Photomicrographs of the fracture surface are useful in determining the type of failure.

In corrosion fatigue experiments, a notch is cut in the specimen at which maximum stress will occur. Crack initiation and growth thus occur at the tip of this notch and may be measured with the aid of a microscope.

5.2.2 Electrochemical testing

Many corrosive environments are sufficiently conducting to allow the use of electrochemical techniques in corrosion testing. The simplest technique is merely to monitor the corrosion potential of the specimen with time but this method may be ambiguous and is generally used in conjunction with other methods. The potential of a corroding metal does not change smoothly or remain constant but consists of random "spikes" or "noise". Analysis of the amplitudes and frequencies of this noise produces "fingerprints" which indicate the type of corrosion occurring and, to some degree, its extent. This technique of corrosion monitoring requires expensive instrumentation for full spectrum analysis but simplified monitors are now available [32].

The a.c. impedance technique [33,34] is used to study the response of the specimen electrode to perturbations in potential. Electrochemical processes occur at finite rates and may thus be out of phase with the oscillating voltage. The frequency response of the electrode may then be represented by an equivalent electrical circuit consisting of capacitances, resistances, and inductors arranged in series and parallel. A simplified circuit is shown in Fig. 16 together with a Nyquist plot which expresses the impedance of the system as a vector quantity. The pattern of such plots indicates the type and magnitude of the components in the equivalent electrical network [35].

Direct current methods are commonly employed in corrosion testing and may be either potentiostatic, where the electrode potential is controlled, and current monitored or galvanostatic (intentiostatic), where the cell current is controlled and the working electrode potential monitored. The potential or current may also be varied with time, producing potentiodynamic or

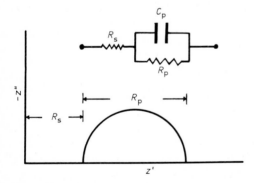

Fig. 16. Nyquist plot of the impedance response of an electrode. The equivalent electrical circuit is shown above the plot. R_s is the solution resistance, C_p the electrode/solution interface capacitance, and R_p the electrode/solution interface polarization resistance.

intentiodynamic plots. Figure 17 shows a typical potentiodynamic plot for stainless steel in an aerated neutral electrolyte.

Each of these techniques may be used with numerous cell designs. The design of cell will depend upon the characteristics of the system to be simulated. Commonly, the flow of electrolyte is important and must be known and in such cases, rotating-disc or ring-disc, flow-in-duct, or rotating-cylinder [36, 37] electrodes would be used.

5.2.3 Atmospheric corrosion testing

If suitable field sites are not available or lack controlled conditions, then corrosion tests must be conducted in the laboratory. Cabinets are constructed in which the atmosphere is controlled and high humidity and temperature can be used to help accelerate the tests. Marine environments are simulated by salt spray and industrial environments by sulphur dioxide or nitrogen dioxide. Figure 18 shows a salt-spray cabinet and the arrangement of test panels. Periodic changes of temperature within the cabinet can be used to simulate night and day. Addition of other aggressive salts or acid into the sprayed solution is further used to accelerate the test.

6. Methods of protection

Metals may be protected from corrosion by using a metal in an environment in which it is immune, by making a physical barrier between the metal and its environment, by means of an electric current, or by changing the environment. All of these methods are used, the choice depending upon the function of the metal, the cost, and the required life of the structure. Ec-

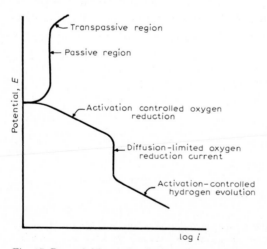

Fig. 17. Potential/log i plot for stainless steel in an aerated neutral solution.

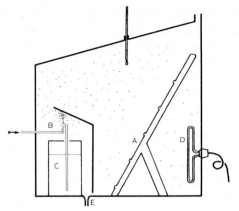

Fig. 18. Salt spray cabinet. A, Rack of test panels; B, mist spray; C, salt solution; D, thermostatic heater; E, drain.

onomics may dictate that more than one method is appropriate, for example the use of paint and cathodic protection on ships' hulls.

6.1 IMPROVED DESIGN

Corrosion protection begins at the design stage. Within the economic constraints, the material should be chosen which fulfils its expected duty. The mechanical properties of a metal are basic to its selection and this includes not only its strengths in the finished structure but also its machinability and ease of fabrication. If costs permit, the most corrosion resistant metal with the appropriate mechanical properties should be chosen; if not, then additional means of protection will be required and their cost taken into account.

In designing a structure, the following points with regard to minimising corrosion should be noted.

(i) Stress should be evenly distributed and well within tolerance limits, effects of SCC and corrosion fatigue on the material should be known and be within design limits.

(ii) Access to the structure for inspection, cleaning, and repair should be incorporated in the design. Sites where debris and moisture may collect should be avoided. Crevices and confined areas where moisture will not evaporate should be avoided and ventilation, if necessary, or insulation included to prevent condensation.

(iii) Use the minimum number of metals; avoid galvanic contact. Join dissimilar pipes by an insulating section or "waster" section.

(iv) Ensure adequate protection, i.e. correct paint scheme, cathodic protection, etc.

(v) Fabrication techniques and handling of components and finished articles should be stipulated. Recommendations should also be made on inspection periods and methods.

(vi) Allowance should be made for the effects of corrosion products. Corrosion products have a larger volume than the metal from which they were derived. The resultant expansion may cause deformation or failure of a component, e.g. corrosion of reinforcing bars in concrete causes cracking and spalling of concrete from the structure, leading to eventual failure.

6.2 PROTECTIVE COATINGS

Before application of protective coatings, the metal surface should be adequately prepared. The degree of cleanliness will depend on the type of coating to be applied. In general, the surface should be free of scale, grease, previous coatings, dirt, and moisture. The surface may be cleaned by wire brush, shot blasting, pickling, chemical degreasing, electrochemical cleaning, or ultrasonic cleaning.

6.2.1 Metallic coatings

Metallic coatings are either more noble than the substrate, as in chromium plate on mild steel, or are base metals which corrode more easily than the substrate. In the former case, the underlying metal is protected by a continuous impervious film of the noble metal which is itself resistant to attack. This method is adequate provided the surface coating contains no holes or flaws and remains intact. Penetration of this layer by the corrosive agent leads to galvanic corrosion at the interface of the two metals. If the object is coated with a more base metal, then protection is by both the physical barrier of the metal film and by cathodic protection at any subsequent defects (Fig. 19).

The more volatile metals may be used as coatings by hot dipping. The article to be coated is immersed in a bath of the molten metal for a short time so that the molten metal thoroughly wets the article and bonds with its surface (alloying occurs). The corrugated iron of shed roofs is coated with zinc in this way (galvanised).

A second method of forming a metallic coating is to heat the article immersed in the powdered metal. This causes diffusion of the coating metal into the substrate. With aluminium, the process is termed calorising, with silicon, siliconising, with chromium, chromising, and with zinc, sherardising. Powdered alloys may also be used for coating.

Coatings may also be formed by spraying the molten metal directly onto the article to be protected. An adaptation of this is to blow the powdered metal into a flame which is played across the article.

(a)   (b)

Fig. 19. Galvanic effect of a metal coating. (a) Chromium plated steel, (b) galvanised steel.

A substrate metal may be clad to improve corrosion resistance, in which case the cladding is applied by hot rolling of the substrate and cladding metal. Aluminium alloys are often clad with high-purity aluminium, the latter having improved corrosion resistance over its alloys.

Electrolytic plating of metals is performed for both protective and decorative reasons. The arrangement and chemistry of the plating bath must be such that a coat of even thickness, non-porous, adherent metal is produced. The "throwing power" is a measure of the ability of the metal ions to reach remote sections of the article to be plated. Accurate control of current, potential, plating bath chemistry, and temperature is essential in order to obtain high-quality results.

6.2.2 Non-metallic coatings

Non-metallic protective coatings include enamels, paints, waxes, and plastics.

Enamelling consists of forming an adherent glass on the metal surface. Surfaces treated in this way are resistant to attack from acids and weak alkalis. The glasses may be coloured and formed into decorative patterns for household ware and jewellery and are prodcued from a mixture of oxides of silicon, boron, sodium, calcium, and lead. Secondary elements are also added, e.g. nitrates and manganese dioxide, which oxidise impurities, other oxides to aid adhesion to the metal surface, and pigments to impart colour and opacity.

Waxes and greases are employed as temporary protectives. Steel sheets, bars, tubes, etc. are transported with a water-repellent film of oil or grease. A secondary method of protecting car bodies, especially door and sill voids, is to spray wax through the small access ports and form an even layer which protects the metal from water, salt, and road dirt.

Application of paint to steel surfaces has the primary purpose of preventing corrosion but can also provide decoration or other specialised effects, e.g. non-slip, anti-condensation, high or low reflectance, etc. Other specialised purposes such as the internal coating of tanks to hold a wide range of liquids, e.g. water, petroleum fuels, foodstuffs, chemicals, etc., may also be fulfilled.

The paint coatings may be either air dried at ambient temperature or heated to provide the requisite film properties. In this discussion, we are mainly concerned with the prevention of corrosion by coatings applied and dried at ambient temperature.

One of the most important aspects of protection of steel by paints is the preparation of the surface. Most steel for heavy industrial use is now shot or grit blasted to remove millscale and corrosion product and this process yields a uniformly grey finely roughened surface virtually free from soluble salts or other contamination. The degree of preparation, judged visually, is illustrated by several widely recognised standards, the most common being the "Pictorial surface preparation standards for painting steel surfaces",

Swedish standard SIS 05 5900–1967. A degree of preparation equal to Sa $2\frac{1}{2}$ is usually employed for most industrial painting projects.

Other methods of preparation are encountered, usually in situations where grit blasting is not suitable because of the large amount of dust which is produced. Acid pickling, often followed by a phosphating process, is used, particularly for light gauge steel plate used in the construction of small naval vessels, and, although less common now than several years ago, it is still carried out on some specialised plants. Flame cleaning, i.e. rapid surface heating by the direct application of gas flames using portable gas lances, is used although the rate is too slow to allow cleaning of very large areas. This process is quite effective at removing millscale but less so for rust scale. It does have a beneficial effect in thoroughly drying the rust layer and volatilising those volatile salts which may be present.

For a number of years, the use of high-pressure water, sometimes with abrasive particles of mineral grit entrained, has increased. Pressures ranging from 1000 p.s.i., usually with abrasive, to 14 000 p.s.i. without abrasive are used. The process, particularly that incorporating abrasive, is very effective both in terms of rate of cleaning and degree of cleanliness achieved. The exposed steel surface will rapidly "flash" rust on drying and this can lead to some difficulty in recognising the standard of cleaning which has been achieved by the jetting process. To overcome this, inhibitors can be incorporated in the water stream, thus maintaining the steel in a "grey" condition for inspection purposes. Removal of wet grit debris from the surface is achieved by water washing with or without inhibitor in the wash water. If inhibitor or fresh water only (leading to flash rusting) is used then the paint supplier will reserve the right to accept or reject the surface for painting. In some areas where flash rusting has occurred, a rapid "sweep blast" using dry mineral grit is carried out at the request of the paint supplier.

Whichever process of preparation is selected, the aim is to produce a clean dry surface to which paint may be applied in a specified number of coats to achieve an adequate film thickness.

There are numerous types of paint employed in the protection of steel and they are designed to meet the conditions imposed by the environment in which they are expected to function. For steel exposed to the atmosphere, the most common type of paint system is based on alkyl resin and this may be mixed with other types or may itself be chemically modified for a specific purpose, e.g. vinyl toluenated or styrenated to give rapid drying. Other generic types are chlorinated rubber, vinyl, acrylic, epoxy, and polyurethane. All have particular attributes and limitations and selection is usually a matter of discussion between user and supplier.

Pigmentation can range from "inert" types, such as barium sulphate, magnesium and calcium silicates together with a primary coloured pigment such as iron oxide, to anticorrosive pigments such as red lead, zinc chromate, zinc or calcium phosphate, etc. Others, which may be included to improve water resistance, are aluminium flake, micaceous iron oxide and mica, the most effective of these lamallae type being aluminium flake. The pigmentation is an important aspect of paint formulations, not only with regard to type but also with respect to total pigment content. Increased pigment loading will generally be beneficial in improving water resistance, drying, adhesion, etc. However, a point will be reached where the pigment content will become excessive and there will be a rapid deterioration of protective properties. This level of pigmentation is referred to as the critical pigment volume concentration.

It can be recognised from the brief description of paint types and the various ingredients that the effectiveness from the point of view of corrosion control will vary appreciably and the suitability of any paint system should be assessed by both laboratory and field testing. Corrosion resistance is not solely dependent on, nor indeed proportional to, water impermeability. The mode of protection varies depending on the type of paint system. Inert pigment types function as barriers to water, oxygen, and ionic transport, those incorporating anticorrosive pigment may function by passivating the steel surface (e.g. chromates) whilst others, such as metallic zinc paints, initially act as an anodic layer in a similar manner to sacrificial anodes used for cathodic protection but later tend to act by passivating the steel surface. These zinc paints remain quite water permeable at all stages of their "lives" compared with the barrier type.

From a practical point of view, the attainment of the specified dry film thickness is important. In many cases, a minimum thickness must be achieved or protection will fail in a short period. This aspect demands skill in application and close inspection and control. For metallic zinc paints, the thickness would normally be of the order of 60–100 μm, for anticorrosive pigmented paints, together with top coats, 100–200 μm, and for inert barrier types 150–300 μm. For special environments or projects, these thicknesses may be varied but it should not be presumed that protection will always be improved or be more effective over a longer period of time by increasing the film thickness.

The most widely used method of application in heavy industrial projects is airless spraying. This utilises high pressure, the liquid paint being ejected through a fine specially designed nozzle which causes the paint stream to break up into fine droplets in the form of a fan. This rapid method of paint deposition also allows application of high dry-film thickness with each coat, 150 μm being not unusual in this respect. A coat of "decorative" paint applied by brush would probably yield a thickness of 30–40 μm.

With the ability and resources to prepare steel surfaces satisfactorily together with equipment and access to apply high thickness in an economical manner, the technology involved in paint has made significant advances in the last several years. It is expected that steel structures would receive an adequate paint system to last for at least five years before any major remedial work is required.

References pp. 277–279

For structures immersed in seawater, the painted steel is often also protected by cathodic protection. With an intact paint system, the demand on the polarising system is very low but, in the case of damage to reveal bare steel or if spontaneous detachment of the paint coatings should occur, demand will be increased. With impressed current systems, monitoring of current demand will give an indication of the integrity of the protective paint system, sudden and sustained demand being indicative of coating failure.

6.3 CATHODIC PROTECTION

Cathodic protection is an electrochemical technique of providing protection from corrosion [38]. The object to be protected is made the cathode of an electrochemical cell and its potential driven negatively to a point where the metal is immune to corrosion. The metal is then completely protected. The reaction at the surface of the object will be oxygen reduction and/or hydrogen evolution. Cathodic protection may be divided into two types, that produced using sacrificial anodes and the second by impressed current from a d.c. generator [39].

Sacrificial anodes for the protection of steel structures have been made from many alloys but for full protection the anode must produce a sufficiently negative potential at the steel surface, thus providing sufficient current to support the cathodic reaction at that potential. Sacrificial anodes do this, as the name implies, by corroding at the required rate when in galvanic contact with the structure to be protected. In order for the anode to be efficient, it must be able to sustain the cathodic reaction at the steel surface, undergo negligible self-corrosion, corrode in a uniform fashion, and have sufficient potential to provide protection at some distance from the structure. Negligible self-corrosion means that portion of dissolution of the anode which is not due to galvanic connection to the structure. The "throwing power" of the anode is also controlled by the resistivity of the medium within which the structure and anode are placed. For example, in seawater an anode may protect several meters of pipe, whereas in a fairly dry soil only the immediate vicinity will be protected.

The metals usually used as anodes in sacrificial cathodic protection are pure zinc, magnesium, and aluminium alloys. Aluminium is alloyed with small amounts of zinc, indium, cadmium, and magnesium to improve the evenness with which it corrodes. Magnesium is alloyed with manganese, zinc, and aluminium; the last two components are varied in order to alter the potential of the anode. Various designs of sacrificial anode are used, ranging from bars of various cross-section, bracelet anodes, circular stubs, hydrodynamically smooth, etc. Some designs of anode allow for the high current demand of the structure for initial polarisation. This is done by casting the anode with large thin fins which produce the high surface area necessary for high current output.

Impressed current systems use anode materials with high corrosion resistance at positive potentials. The current supply is derived from a d.c. generator, the negative terminal being connected to the structure to be protected and the positive terminal to the anode. Anodes may be composed of magnetite, lead, platinised titanium, silicon iron, or graphite. The anode material should be sufficiently robust to withstand the shocks and abrasions encountered in use. With impressed current systems, it is necessary to control the potential of the structure within certain limits. At too high a potential, insufficient protection is obtained and at too low a potential, excessive hydrogen evolution occurs [40] which may lead to hydrogen embrittlement of the metal. In flowing seawater, the optimum potential is considered to be about -0.8 V (relative to S.C.E.). In order to control the potential, a reference electrode is necessary and must itself be robust and provide a reliable potential. Commonly used reference electrodes are a piece of metal of the same composition as the structure to be protected, pure zinc, copper/copper sulphate, or silver/silver chloride.

Cathodic protection has many applications, e.g. in refineries, power stations, gas, water, and oil utilities; on marine structures, e.g. jetties, piers, locks, offshore platforms, pipelines, ships' hulls, etc.; and on land structures, e.g. buried pipeline, storage tanks, cables, etc. For each use, the cathodic protection system requires careful design, either impressed current, sacrificial anodes, or a combination of both may be chosen. There may also be other protection systems, e.g. paint, the nature of which will affect the design parameters and must be taken into consideration.

Marine conditions are ideal for the use of cathodic protection systems. The seawater has low electrical resistivity. Thus, lower driving potentials may be used on impressed current systems and sacrificial anodes are capable of protecting at greater distances along the structure. A secondary beneficial effect is the production of a calcareous deposit upon the protected steel surface. This is formed as a result of the production of hydroxyl ions at the steel surface from the reduction of oxygen. The resultant high surface pH causes precipitation of calcium carbonate (aragonite) and magnesium hydroxide (brucite). This layer of salt acts as a barrier to ion and oxygen transport, thus reducing the current required for polarisation. The amount of current reduction caused by the calcareous deposit depends upon the conditions under which it was formed. Temperature, flow rate, and potential have a controlling effect upon the properties of the scale.

Care must be taken when designing cathodic protection systems in built-up areas as electrical interference from transmission lines or electrical rail systems may cause stray current corrosion at points in the structure close to the tramway or transmissions line [41].

6.4 INHIBITORS

Inhibitors are substances which, when added to the corrosive medium, reduce the rate of attack on a metal [42, 43]. Inhibitors may be used in acids, alkalis, impregnated into wrapping material, in cooling water, cutting fluids, lubricating oils, and many other applications. The inhibitor chosen should not interfere with other functions of the fluid, e.g. should not cause breakdown of oil or demulsification in cutting fluids. Inhibitors may be specific to one environment and type of metal. They function by reducing the rate of either the anodic or cathodic reaction or a combination of both. This can be brought about by increasing the activation energy for the anodic or cathodic reaction, preventing or reducing the escape of metal ions by inducing the growth of a protective oxide, reducing access of depolarising substances such as oxygen to the metal surface by forming a deposit on the surface, or direct adsorption of the inhibitor to form an insulating film on the metal surface. Inhibitors are thus generally classified, according to their type of effect, anodic, cathodic, adsorption, etc. This, however, infers that the inhibitor is working by one mode only, which is not generally the case. For example, the precipitation of a sparingly soluble salt on a metal surface forms a partial barrier to oxygen diffusion, thus reducing the cathodic reaction. However, the migration of metal ions is also hindered and an oxide may be formed at the metal/precipitate interface which is different from that formed in the absence of precipitate; hence the anodic reaction is also reduced. A total description of the mode of action, from which a better idea of the effects of change in environment can be assessed, is therefore to be preferred.

The use of inhibitors in cooling waters may be taken as an example of their application [44, 45]. The quality of water used for cooling purposes has a very broad range and depends upon availability, geographical location, and geology of the source area. In arid regions, bore hole water with a high solids content may be used and reservoir waters may vary from very soft acid water in granite regions to hard waters from limestone regions. Water may be pretreated before use by lime softening, in which case alkalinity is increased and hardness reduced. Water for cooling purposes is used in large quantities and is therefore an expensive commodity. The greatest use of available water is made by recycling before it is sent to drain. The larger water cooling plants achieve heat loss by evaporation of a portion of the cooling water from within the cooling tower. This evaporation results in the requisite loss of heat but also a loss of water and subsequent increase in dissolved solids. The build-up of salts in the water must be limited by "blowing down", i.e. removing a portion of the system water. The concentration factor to which the water may be taken will depend upon the quality of make-up water and the ability of the chemical additives to prevent scale build-up and corrosion problems. Additives to water therefore consist of a formulation which controls scale, biological fouling, corrosion, and general dispersal of solid material. Some of these elements may be omitted or more than one additive for one purpose may be included. Generally, more than one corrosion inhibitor is incorporated into the formulation, one for yellow metals (brass etc.) [46] and one for ferrous metals. Two corrosion inhibitors may be used for the same metal as they "synergise" with one another, i.e. they produce improved corrosion protection together than when used alone.

The technology of corrosion inhibition is in rapid development due to environmental concerns and the need to make greater use of available water supplies of poorer quality. An early inhibitor for ferrous metal consisted of chromate, which produced a protective oxide on the steel surface. The oxide layer prevents anodic dissolution but is conductive enough to allow electron transfer for oxygen reduction; thus, insufficient chromate could lead to localised attack on insufficiently inhibited surfaces. Improved performance was obtained by the addition of zinc, which forms an insoluble hydroxide at cathodic sites [47]. Environmentally, chromate is unacceptable due to its high toxicity. Zinc salts alone are not efficient inhibitors and require careful formulation with polymeric stabilisers, which hold sufficient zinc in solution and help form a more protective deposit on the steel surface. Zinc formulations [48] are in common use but, although less toxic than chromate, they are not environmentally acceptable. Newer formulations with lower levels of zinc (1–2 ppm in the system water) are now being used and are efficient in waters of higher alkalinity with little or no acid dosing. Soft waters may be treated with nitrite [49], silicates, polyphosphates [49], and molybdates. These may be used in combination with stabilisers such as phosphonates, polyphosphino carboxylic acids [49], polymaleates [49], and polyacrylates. Nitrites are generally used in closed systems. In open systems, there is a danger of oxidation to nitrate and biofouling. Silicates require careful formulation in order to maintain a sufficient concentration in solution; they should not be overdosed as fouling by precipitated SiO_2 may become a problem. Molybdates [50, 51] are increasing in use and have the advantage over chromates of low toxicity but they are less efficient and require an oxidising environment for good corrosion inhibition. Polyphosphate and orthophosphate are formulated together with stabilisers against precipitation of calcium salts. Such formulations [49] can give excellent results provided dosing procedures are accurately followed. Several countries have legislation which limits the amount of phosphorus that may be discharged due to it acting as a food source for algae, etc. All organic corrosion inhibitors [52] are the most recent advances in corrosion control and contain very low or no phosphorus compounds. The non-phosphorus formulations consist of polymers which inhibit the precipitation of calcium carbonate and may only be used in alkaline waters with appreciable calcium content. They function by allowing a thin film of calcium carbonate to be deposited on the metal surface while preventing any bulk precipitation of calcium carbonate. Recent low phosphorus-containing formulations are applicable in waters of a broader range and function by the formation of a protective calcium phosphonate film on the metal surface in hard waters and by a mixed calcium/iron phosphonate film in softer waters where, due to the more aggressive nature, slightly higher doses are required. Thus, the ultimate aim of the water treatment industry is to produce a number of environmentally acceptable corrosion inhibitor formulations which will provide protection over the whole range of water qualities found in industrial use.

References pp. 277–279

The way to produce improved corrosion inhibitors is through an understanding of their mechanism of controlling the corrosion process. From the understanding, a picture can be built up of the role of functional groups and the stereochemistry of the whole molecule at the metal or oxide surface.

Lorenz and Mansfeld [53] and others [54] have considered inhibitors to be effective in either the interface or interphase of a metal and its environment. In the case of interface action, the inhibitor absorbs directly at the corroding surface forming a two-dimensional layer. This layer can affect the corrosion reaction in different ways, either by geometrical blocking, blocking of active surface sites, or by the layer itself being reactive [55–57]. The reactive adsorbate can either stimulate or inhibit corrosion by acting as an electrocatalyst on the corrosion reaction and/or itself undergoing an electrochemical redox process, the product of which may also be active. Interface inhibition generally occurs in systems where the bare metal surface is in contact with the corrosive medium, for example in dissolution of metal in acid [58, 59]. Mukherjee et al. [60] have correlated corrosion inhibition efficiency with molecular orbital energy levels of functional groups. This type of approach may be used to indicate possible interface active molecules. Wang Nai-Long et al. [61] have used quantum mechanics together with other structural parameters to devise a model for corrosion inhibitors in acid media.

Interphase inhibition [52] occurs where the inhibitive layer has a three-dimensional structure situated between the corroding metal and the electrolyte. The interphase layers generally consist of weakly soluble compounds such as oxides, hydroxides, carbonates, inhibitors, etc. and are considered to be porous. Non-porous three-dimensional layers are characteristic of passivated metals. The inhibitive efficiency depends on the properties of the three-dimensional layer, especially on porosity and stability. Interphase inhibition is generally encountered in neutral media, either in the presence or absence of oxygen. In aerated solutions, the inhibitor efficiency may be correlated with the reduction in the oxygen transport limited current at the metal surface.

Modelling of the interphase region [62] is possible with the aid of a computer. With this technique, the "lattice fit" of prospective molecules may be determined. This "fit" would include both spatial configuration and energy considerations.

Because of the complexity of aqueous corrosion processes, it is not always a simple matter of using one corrosion inhibitor. As mentioned earlier, better performance can be achieved using more than one inhibitor. Such "synergistic" formulations work by affecting the corrosion process by different mechanisms. For example, some organic compounds form insoluble calcium salts on cathodic sites on the steel surface, thus reducing the access of oxygen and lowering the surface potential. This allows increased absorption of cationic polymers, thus blocking more of the metal surface and forming a more adherent protective film. Other additives are used in formulations to stabilise bulk water chemistry and prevent excessive precipitation of sparingly soluble salts which would lead to under-deposit corrosion.

A successful formulation would provide good corrosion protection over a broad range of water chemistry, would prevent fouling by sparingly soluble salts, would be easily dosed to a system, would have a method of detection, would be stable on storage over a broad temperature range, would not be affected by chlorine, high levels of calcium or iron, would have low toxicity and would be inexpensive. Thus, the search for corrosion inhibitors is not a straightforward exercise but is complicated by financial and environmental factors.

7. Future needs

Corrosion is an everyday part of life and generally tends to be taken for granted or overlooked altogether. The first step in combating corrosion, therefore, is to increase the awareness of those people involved in design, construction, and the use of metal-containing items. Design is the first stage of a project and proper consultation at this stage may save time and money later. Computers are now being used as "experts" in corrosion [63]. The knowledge of experts is being stored on computer in a form which is able to give answers to questions by use of decision trees.

As technology advances, the need for improved performance from materials increases. There is also a continuing economic pressure to minimise costs, which will become increasingly difficult as resources become depleted. Thus, research is aiming at more efficient corrosion protection methods such as improved formulations for paints and corrosion inhibitors, improved metal alloys, better design, and the use of novel materials such as ceramics and plastics.

References

1　Natl. Bur. Stand. (U.S.) Spec. Publ., 511-1
2　R.C. Weast (ed.), Handbook of Chemistry and Physics, CRC Press, Boca Raton, FL, 61st edn., 1980.
3　F.L. LaQue, Marine Corrosion, 3 Causes and Prevention, Wiley, New York, 1975.
4　R.J. O'Halloran, L.F.G. Williams and C.P. Lloyd, Corrosion (NACE), 40 (7) (1984) 344.
5　M. Pourbaix, Atlas of Electrochemical Equilibria in Aqueous Solutions, Pergamon Press, Oxford, 1966.
6　D.C. Silverman, Corrosion (NACE), 38 (8) (1982) 453.
7　J. Tafel, Z. Phys. Chem., 50A (1905) 641.
8　J.A.V. Butler, Trans. Faraday Soc., 19 (1934) 729.
9　T. Erdey-Gruz and M. Volmer, Z. Phys. Chem., 150A (1930) 203.
10　M. Stern and A.L. Geary, J. Electrochem. Soc., 104 (1957) 56.
11　M.J. Graham, R.J. Hussey and D.F. Mitchell, 24th Corrosion Science Symposium, Cambridge, 1983, paper 26.

278

12   J.B. Johnson, B.S.S. Kerry and G.C. Wood, 24th Corrosion Science Symposium, Cambridge, 1983, paper 32.
13   K.C. Waterton, G.E. Thompson and G.C. Wood, 24th Corrosion Science Symposium, Cambridge, 1983, paper 17.
14   H.M. Seip, Chem. Br., 20 (1984) 791.
15   M. Cross, New Sci., 103 (1984) 10.
16   P.A. Farinha, PhD Thesis, Manchester University (UMIST), 1982.
17   S.C. Dexter and C.M. Culberson, Corrosion/79, Atlanta, GA, 1979, paper 227.
18   S. Elbeik, A.C.C. Tseung and A.L. Mackay, Corros. Sci., 26 (1986) 669.
19   C.R. Vanentini and C.A. Moina, Corros. Sci., 25 (1981) 317.
20   J.G.N. Thomas and J.D. Davies, Br.Corros. J., 12 (2) (1977) 108.
21   T.E. Castle and A.R. Daud, 24th Corrosion Science Symposium, Cambridge, 1983, paper 61.
22   R.P.M. Procter, Cathodic Protection, Theory and Practice. The Present Status, Conference held at Coventry, 1982, paper 18.
23   M. Sakashita and N. Sato, Corrosion (NACE), 35 (8) (1979) 351.
24   H.H. Strehblow and H.D. Speckmann, Werkst. Korros., 35 (1984) 512.
25   R.D. Misra and T. Burstein, 24th Corrosion Science Symposium, Cambridge, 1983, paper 9.
26   S.W. Smith, Jr., N.D. Kackrey and R.M. Latanision, Corrosion/83, Anakeim, CA, 1983.
27   J.C. Rowlands and M.N. Bentley, Br.Corros.J., 7 (1) (1972) 42.
28   M. Metikos-Huckovic and C. Zernick, Werkst Korros., 33 (1982) 361.
29   Betz Laboratories, U.S. Pat. No. 1 357 573, 1974.
30   J. Asher, J.W. Webb, N.J.M. Wilkins and P.F. Lawrence, UKAERE Rep. 10574, AERE Harwell, 1982.
31   L. Clerbois, E. Heitz, F.P. Igsseling, J.C. Rowlands and J.P. Simpson, Br.Corros.J., 20 (3) (1985) 107.
32   K. Hladky and J.L. Dawson, Corros. Sci., 21 (1981) 317.
33   K. Hladky, L.M. Callow and J.L. Dawson, Br.Corros.J., 15 (1) (1980) 20.
34   D.E. Williams, C.C. Naish, Introduction to the A.C. Impedence Technique and its Application to Corrosion Problems, HL 85/1066 (C14), Materials Development Division, AERE Harwell, 1985.
35   D.E. Williams and C.C. Naish, UK AERE Memo 3461, AERE, Harwell 1985
36   D.R. Grabe and F.C. Walsh, J. Appl. Electrochem., 13 (1983) 3.
37   D.C. Silverman, Corrosion (NACE), 40 (5) (1984) 220.
38   V. Ashworth, Cathodic Protection, Theory and Practice. The Present Status, Coventry, 1982, paper 1.
39   K.G.C. Berkeley, Corrosion/84, New Orleans, LA, 1984, paper 48.
40   R.N. Perkins, Corrosion/84, New Orleans, LA, 1984, paper 47.
41   M .Allen and D. Ames, Cathodic Protection, Theory and Practice. The Present Status, Coventry, 1982, paper 20.
42   A.D. Mercer, Br.Corros.J., 20, (2) (1985) 61.
43   R.T. White, Mintek Rep. No. 66, 1983.
44   B.P. Boffardi, Mater. Perform., 23 (11) (1984) 17.
45   A.J. Freedman, Corrosion/83, Anakeim, CA, 1983, paper 273.
46   O. Hollander and R.C. May, Corrosion (NACE), 41(1) (1985) 39.
47   R.V. Comeaux, Hydrocarbon Process., 46 (12) (1967) 129.
48   A. Marshall, Corrosion/81, Toronto, Ont., 1981.
49   Product literature, Ciba-Geigy Industrial Chemicals, Trafford Park, Manchester, Gt. Britain.
50   M.A. Stranick, Corrosion (NACE), 40 (6) (1984) 296.
51   M.S. Vukasovich and J.P.G. Farr, Mater. Perform., 25 (5) (1986) 9.
52   G. Bohnsack, Corrosion/85, Boston, MA, 1985, paper 379.
53   W.J. Lorenz and F. Mansfeld, Proceedings of the 6th European Symposium on Corrosion Inhibition, 1985, Ferrara, Italy, Vol.1, p.23.

54 S.W. Dean, R. Derby and G.T. Von Dem Bussche, Corrosion/81, Toronto, Ont. 1981, paper 253.

55 W.J. Lorenz, Corros. Sci., 5 (1965) 121.

56 E.J. Kelly, J. Electrochem. Soc., 115 (1968) 1111.

57 H. Vaidyanaathan and N. Hackerman, Corros. Sci., 11 (1971) 737.

58 R. Driver and R.J. Meakins, Br.Corros. J., 9 (1974) 233.

59 R.K. Dinnappa and S.M. Mayanna, Corrosion (NACE), 38 (10) (1982) 525.

60 D.C. Mukherjee, T. Kar and A. Chakrabarti, Proceedings of the 6th European Symposium on Corrosion Inhibition, 1985, Ferrara, Italy, Vol.1, p.465.

61 Wang Hai-Long, Yao Zhan-hua and Wang Wen-Zhen, Proceedings of the 6th European Symposium on Corrosion Inhibition, 1985, Ferrara, Italy, Vol.1, p.199.

62 M. Lees, personal communication, 1987.

63 J.G. Hines, Br.Corros. J., 21 (2) (1986) 81.

64 T.R.S. Wilson, in J.P. Riley and G. Skirrow (Eds.) Chemical Oceanography, Vol.1, Academic Press, New York, 1965.

65 F. Culkin, in J.P. Riley and G. Skirrow, (Eds.), Chemical Oceanography, Vol.1, Academic Press, New York, 1965.

Index